Highlights in Computational Chemistry II

Tim Clark (Ed.)

Highlights in Computational Chemistry II

Special reprint edition of selected papers
published in the *Journal of Molecular Modeling*
on the occassion of Professor
Paul von Ragué Schleyer's 75th Birthday

With contributions of leading experts
in the field of computational chemistry

Springer

Editor:
Professor Dr. Tim Clark
Journal of Molecular Modeling
Computer-Chemie-Centrum
Universität Erlangen-Nürnberg
Nägelsbachstraße 25
91052 Erlangen
Germany

Library of Congress Control Number: 2001042000

ISSN 0948-5023
ISBN-10 3-540-37592-9 Springer Berlin Heidelberg New York
ISBN-13 978-3-540-37592-0 Springer Berlin Heidelberg New York

Springer is a part of Springer Science+Business Media
springer.com

Cover design: WMX-Design, Heidelberg
Printing and Binding: Druckerei Stürtz, Würzburg
Printed on acid-free paper SPIN 11780618 52/3141/ba 5 4 3 2 1 0

Paul von Ragué Schleyer

Five years ago, we published the papers submitted to the *Journal of Molecular Modeling* Festschrift for Paul Schleyer's 70th Birthday as *Highlights in Computational Chemistry*. [1] The positive response to this volume has prompted us to repeat the exercise for Paul's 75th birthday, which was celebrated by a small symposium in Athens, Georgia in February 2005.

The similarities and differences between the two volumes are fascinating and provide some insight into Paul's continuing impact on computational chemistry. Once again, the response to invitations to contribute to a Festschrift honoring Paul was overwhelming. However, the subjects of the contributions are more diverse than five years ago, indicating the development of the "Schleyer School" into areas that Paul himself has not considered. His scientific style, which has influenced us all to a greater or lesser extent, is still evident in many contributions from areas far removed from his own research. His unique way of looking at problems and formulating their solutions has formed an entire generation of computational chemists. I am not sure that he himself appreciates that this is also true for areas such as life-science modeling that have never featured in his own research.

Perhaps this is the real mark of a real scientific personality. It is comparatively easy to influence a field in which you have worked for thirty years. It is far more difficult to shape a generation of scientists who have spread from Paul's own areas into very different scientific subjects and disciplines. This Festschrift provides the evidence that Paul has done exactly that.

Once again, it was a real pleasure to edit and produce this Festschrift. All of us in Erlangen and at Springer wish Paul good health and many more years of productive research. Nothing would please me more than to edit the third volume for his 80th birthday.

Erlangen, 13th July 2006 *Tim Clark*

TABLE OF CONTENTS

J Mol Model (2006) 12: 529
DOI 10.1007/s00894-006-0119-x

EDITORIAL

Tim Clark

Paul von Ragué Schleyer 75th birthday Festschrift

Published online: 25 May 2006

It seems not long ago that I wrote the introduction to the Festschrift for Paul's 70th birthday. Happily, Paul is still well and active (active means that he still publishes as much as about any three of us put together). Maybe he is not quite the fireball he used to be, but I will attribute that to the fact that his child, computational chemistry, has come of age and is accepted as essential for modern research, so that the proud father can take it easy and no longer has to look out for its interests. Just how much the attitude to computational chemistry has changed, even in the 5 years since the last Festschrift, became clear to me in the last few weeks as we were writing a major cooperative grant proposal for the development of new materials. Modeling and simulation was assumed by all concerned to be an integral part of the concept from the beginning.

Paul is responsible for much of this change. At Gernot Frenking's recent 60th birthday symposium, Paul suggested that Gernot had achieved more in gaining acceptance for computational chemistry in his diplomatic art than Paul ever had with his frontal attacks. This may be false modesty. The young people who were in awe of Paul and relished his attacks on their elders are now the decision makers. They are the ones who are now using modeling and simulation (in the widest sense) in their chemistry.

Much of this was achieved in the "golden years" in Erlangen. Paradoxically, this was the low point in the acceptance of computational chemistry, at least in Germany. I have also only recently come to appreciate just how important and stimulating those years were. Many of the authors who have contributed to this festschrift also experienced the atmosphere in the *Blaue Saal* back then. In this respect, this volume documents Paul's seminal effect on a whole generation of computational chemists. As always, the response to my invitation to contribute papers to the Festschrift was overwhelming. This is also a measure of Paul's stature and standing in our community.

Let me close this with the wish that we can continue to produce a Paul Schleyer Festschrift every 5 years for a very long time and a wish for Paul and Inge's health and happiness.

Tim Clark, Erlangen, April 26th 2006

T. Clark (✉)
Journal of Molecular Modeling,
Erlangen, Germany
e-mail: Journal.of.Molecular.Modeling@chemie.uni-erlangen.de

J Mol Model (2006) 12: 531–536
DOI 10.1007/s00894-005-0041-7

ORIGINAL PAPER

J. Chris Slootweg · Andreas W. Ehlers
Koop Lammertsma

Valence isomerization of 2-phospha-4-silabicyclo[1.1.0]butane: a high-level ab initio study

Received: 23 February 2005 / Accepted: 23 June 2005 / Published online: 27 April 2006

Abstract The rearrangements for 2-phospha-4-silabicyclo[1.1.0]butane, analogous to the valence isomerization of the hydrocarbons bicyclobutane, 1,3-butadiene, and cyclobutene, were studied at the (U)QCISD(T)/6-311+G**//(U)QCISD/6-31G* level of theory. The monocyclic 1,2-dihydro-1,2-phosphasiletes are shown to be the thermodynamically preferred product, in contrast to the isomerization of the hydrocarbons, which favors the 1,3-butadiene structure. Furthermore, an unprecedented direct isomerization pathway to the 1,2-dihydro-1,2-phosphasiletes was identified. This pathway is competitive with the isomerization via the open-chain butadienes and becomes favorable when electron-donating substituents are present on silicon.

Keywords Heterobicyclobutanes · Valence isomerization · Ab initio theory

Introduction

Bicyclo[1.1.0]butane with its strain energy of over 60 kcal mol^{-1} is a fascinating compound that has attracted the interest of both experimental and theoretical chemists [1]. It is now well established that bicyclo[1.1.0]butane (**1**) opens to the more stable valence isomer *gauche*-butadiene (**2**) by a pericyclic rearrangement, which is characterized by a concerted, asynchronous conrotatory ring opening where the central C–C bond remains intact [2, 3]. This is an allowed [σ2s+σ2a] conrotatory rearrangement according to the Woodward–Hoffmann (W–H) orbital-symmetry rules [4–6], affording kinetic intermediate **2** that can easily rotate to *s-trans*-1,3-butadiene (**3**). The activation barrier of 41.5 kcal mol^{-1} calculated at the multiconfiguration self-consistent field level of theory [2] agrees closely with the experimental value of 40.6 kcal mol^{-1} [7, 8]. The disrotatory, W–H forbidden, thermal ring opening of **1** is less favorable, and was calculated to be about 15 kcal mol^{-1} higher in energy [2]. Another rearrangement is also feasible; stretching of the central C–C bond leads to a planar singlet diradical transition structure for inversion, which is also a higher energy process with a barrier of 47.4 kcal mol^{-1} [9].

1 **2** **3** **4**

Valence isomer cyclobutene (**4**) is of intermediate stability between **1** and **3** and converts thermally to *gauche*-butadiene **2** by an electrocyclic ring opening [10, 11]. This pericyclic rearrangement follows a W–H allowed concerted, conrotatory pathway. The calculated activation barrier at the MP2/6-311G** level of theory of 33.7 kcal mol^{-1} [12–14] for this process is in agreement with the experimental value of 32.9±0.5 kcal mol^{-1} [10, 11]. Usually for the ring opening of cyclobutenes, steric effects dominate the preference for inward versus outward rotation [15]. However, electronic effects can also dictate this rearrangement, as was reported very recently for the sterically hindered substrate **5**, which prefers to react via the more crowded inward rotatory pathway, leading mainly to butadiene **6** (Scheme 1) [16, 17].

Electronic Supplementary Material Supplementary material is available for this article at http://dx.doi.org/10.1007/s00894-005-0041-7 and is accessible for authorized users.

Dedicated to Professor Dr. Paul von Ragué Schleyer on the occasion of his 75th birthday.

J. C. Slootweg · A. W. Ehlers · K. Lammertsma (✉)
Department of Organic and Inorganic Chemistry,
Faculty of Sciences, Vrije Universiteit, De Boelelaan 1083,
1081 HV Amsterdam, The Netherlands
E-mail: lammert@chem.vu.nl
Fax: +31-20-5987488

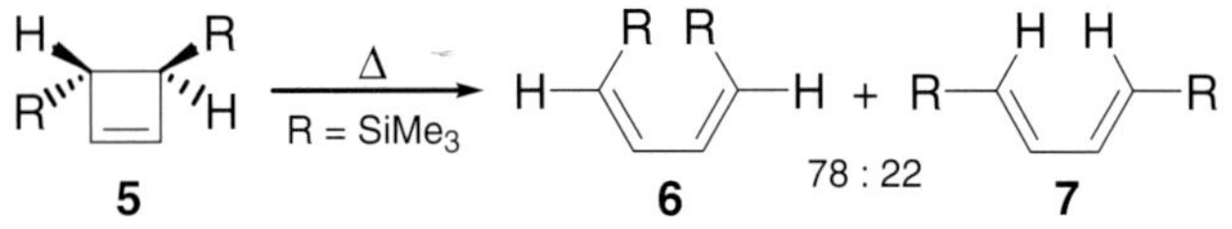

Scheme 1 Ring opening of cyclobutene 5

Bicyclo[1.1.0]butanes with main-group hetero-elements in the ring have also received considerable attention [18]. However, little is known about the phosphorus-containing analogues [19–22]. In our ongoing research on small strained organophosphorus ring systems, we became interested in the yet unknown 2-phospha-4-silabicyclo[1.1.0]butanes, whose occurrence we reported as a reactive intermediate recently [23, 24]. Valence isomerization of the 2-phospha-4-silabicyclo[1.1.0]butane **9** to the 1,2-dihydro-1,2-phosphasiletes **10a**,**b** was indicated by reacting 1H-phosphirene **8** with silylene $Si[(NCH_2^tBu)_2C_6H_4\text{-}1,2]$ [≡ Si(NN)] (Scheme 2).

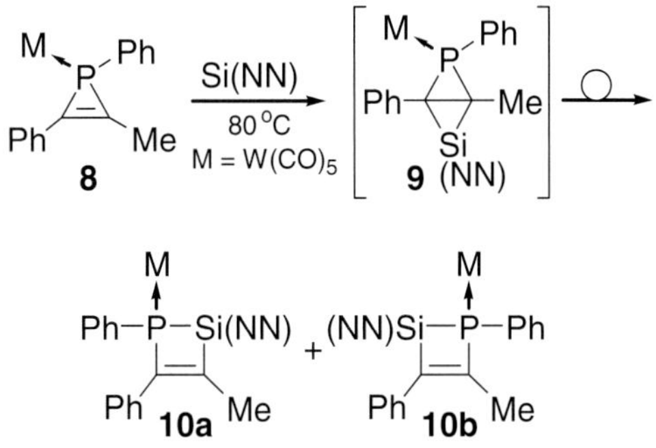

Scheme 2 Isomerization of bicyclo[1.1.0]butane 9

SCS-MP2/6-311+G** calculations on B3LYP/6-31G* model structures show that the intermediate 2-phospha-4-silabicyclo[1.1.0]butane isomerizes directly, via an unprecedented W–H allowed [σ2s + σ2a] process, to the thermodynamically preferred 1,2-dihydro-1,2-phosphasilete [23, 24]. This pathway is favored over the concerted, asynchronous conrotatory ring opening leading to *s-trans*-1-phospha-4-sila-1,3-butadiene [25].

Here, we report on the isomerization of 2-phospha-4-silabicyclo[1.1.0]butane **A** to its valence isomers 1-phospha-4-sila-1,3-butadiene **B** and 1,2-dihydro-1,2-phosphasilete **C** (only one other synthesis of 1,2-dihydro-1,2-phosphasiletes was reported: [26–28]), using high-level ab initio calculations at the (U)QCISD(T)/6-311+G**//(U)QCISD/6-31G* level of theory. We will compare the differences between a direct **A** → **C** pathway versus the isomerization via butadiene **B**. In addition, the influence of substituents on silicon on the rearrangements will also be discussed.

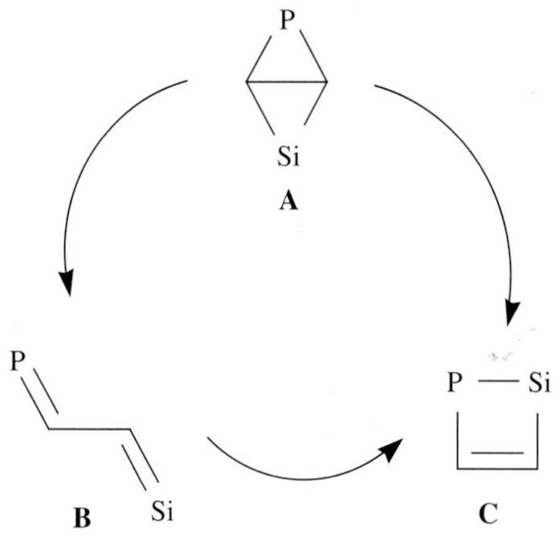

Computational details

All calculations were performed using the GAUSSIAN 98 [29] suite of programs. Geometries were optimized using the standard 6-31G* basis set at the (U)MP2 and (U)QCISD [30, 31] level of theory, while single-point calculations were preformed at the (U)QCISD(T)/6-311+G** level using the (U)QCISD/6-31G* geometries. First and second order energy derivatives were computed to confirm that minima or transition structures had been located at the (U)MP2/6-31G* level. Intrinsic reaction coordinate driving calculations were performed at the (U)MP2/6-31G* level to establish the connections between transition structures and minima. The total energies calculated at the (U)MP2, (U)QCISD, and (U)QCISD(T) levels were corrected for the (U)MP2/6-31G* level zero-point energies scaled by a factor of 0.967 [32].

Results and discussion

First, we investigated the rearrangements of bicyclo[1.1.0]butane (**1**) and cyclobutene (**4**) into the more stable *s-trans*-1,3-butadiene (**3**) at the (U)QCISD(T)/6-311+G**//(U)QCISD/6-31G* level of theory (this method gives similar energies when compared to the CASSCF(10,10)/6-31G* level of theory as was reported for the isomerization of 2-oxabicyclo[1.1.0]butane: [33]), since no complete study of the valence isomerizations of all C_4H_6 isomers at the same level of theory were reported to date. Subsequently, we investigated the rearrangements of the 2-phospha-4-silabicyclo[1.1.0]butanes, where the effects of heteroatom substitution on the characteristics of the rearrangements become apparent.

Bicyclo[1.1.0]butane (**1**) leads to *gauche*-butadiene **2** via a concerted, asynchronous conrotatory ring opening [2, 3], which has a barrier of 39.2 kcal mol^{-1}, and is exothermic by 26.0 kcal mol^{-1} (Fig. 1). This closed-shell rearrangement is favored over the corresponding

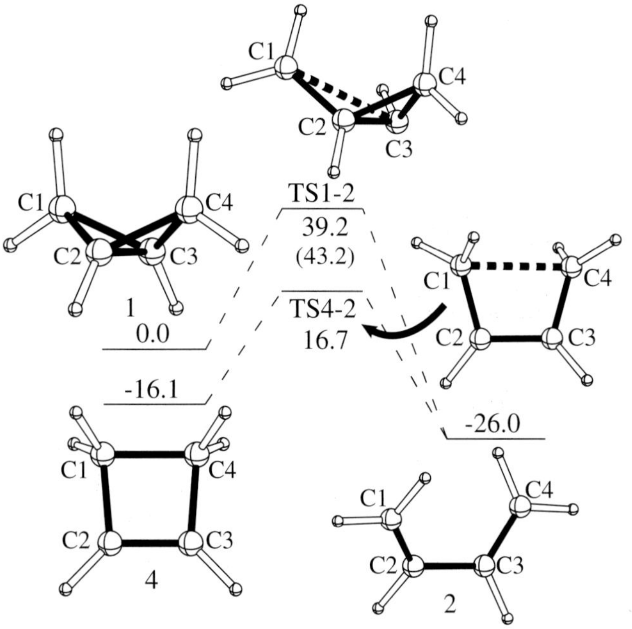

Fig. 1 Relative QCISD(T)/6-311+G**//QCISD/6-31G* (UQCISD(T)/6-311+G**//UQCISD/6-31G* in parenthesis) energies (ZPE corrected, in kcal mol^{-1}) for the rearrangements of **1** and **4** into **2**. Selected bond lengths [Å], angles and torsion angles [°] of **1** (C_{2v}): C1–C2 1.498, C2–C3 1.494, C2–C1–C3 59.8, C1–C2–C3–C4 121.9; **2** (C_2): C1–C2 1.342, C2–C3 1.479, C3–C4 1.342, C1–C2–C3–C4 37.9; **4** (C_{2v}): C1–C2 1.520, C1–C4 1.570, C2–C3 1.346, C1–C2–C3 94.2; **TS1–2** (closed-shell): C1–C2 1.403, C1–C3 2.344, C2–C3 1.542, C2–C4 1.569, C2–C1–C3 39.4; **TS4–2** (C_2): C1–C2 1.430, C1–C4 2.150, C2–C3 1.379, C1–C2–C3–C4 21.7

diradical open-shell pathway ($\Delta E^{\ddagger}=43.2$ kcal mol^{-1}, $<S^2>=0.85$). In addition, cyclobutene (**4**) also gives **2** via a synchronous (C_s symmetry) conrotatory ring opening [12–14] that requires 32.8 kcal mol^{-1}, and is exothermic by 9.9 kcal mol^{-1}. Both calculated reaction barriers are in excellent agreement with the experimental values of 40.6 kcal mol^{-1} [7, 8] and 32.9 kcal mol^{-1} [10, 11], respectively.

The kinetic *gauche*-butadiene **2** can easily transform into its enantiomer **2′** via the planar *s-cis*-1,3-butadiene (**TS2-2′**) [2, 34] with a barrier of only 0.7 kcal mol^{-1}, or can rotate to the more stable *trans*-butadiene **3** ($\Delta E^{\ddagger}=2.5$ kcal mol^{-1}) with an exothermicity of 2.6 kcal mol^{-1} (Fig. 2) [12–14]. The geometrical parameters of the optimized structures **1**, **3**, and **4** at the QCISD/6-31G* level of theory are in excellent agreement with the experimental estimates (experimental structures—**1**, **3**, and **4**: [35–37]).

Incorporating heteroatoms into the bicyclo[1.1.0]butane framework has a profound impact. We found that 2-phospha-4-silabicyclo[1.1.0]butane (**11**) opens with a modest exothermicity (0.4 kcal mol^{-1}) directly to valence isomer *s*-1-phospha-4-sila-1,3-butadiene (**12**) in its *trans* configuration via a concerted, asynchronous conrotatory ring opening. In this process, the P–C2 bond becomes elongated well before that of the Si–C1 bond (Fig. 3). The activation barrier of 38.8 kcal mol^{-1} is very similar to the calculated activation barrier of

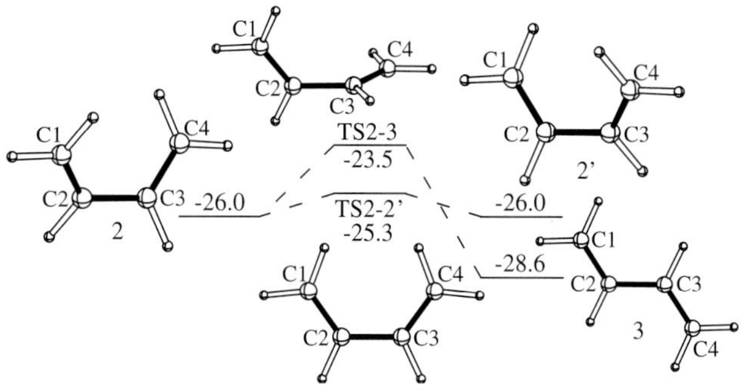

Fig. 2 Relative QCISD(T)/6-311+G**//QCISD/6-31G* energies (ZPE corrected, in kcal mol^{-1}) for the rearrangements of **2**. Selected bond lengths [Å], angles and torsion angles [°] of **3** (C_{2h}): C1–C2 1.343, C2–C3 1.467, C1–C2–C3 123.8; **TS2–3** (C_2): C1–C2 1.340, C2–C3 1.490, C1–C2–C3–C4 101.9; **TS2–2′** (C_{2v}): C1–C2 1.430, C2–C3 1.379, C1–C2–C3–C4 0.0

39.2 kcal mol^{-1} for the [σ2s+σ2a] process in bicyclo[1.1.0]butane (**1**). The closed-shell rearrangement **11** → **12** is favored over the corresponding diradical open-shell pathway ($\Delta E^{\ddagger}=$ 41.3 kcal mol^{-1}, $<S^2>=0.97$).

s-Trans-butadiene **12** can transform into the slightly less stable *gauche*-butadiene **13** ($\Delta E=2.6$ kcal mol^{-1}) with an energy barrier of 7.5 kcal mol^{-1}. Subsequently, butadiene **13** can isomerize via a conrotatory electrocyclic ring closure to the much more stable 1,2-dihydro-1,2-phosphasilete (**14**) ($\Delta E=-23.9$ kcal mol^{-1}), with a rearrangement barrier of only 3.2 kcal mol^{-1}. Clearly, if a 1-phospha-4-sila-butadiene is to be formed from **11**, it will rearrange to the four-membered ring structure **14**. We conclude that in contrast to the hydrocarbons, where butadiene **3** is the favored product, the P,Si-derivatives **12** and **13** are not likely candidates to be observed on rearranging bicyclic compound **11**.

As **14** is thermodynamically the preferred valence isomer, we also explored whether it could be formed directly from bicyclic **11**. Indeed, forcing an asynchronous conrotatory ring opening with an initial SiH_2-group rotation resulted in a transition structure **TS11–14** for the direct rearrangement of **11** into **14** (Fig. 4). The barrier of 39.0 kcal mol^{-1} for this closed-shell process is similar to the conversion via the P,Si-butadienes ($\Delta E^{\ddagger}=38.8$ kcal mol^{-1}, Fig. 3)[1]. The rearrangement via **TS11–14** obeys the orbital symmetry rules and can be described as a [σ2s+σ2a] process. Such a pathway is unprecedented for the isomerization of the carbon analogue bicyclo[1.1.0]butane (**1**) [2], for which *s-trans*-1,3-butadiene is the favored product.

Due to the similarities in activation energy for the conversions **11** → **12** and **11** → **14** at the QCISD(T)/6-311+G**//QCISD/6-31G* level of theory, we have also incorporated in our computational model the cyclic diamine HN-C=C-NH as substituent on silicon to investigate the effect of donating N atoms, which are also present in our experimental system [23, 24] on the rearrangements.

[1]No competitive open-shell rearrangement is present for **TS11–14**.

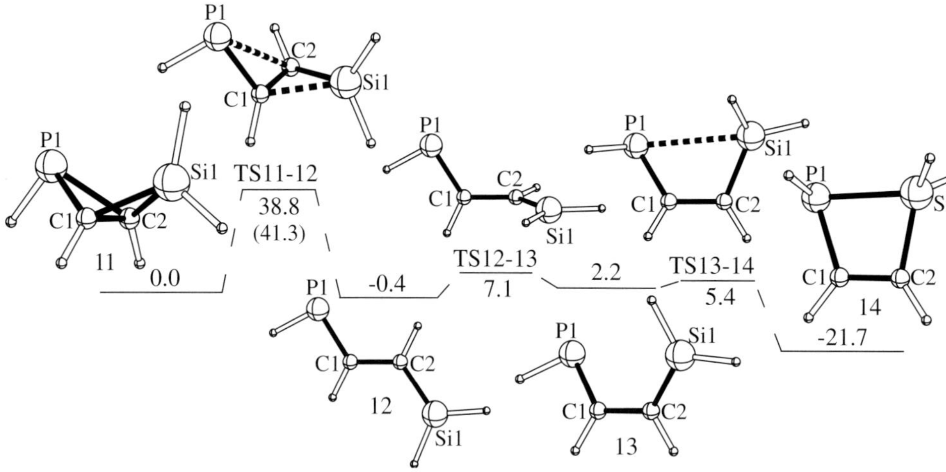

Fig. 3 Relative QCISD(T)/6-311 + G**//QCISD/6-31G* (UQCISD(T)/6-311 + G**//UQCISD/6-31G* in parenthesis) energies (ZPE corrected, in kcal mol^{-1}) for the rearrangements of **11** into **14**. Selected bond lengths [Å], angles and torsion angles [°] of **11** (C_s): P1–C1 1.852, Si1–C1 1.840, C1–C2 1.548, C1–P1–C2 49.4, C1–Si1–C2 49.7, P1–C1–C2–Si1 119.0; **TS11–12**: P1–C1 1.782, P1–C2 2.664, Si1–C1 1.982, Si1–C2 1.785; **12** (C_s): P1–C1 1.708, Si1–C2 1.741, C1–C2 1.443; **TS12–13**: P1–C1–C2–Si1 103.3; **13**: P1–C1–C2–Si1 36.3; **TS13–14**: P1–C1 1.736, P1–Si1 3.001, Si1–C2 1.774, C1–C2 1.414, P1–C1–C2–Si1 34.1; **14**: P1–C1 1.869, P1–Si1 2.290, Si1–C2 1.872, C1–C2 1.354

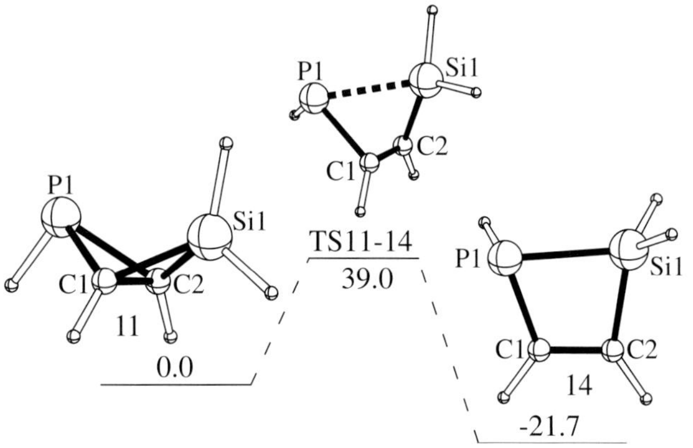

Fig. 4 Relative QCISD(T)/6-311 + G**//QCISD/6-31G* energies (ZPE corrected, in kcal mol^{-1}) for the direct rearrangement of **11** into **14**. Selected bond lengths [Å] and torsion angles [°] of **TS11–14**: P1–C1 1.834, P1–Si1 2.431, Si1–C2 1.800, C1–C2 1.422, P1–C1–C2–Si1 76.0

Substituted 2-phospha-4-silabicyclo[1.1.0]butane **15** leads to its valence isomer *s-trans*-1-phospha-4-sila-1,3-butadiene **16** via a concerted, asynchronous conrotatory ring opening ($\Delta E^{\ddagger} = 39.0$ kcal mol^{-1}), with a modest endothermicity of 1.5 kcal mol^{-1} (Fig. 5). The associated transition structure **TS15–16** shows features similar to the parent analogue **TS11–12**, and the closed-shell rearrangement **15** → **16** is favored over the corresponding diradical open-shell pathway ($\Delta E^{\ddagger} =$ 46.1 kcal mol^{-1}, $<S^2> = 0.97$)[2]. In addition, *s-trans*-butadiene **16** can transform into the slightly more stable planar *cis*-butadiene **17** ($\Delta E = -1.5$ kcal mol^{-1}), which is now an energy minimum, with an energy barrier of only 8.2 kcal mol^{-1}.

[2] Open-shell **TS15–16** was calculated at the UQCISD(T)/6-311 + G**//UMP2/6-31G* level of theory.

Subsequently, **17** can isomerize via a conrotatory electrocyclic ring closure to the much more stable 1,2-dihydro-1,2-phosphasilete **18** ($\Delta E = -25.6$ kcal mol^{-1}) with a minute barrier of only 1.3 kcal mol^{-1}. The geometrical parameters of the optimized **18** are in good agreement with the single-crystal X-ray analysis of **10a** [23, 24].

Interestingly, the direct valence isomerization now becomes favorable, and 2-phospha-4-silabicyclo[1.1.0]butane **15** gives cyclobutene derivative **18** ($\Delta E^{\ddagger} =$ 27.7 kcal mol^{-1}) via a W–H allowed [$\sigma 2s + \sigma 2a$] process, with an exothermicity of 25.6 kcal mol^{-1} (Fig. 6).

The lower barrier for the direct conversion **15** → **18** compared to that of the parent **11** → **14** can be ascribed to the presence of the donating amino groups on silicon. Generally, π-donor (e.g., NH_2) and σ-acceptor (e.g., F) substituents destabilize three-membered rings, making them more reactive, as indicated by their increased ring strain [38, 39]. This is also evident for the **15** → **18** conversion by an increased exothermicity ($\Delta E_{\mathbf{11} \to \mathbf{14}} =$ 21.7 kcal mol^{-1}; $\Delta E_{\mathbf{15} \to \mathbf{18}} = 25.6$ kcal mol^{-1}). Additionally, the analogous rearrangement for the fluoro-substituted 2-phospha-4-silabicyclo[1.1.0]butane **19** confirms this trend ($\Delta E_{\mathbf{19} \to \mathbf{20}} = 28.1$ kcal mol^{-1}, Fig. 7). Furthermore, the associated transition state of this novel pathway is stabilized by the electron-donating *N*-heterocyclic substituent on silicon ($\Delta E^{\ddagger}_{\mathbf{11} \to \mathbf{14}} = 39.0$ kcal mol^{-1}; $\Delta E^{\ddagger}_{\mathbf{15} \to \mathbf{18}} = 27.7$ kcal mol^{-1}; $\Delta E^{\ddagger}_{\mathbf{19} \to \mathbf{20}} = 35.0$ kcal mol^{-1}).

Conclusions

Hetero substitution changes the stability of the valence isomers of bicyclo[1.1.0]butane (**1**). 2-Phospha-4-silabicyclo[1.1.0]butane (**11**) is the least stable isomer and 1,2-dihydro-1,2-phosphasilete (**14**) the most stable one at the

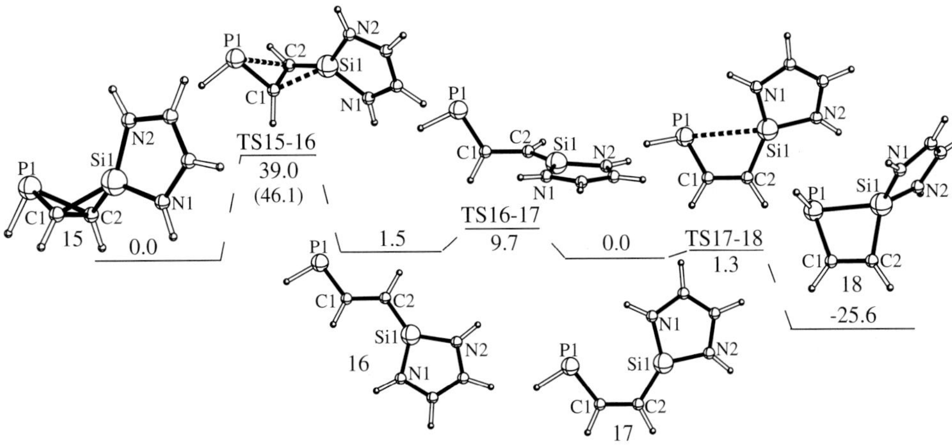

Fig. 5 Relative QCISD(T)/6-311+G**//QCISD/6-31G* (UQCISD(T)/6-311+G**//UMP2/6-31G* in parenthesis) energies (ZPE corrected, in kcal mol^{-1}) for the rearrangements of **15** into **18**. Selected bond lengths [Å], angles and torsion angles [°] of **15** (C_s): P1–C1 1.852, Si1–C1 1.823, Si1–N1 1.730, C1–C2 1.613, C1–P1–C2 51.6, C1–Si1–C2 52.5, P1–C1–C2–Si1 122.1; **TS15–16**: P1–C1 1.768, P1–C2 2.590, Si1–C1 1.977, Si1–C2 1.748, Si1–N1 1.726; **16** (C_s): P1–C1 1.718, Si1–C2 1.724, Si1–N1 1.717, C1–C2 1.434; **TS16–17**: P1–C1–C2–Si1 98.2; **17** (C_s): P1–C1 1.727, Si1–C2 1.732, Si1–N1 1.710, Si1–N2 1.717; **TS17–18**: P1–C1 1.743, P1–Si1 3.103, Si1–C2 1.765, Si1–N1 1.719, C1–C2 1.406, P1–C1–C2–Si1 24.8; **18**: P1–C1 1.867, P1–Si1 2.309, Si1–C2 1.867, Si1–N1 1.741, C1–C2 1.356

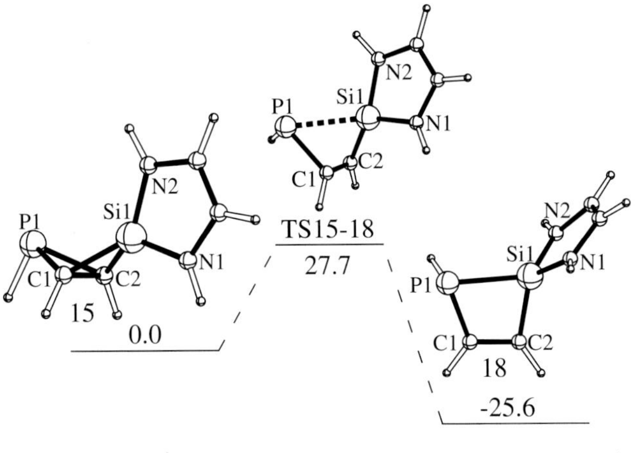

Fig. 6 Relative QCISD(T)/6-311+G**//QCISD/6-31G* energies (ZPE corrected, in kcal mol^{-1}) for the direct rearrangement of **15** into **18**. Selected bond lengths [Å] and torsion angles [°] of **TS15–18**: P1–C1 1.841, P1–Si1 2.458, Si1–C2 1.777, Si1–N1 1.742, Si1–N2 1.736, C1–C2 1.444, P1–C1–C2–Si1 78.0

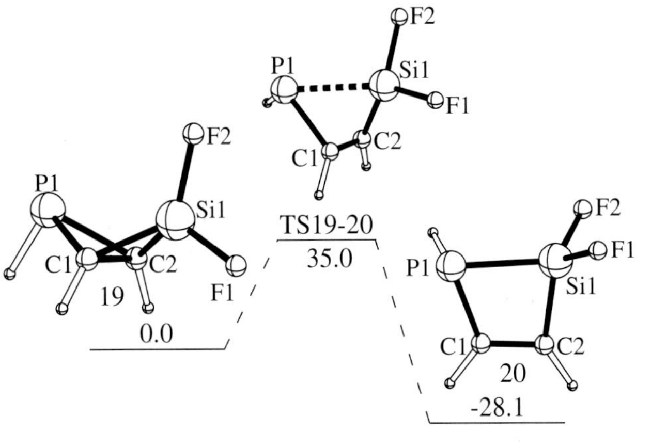

Fig. 7 Relative QCISD(T)/6-311+G**//QCISD/6-31G* energies (ZPE corrected, in kcal mol^{-1}) for the direct rearrangement of **19** into **20**. Selected bond lengths [Å] and torsion angles [°] of **19** (C_s): P1–C1 1.852, Si1–C1 1.797, Si1–F1 1.600, C1–C2 1.631, C1–P1–C2 52.2, C1–Si1–C2 54.0, P1–C1–C2–Si1 122.0; **TS19–20**: P1–C1 1.834, P1–Si1 2.383, Si1–C2 1.758, Si1–F1 1.613, Si1–F2 1.610, C1–C2 1.457, P1–C1–C2–Si1 77.5; **20**: P1–C1 1.879, P1–Si1 2.252, Si1–C2 1.841, Si1–F1 1.607, C1–C2 1.357

QCISD(T)/6-311+G**//QCISD/6-31G* level of theory [40]. Two reaction pathways for the thermal isomerization of 2-phospha-4-silabicyclo[1.1.0]butane (**11**) have been found: (a) a three-step process starting with a barrier of 38.8 kcal mol^{-1} for the concerted, asynchronous conrotatory ring opening of **11** to *s-trans*-1-phospha-4-sila-1,3-butadiene (**12**), followed by a conformational change to the *gauche* isomer **13** and a subsequent conrotatory electrocyclic ring closure to **14**, and (b) a direct transformation of **11** into **14** via a $[\sigma 2s+\sigma 2a]$ process with a barrier of 39.0 kcal mol^{-1} which becomes favorable when electron-donating substituents are present on silicon. This latter path is unprecedented for the analogous isomerization of bicyclo[1.1.0]butane.

Cartesian coordinates and energies of all stationary points are available in the electronic supplementary material.

References

1. Hoz S (1987) Bicyclo[1.1.0]butane. In: Rappoport Z (ed) The chemistry of the cyclopropyl group, Part 2, chapter 19. Wiley, Chichester
2. Nguyen KA, Gordon MS (1995) J Am Chem Soc 117:3835–3847

3. Shevlin PB, McKee ML (1988) J Am Chem Soc 110:1666–1671
4. Woodward RB, Hoffmann R (1969) Angew Chem 81:797–870
5. Woodward RB, Hoffmann R (1969) Angew Chem Int Ed 8:781–853
6. Woodward RB, Hoffmann R (1970) The conservation of orbital symmetry. Academic, New York
7. Frey HM, Stevens IDR (1965) Trans Faraday Soc 61:90–94
8. Srinivasan R, Levi AA, Haller I (1965) J Phys Chem 69:1775–1777
9. Nguyen KA, Gordon MS, Boatz JA (1994) J Am Chem Soc 116:9241–9249
10. Cooper W, Walters WD (1958) J Am Chem Soc 80:4220–4224
11. Carr RW Jr, Walters WD (1965) J Phys Chem 69:1073–1075
12. Deng L, Ziegler T (1995) J Phys Chem 99:612–618
13. Wiest O, Houk KN, Black KA, Thomas BE IV (1995) J Am Chem Soc 117:8594–8599
14. Spellmeyer DC, Houk KN (1988) J Am Chem Soc 110:3412–3416
15. Niwayama S, Kallel EA, Spellmeyer DC, Sheu C, Houk KN (1996) J Org Chem 61:2813–2825
16. Murakami M, Hasegawa M (2004) Angew Chem 116:4981–4984
17. Murakami M, Hasegawa M (2004) Angew Chem Int Ed 43:4873–4876
18. Iwamoto T, Yin D, Kabuto C, Kira M (2001) J Am Chem Soc and references therein 123:12730–12731
19. Tebby JC (2001) Bicyclic and polycyclic systems with a ring junction phosphorus atom. In: Mathey F (ed) Phosphorus-carbon heterocyclic chemistry: the rise of a new domain. Pergamon, Amsterdam, pp 683
20. Niecke E, Fuchs A, Nieger M (1999) Angew Chem 111:3213–3216
21. Niecke E, Fuchs A, Nieger M (1999) Angew Chem Int Ed 38:3028–3031
22. Jones C, Platts JA, Richards AF (2001) Chem Commun 663–664
23. Slootweg JC, de Kanter FJJ, Schakel M, Ehlers AW, Gehrhus B, Lutz M, Mills AM, Spek AL, Lammertsma K (2004) Angew Chem 116:3556–3559
24. Slootweg JC, de Kanter FJJ, Schakel M, Ehlers AW, Gehrhus B, Lutz M, Mills AM, Spek AL, Lammertsma K (2004) Angew Chem Int Ed 43:3474–3477
25. Slootweg JC, Ehlers AW, Lammertsma K (2004) Phosphorus, sulfur and silicon 179:803–807
26. Haber S, Boese R, Regitz M (1990) Angew Chem 102:1523–1525
27. Haber S, Boese R, Regitz M (1990) Angew Chem Int Ed 29:1436–1438
28. Haber S, Schmitz M, Bergsträßer U, Hoffmann J, Regitz M (1999) Chem Eur J 5:1581–1589
29. Frisch MJ, Trucks GW, Schlegel HB, Scuseria GE, Robb MA, Cheeseman JR, Zakrzewski VG, Montgomery JA Jr, Stratmann RE, Burant JC, Dapprich S, Millam JM, Daniels AD, Kudin KN, Strain MC, Farkas O, Tomasi J, Barone V, Cossi M, Cammi R, Mennucci B, Pomelli C, Adamo C, Clifford S, Ochterski J, Petersson GA, Ayala PY, Cui Q, Morokuma K, Rega N, Salvador P, Dannenberg JJ, Malick DK, Rabuck AD, Raghavachari K, Foresman JB, Cioslowski J, Ortiz JV, Baboul AG, Stefanov BB, Liu G, Liashenko A, Piskorz P, Komaromi I, Gomperts R, Martin RL, Fox DJ, Keith T, Al-Laham MA, Peng CY, Nanayakkara A, Challacombe M, Gill PMW, Johnson B, Chen W, Wong MW, Andres JL, Gonzalez C, Head-Gordon M, Replogle ES, Pople JA (2002) Gaussian 98 (Revision A.11.4). Gaussian Inc., Pittsburgh
30. Gauss J, Cremer C (1988) Chem Phys Lett 150:280–286
31. Lee TJ, Rendell AP, Taylor PR (1990) J Phys Chem 94:5463–5468
32. Scott AP, Radom L (1996) J Phys Chem 100:16502–16513
33. Okovytyy S, Gorb L, Leszczynski J (2001) Tetrahedron 57:1509–1513
34. Breulet J, Lee TJ, Schaefer HF III (1984) J Am Chem Soc 106:6250–6253
35. Bock CW, Panchenko YN (1989) J Mol Struct 187:69–82
36. Kuchitsu K, Fukuyama T, Morino Y (1968) J Mol Struct 1:463–479
37. Bak B, Led JJ, Nygaard L, Rastrup-Andersen J, Sørensen GO (1969) J Mol Struct 3:369–378
38. Bach RD, Dmitrenko O (2002) J Org Chem 67:2588–2599
39. Cremer D, Kraka E (1985) J Am Chem Soc 107:3811–3819
40. Driess M, Pritzkow H, Rell S, Janoschek R (1997) For $P_2Si_2H_4$ the diphospha-disilabicyclo[1.1.0]butane isomers are the most stable ones. Inorg Chem 36:5212–5217

J Mol Model (2006) 12: 537–542
DOI 10.1007/s00894-005-0042-6

ORIGINAL PAPER

Hai-Shun Wu · Haijun Jiao

Boron nitride cages from $B_{12}N_{12}$ to $B_{36}N_{36}$: square–hexagon alternants vs boron nitride tubes

Received: 1 March 2005 / Accepted: 27 July 2005 / Published online: 29 April 2006

Abstract The structures and stabilities of square–hexagon alternant boron nitrides (B_xN_x, x = 12–36) vs their tube isomers containing octagons, decagons and dodecagons have been computed at the B3LYP density functional level of theory with the correlation-consistent cc-pVDZ basis set of Dunning. It is found that octagonal $B_{20}N_{20}$ and $B_{24}N_{24}$ tube structures are more stable than their square–hexagon alternants by 18.6 and 2.4 kcal mol^{-1}, respectively, while the square–hexagon alternants of other cages are more stable. Trends in stability as a function of cluster size are discussed.

Keywords DFT · Boron nitrides · Fullerene · Structure · Stability

Introduction

Boron nitrides B_xN_x have been studied extensively since they are the isoelectronic analogs of the carbon fullerenes. Boron nitride molecules [1, 2] and nanotubes [3–6] have been synthesized and/or characterized. For small molecules (x = 3–10), ring isomers are the most stable form of boron nitride [7, 8]. For $x > 10$, cages consisting of three-coordinate networks of boron and nitrogen are the most stable species [9, 10]. A wide variety of cage isomers exists, and the various classes of cages have been studied to determine which cages are the most stable. For example, fullerene-like cages with pentagons and hexagons in the cage network have been compared to cages with squares and hexagons containing alternate B–N bonds (alternant cage). Theoretical calculations comparing these two classes of molecules predict that the square–hexagon alternants are more stable than fullerene-like molecules for $x = 12$ [11], x = 13, 14 and 16 [12] and $x = 24$ [13]. The energetic penalty associated with the pentagons and their inevitable non-alternate B–B and N–N bonds is the determining factor in the relative stability of these two classes. The most stable cages should have only polygons with an even number of atoms, so that full alternation of the boron and nitrogen atoms can take place, resulting in all the bonds being B–N bonds.

However, even-sided polygons include polygons larger than squares and hexagons, and molecules with octagons and larger even-sided polygons have been studied as well. For $x = 24$, Pokropivny et al. [14] predicted that the most stable spheroidal structure has 12 squares, 8 hexagons and 6 octagons. More recently, a $B_{24}N_{24}$ cage with 8 squares, 16 hexagons and 2 octagons was shown [13] to be the most stable, but only slightly more stable than a square–hexagon alternant (6 squares and 20 hexagons).

An interesting result from the most recent study on $B_{24}N_{24}$ [13] is that the most stable octagon-containing molecule is an octagonal tube with exactly two octagons, more specifically with the octagons at each end of a tube consisting of squares and hexagons. Each such tube has a C_4 axis of symmetry down the center of the molecule. In the current study, a stability comparison is made between these tube structures and alternant cages composed entirely of squares and hexagons. However, such tubes exist only if $x = 4n$, where n is a positive integer. For molecule sizes where $x = 4n+2$, the tube will be capped on one end by an additional ring of four alternating B and N atoms. Over a range of molecule sizes from $x = 12$ to 36, calculations are carried out to determine how the stability relationship between these two

Dedicated to Professor Dr. Paul von Ragué Schleyer on the occasion of his 75th birthday.

H.-S. Wu (✉) · X.-Y. Cui · X.-F. Qin
Department of Chemistry, Shanxi Normal University,
Linfen, 041004, China
E-mail: wuhs@dns.sxtu.edu.cn

D. L. Strout
Department of Physical Sciences, Alabama State University,
Montgomery, AL 36101, USA
E-mail: dstrout@alasu.edu

H. Jiao
Leibniz-Institut für Organische Katalyse, Universität Rostock,
Albert-Einstein-Strasse 29a, 18059 Rostock, Germany
E-mail: haijun.jiao@ifok-rostock.de

classes of molecules varies with the size of the molecules. For each molecule size x, if F_4, F_6 and F_8 represent the number of squares, hexagons and octagons, respectively, the molecules are subject to the following mathematical bounds:

$$F_4 - F_8 = 6 \text{ (Euler's theorem)} \quad (1)$$

$$F_4 + F_6 + F_8 = x + 2 \quad (2)$$

The alternant cages have $F_4 = 6$ and $F_8 = 0$, whereas the fully open tube structures have $F_4 = 8$ and $F_8 = 2$. The capped tube structures have $F_4 = 7$ and $F_8 = 1$, the one octagon on the uncapped end. Further, tube structures are chosen such that each octagon is adjacent to four squares that surround it in a fourfold symmetry (capped tubes only have a twofold symmetry overall because of the structure of the cap). Such square–octagon adjacency has been shown [13] to have a stabilizing influence on the molecules. Trends in stability will be discussed over the range of all even x from 12 to 36. The molecule size $x = 14$ is excluded because no square–hexagon isomer exists for $B_{14}N_{14}$ without edge-sharing squares.

Also, the square–hexagon alternants will be compared to decagonal and dodecagonal tube-shaped molecules, at molecule sizes where such tube molecules exist. The decagonal tube has a ten-membered ring and five adjacent squares at each end, and the structure and energy of decagonal tubes will be calculated for $x = 20$ and 30 (sizes at which the number of atoms is a multiple of ten). Likewise, the structure and energy of dodecagonal tubes, which have a 12-membered ring and six adjacent squares at each end, are calculated for $x = 18$, 24, 30 and 36 (sizes at which the number of atoms is a multiple of 12). Trends of stability for decagonal and dodecagonal tubes will be discussed as a function of cluster size.

Computational details

Geometries of all the molecules in this study were optimized with the B3LYP density functional method [15, 16]. The basis set [17] was the correlation-consistent double-zeta (cc-pVDZ) set of Dunning. All optimized geometries were characterized as energy- minimum structures by frequency analysis. All calculations were carried out with the Gaussian 98 quantum chemistry software package [18].

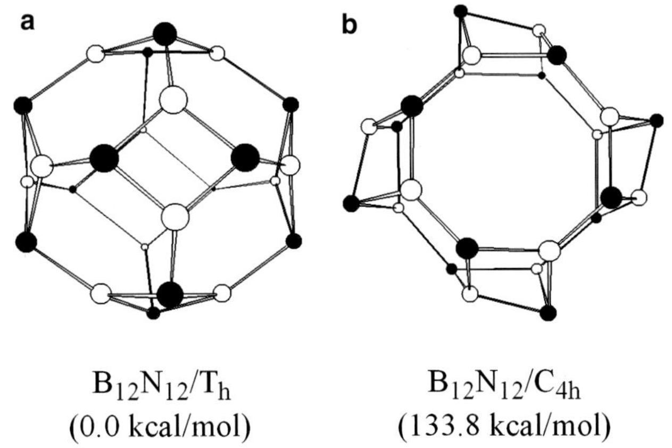

Fig. 1 $B_{12}N_{12}$ molecules: (a) square–hexagon alternant; (b) octagonal tube

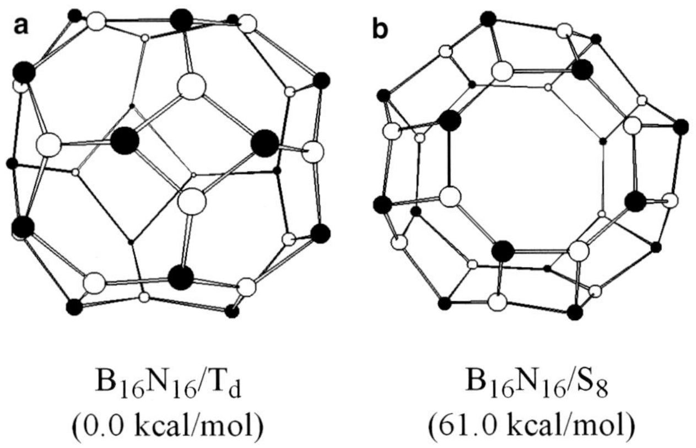

Fig. 2 $B_{16}N_{16}$ molecules: (a) square–hexagon alternant; (b) octagonal tube

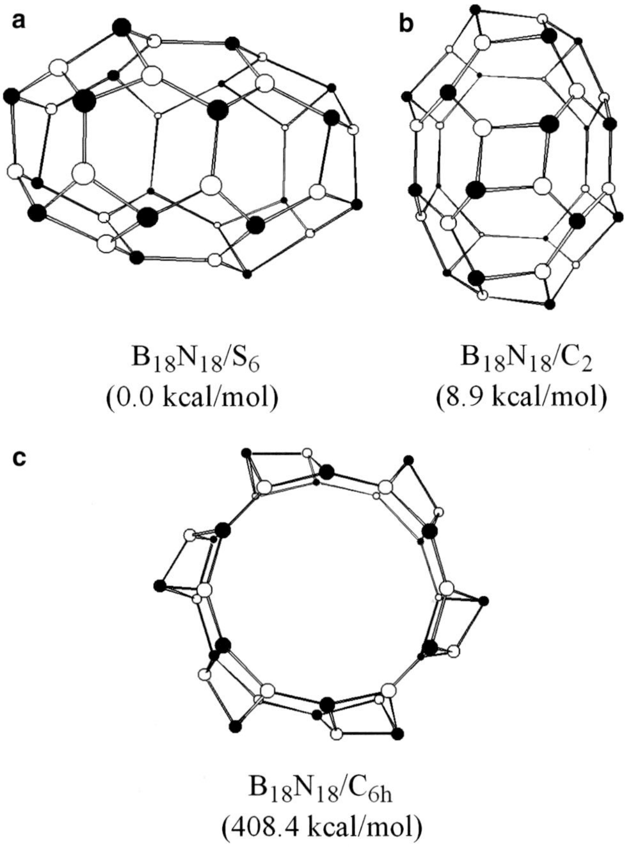

Fig. 3 $B_{18}N_{18}$ molecules: (a) square–hexagon alternant; (b) capped octagonal tube; (c) dodecagonal tube

Results and discussion

Twelve different molecule sizes were considered in this study, and the structures of the molecules are shown in Figs. 1, 2, 3, 4, 5, 6, 7, 8, 9, 10, 11 and 12, respectively, for B_xN_x molecules with $x = 12$, 16, 18, 20, 22, 24, 26, 28,

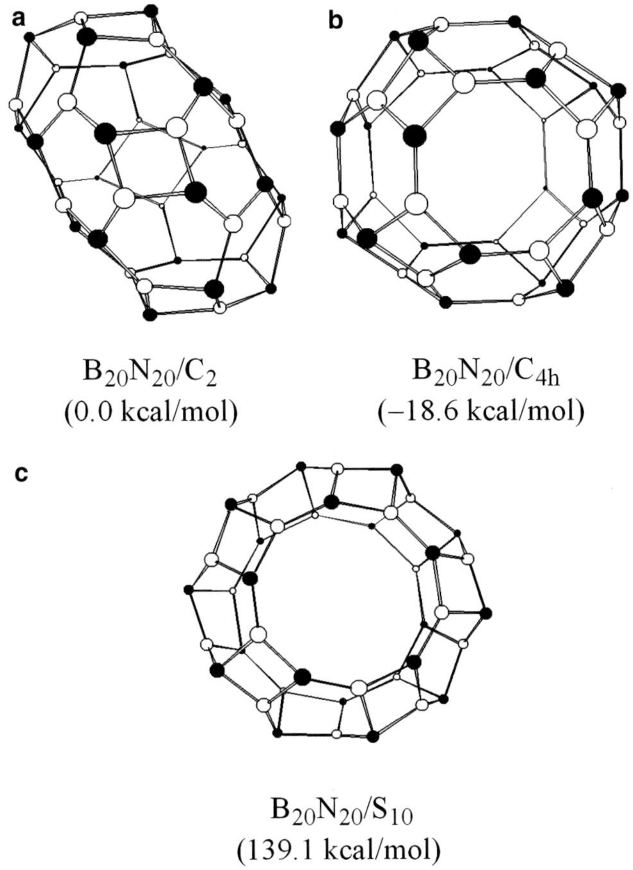

Fig. 4 $B_{20}N_{20}$ molecules: (a) square–hexagon alternant; (b) octagonal tube; (c) decagonal tube

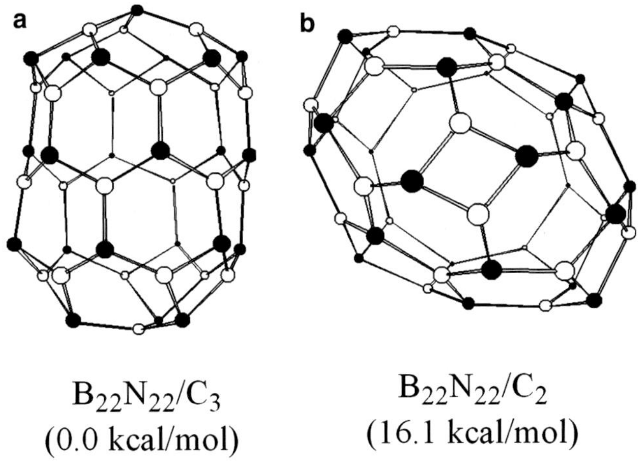

Fig. 5 $B_{22}N_{22}$ molecules: (a) square–hexagon alternant; (b) capped octagonal tube

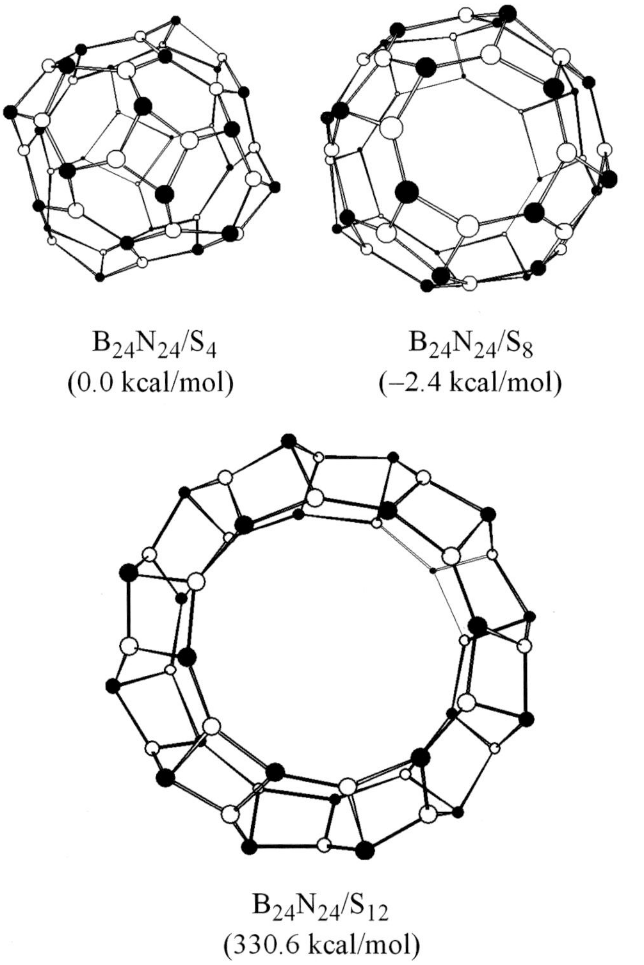

Fig. 6 $B_{24}N_{24}$ molecules: (a) square–hexagon alternant; (b) octagonal tube; (c) dodecagonal tube

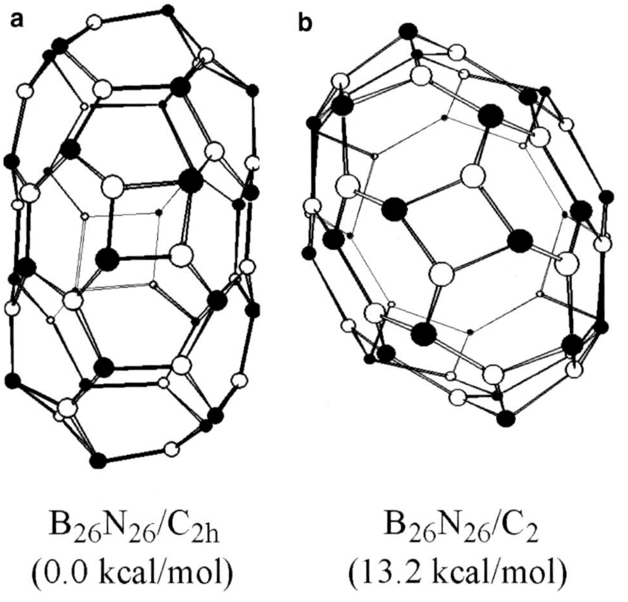

Fig. 7 $B_{26}N_{26}$ molecules: (a) square–hexagon alternant; (b) capped octagonal tube

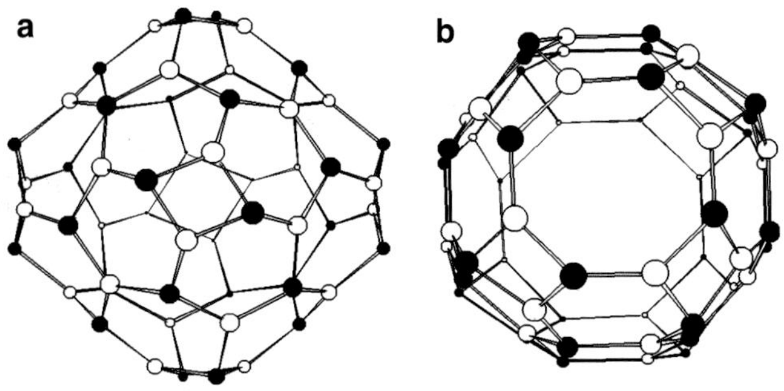

$B_{28}N_{28}/T$
(0.0 kcal/mol)

$B_{28}N_{28}/C_{4h}$
(28.2 kcal/mol)

Fig. 8 $B_{28}N_{28}$ molecules: (a) square–hexagon alternant; (b) octagonal tube

30, 32, 34 and 36. Total energies for the square–hexagon alternants and tubes are given in Table 1, and their relative energies in Table 2. From the data, the following general trends are evident.

Octagonal tubes

For $x=12$, the octagonal tube is unstable by 133.8 kcal mol^{-1} above the square–hexagon alternant because the eight squares form four edge-sharing pairs (Fig. 1), which is a destabilizing feature that is inevitable for very short tubes. The separation of squares first occurs for $x=16$, but the octagonal tube is still 61.0 kcal mol^{-1} less stable than the square–hexagon alternant (Fig. 2). As the tube is lengthened, however, the octagonal tube becomes much more energetically competitive ($x=18$, Fig. 3), becoming more stable than the alternant at the intermediate sizes ($x=20$, 24, Figs. 4, 6). Toward the upper end of molecule sizes in this study, the octagonal tubes show signs of becoming less stable, but they are still close in energy to their square–hexagon alternants. It may be the case, however, that the gap widens again for very large sizes. The octagons do not incur a substantial energy penalty in the molecules in this study, and octagon-containing molecules are close in energy to alternants over a range of sizes.

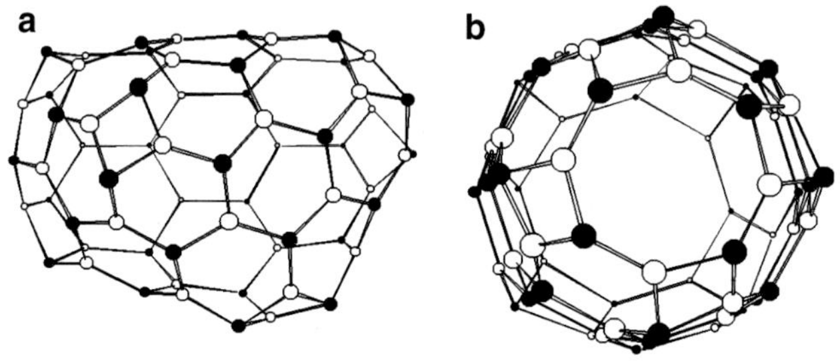

$B_{32}N_{32}/C_1$
(0.0 kcal/mol)

$B_{32}N_{32}/S_8$
(7.7 kcal/mol)

Fig. 10 $B_{32}N_{32}$ molecules: (a) square–hexagon alternant; (b) octagonal tube

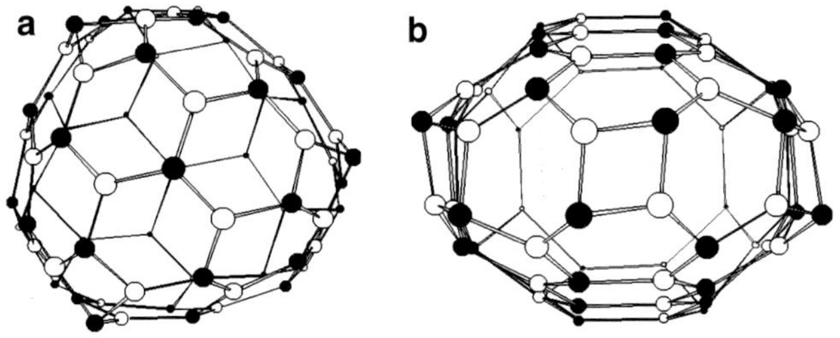

$B_{34}N_{34}/C_3$
(0.0 kcal/mol)

$B_{34}N_{34}/C_2$
(49.0 kcal/mol)

Fig. 11 $B_{34}N_{34}$ molecules: (a) square–hexagon alternant; (b) capped octagonal tube

Capped octagonal tubes

Capping one end of the octagonal tube with a square (to produce molecules with $x=4n+2$ as opposed to $4n$) also does not incur an energetic penalty for the molecules. Capped tubes, with a single octagon, are very close in energy to the square–hexagon alternants over a wide

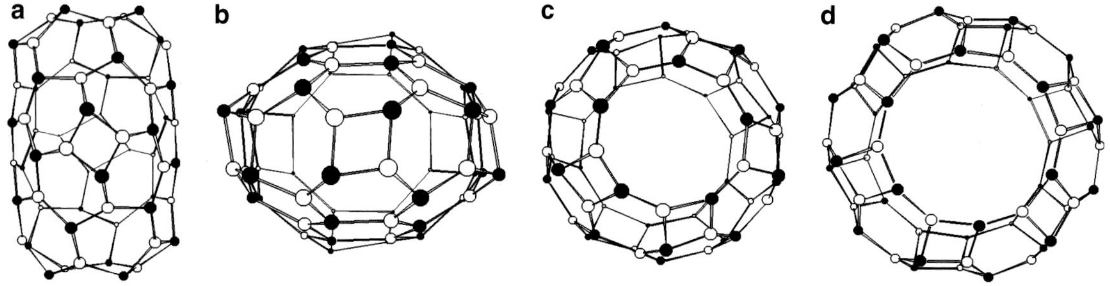

$B_{30}N_{30}/C_2$
(0.0 kcal/mol)

$B_{30}N_{30}/C_2$
(21.3 kcal/mol)

$B_{30}N_{30}/S_{10}$
(54.0 kcal/mol)

$B_{30}N_{30}/C_{6h}$
(226.0 kcal/mol)

Fig. 9 $B_{30}N_{30}$ molecules: (a) square–hexagon alternant; (b) capped octagonal tube; (c) decagonal tube; (d) dodecagonal tube

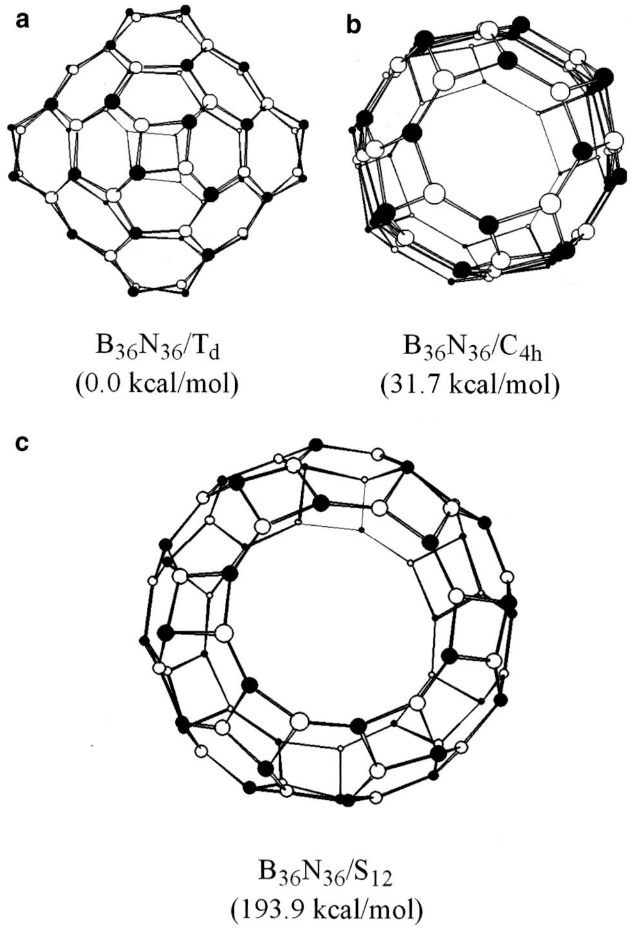

Fig. 12 $B_{36}N_{36}$ molecules: (a) square–hexagon alternant; (b) octagonal tube; (c) dodecagonal tube

range. For $x = 18$, 22, 26 and 30, the capped octagonal tubes are within 9–22 kcal mol^{-1} of the square–hexagon alternants. The capped octagonal tube is not as stable for $x = 34$, which may be a special case since the smaller sizes do not show a strong trend upward in energy for the capped octagonal tubes. For $B_{28}N_{28}$, the square–hexagon alternant structure (Fig. 8) is the most stable isomer, and structures with one or more octagons are higher in energy among ten isomers containing squares, hexagons and octagons [19]. For $B_{32}N_{32}$, the square–hexagon alternant structure (Fig. 10) is the most stable isomer, and other isomers are higher in energy [20].

Decagonal and dodecagonal tubes

The decagonal tubes for $x = 20$ and 30 show much more favorable stability for the longer $x = 30$ tube relative to the respective square–hexagon alternants as these molecule sizes. However, the $x = 30$ decagonal tube is still over 50 kcal mol^{-1} higher in energy than the corresponding square–hexagon alternant, so that studies of $x = 40$ and 50 would probably be required to establish whether these tubes become energetically competitive with molecules composed entirely of squares and hexagons. The dodecagonal tubes also show a trend of increasing stability with increasing length, and are very much higher in energy at all molecule sizes relative to the square–hexagon alternants. Even for $x = 30$ and 36, these tubes are about 200 kcal mol^{-1} above the square–hexagon molecules, with no sign of significant stabilization. The dodecagonal tubes would likely remain high in energy even at sizes larger than the range studied here.

Conclusion

Boron nitride cages are the most stable if all the polygons have an even number of atoms, but even within that framework, several types of molecules exist with comparable stability, as found previously. Squares and hexagons are not the only structural features that lead to stable molecules. A systematic investigation of square–hexagon alternants and tubes of various diameters reveals that octagonal tubes are comparable in stability

Table 1 B3LYP/cc-pvDZ total energies (E_{tot}, au) for B_xN_x ($x = 12$–36)

B_xN_x	Symmetry	E_{tot}	B_xN_x	Symmetry	E_{tot}
$B_{12}N_{12}$ (1a)	T_h	−956.18558	$B_{26}N_{26}$ (7a)	C_{2h}	−2072.26843
$B_{12}N_{12}$ (1b)	C_{4h}	−955.97237	$B_{26}N_{26}$ (7b)	C_2	−2072.24745
$B_{16}N_{16}$ (2a)	T_d	−1275.08805	$B_{28}N_{28}$ (8a)	T	−2231.77243
$B_{16}N_{16}$ (2b)	S_8	−1274.99081	$B_{28}N_{28}$ (8b)	C_{4h}	−2231.72748
$B_{18}N_{18}$ (3a)	S_6	−1434.47716	$B_{30}N_{30}$ (9a)	C_2	−2391.17447
$B_{18}N_{18}$ (3b)	C_2	−1434.46296	$B_{30}N_{30}$ (9b)	C_2	−2391.14047
$B_{18}N_{18}$ (3c)	C_{6h}	−1433.82626	$B_{30}N_{30}$ (9c)	S_{10}	−2391.08837
$B_{20}N_{20}$ (4a)	C_2	−1593.90192	$B_{30}N_{30}$ (9d)	C_{6h}	−2390.81435
$B_{20}N_{20}$ (4b)	C_{4h}	−1593.93149	$B_{32}N_{32}$ (10a)	C_1	−2550.63859
$B_{20}N_{20}$ (4c)	S_{10}	−1593.68023	$B_{32}N_{32}$ (10b)	S_8	−2550.62638
$B_{22}N_{22}$ (5a)	C_3	−1753.38576	$B_{34}N_{34}$ (11a)	C_3	−2710.11515
$B_{22}N_{22}$ (5b)	C_2	−1753.36011	$B_{34}N_{34}$ (11b)	C_2	−2710.03702
$B_{24}N_{24}$ (6a)	S_4	−1912.83071	$B_{36}N_{36}$ (12a)	T_d	−2869.57670
$B_{24}N_{24}$ (6b)	S_8	−1912.83454	$B_{36}N_{36}$ (12b)	C_{4h}	−2869.52613
$B_{24}N_{24}$ (6c)	S_{12}	−1912.29521	$B_{36}N_{36}$ (12c)	S_{12}	−2869.26770

Numbers given in the parentheses denote figure numbers

Table 2 Relative energies (kcal mol^{-1}) of the boron nitride molecules, B_xN_x at the B3LYP/cc-pvDZ level of theory

B_xN_x	Square–hexagon alternant	Octagonal tube	Capped octagonal tube	Decagonal tube	Dodecagonal tube
$B_{12}N_{12}$	0.0 (1a)	+133.8 (1b)			
$B_{16}N_{16}$	0.0 (2a)	+61.0 (2b)			
$B_{18}N_{18}$	0.0 (3a)		+8.9 (3b)		+408.4 (3c)
$B_{20}N_{20}$	0.0 (4a)	-18.6 (4b)		+139.1 (4c)	
$B_{22}N_{22}$	0.0 (5a)		+16.1 (5b)		
$B_{24}N_{24}$	0.0 (6a)	-2.4 (6b)			+336.0 (6c)
$B_{26}N_{26}$	0.0 (7a)		+13.2 (7b)		
$B_{28}N_{28}$	0.0 (8a)	+28.2 (8b)			
$B_{30}N_{30}$	0.0 (9a)		+21.3 (9b)	+54.0 (9c)	+226.0 (9d)
$B_{32}N_{32}$	0.0 (10a)	+7.7 (10b)			
$B_{34}N_{34}$	0.0 (11a)		+49.0 (11b)		
$B_{36}N_{36}$	0.0 (12a)	+31.7 (12b)			+193.9 (12c)

Numbers given in the parentheses denote figure numbers

to square–hexagon alternant over a wide range of molecule sizes. Larger tubes are not as stable, however. Tubes with a decagonal structure are less stable than octagonal tubes, and the decagonal tubes will be comparable in stability to alternants only at large sizes or perhaps not at all. The dodecagonal tubes are very unstable relative to square–hexagon alternants and octagonal tubes across the range of sizes in this study and, in all likelihood, at larger sizes as well. The general trend is toward decreasing tube stability with increasing tube diameter. Therefore, tubes based on openings larger than 12-membered rings are likely even less stable.

Acknowledgments D.L.S. acknowledges the Alabama Supercomputer Authority for a grant of computer time on the Cray SV1 and SGI Altix operated in Huntsville, AL and the National Institutes of Health (NIH/NCMHD 1P20MD000547-01) for support. We also thank the Natural Science Foundation of China (20471034) for support of this work.

References

1. Jensen F (1993) Chem Phys Lett 209:417–422
2. Silaghi-Dumitrescu I, Lara-Ochoa F, Bishof P, Haiduc I (1996) J Mol Struct (Theochem) 367:47–54
3. Chopra NG, Luyken RJ, Cherrey K, Crespi VH, Cohen ML, Louie SG, Zettl A (1995) Science 269:966–967
4. Terrones M, Hsu WK, Terrones H, Zhang JP, Ramos S, Hare JP, Castillo R, Prassides K, Cheetham AK, Kroto HW, Walton DRM (1996) Chem Phys Lett 259:568–573
5. Loiseau A, Williame F, Demoncy N, Hug G, Pascard H (1996) Phys Rev Lett 76:4733–4740
6. Saito Y, Maida M (1999) J Phys Chem A 103:1291–1293
7. Martin JML, El-Yazal J, Francois JP, Gijbels R (1995) Chem Phys Lett 232:289–294
8. Martin JML, El-Yazal J, Francois JP (1996) Chem Phys Lett 248:95–101
9. Strout DL (2000) J Phys Chem A 104:3364–3366
10. Strout DL (2001) J Phys Chem A 105:261–263
11. Wu HS, Xu XH, Zhang FQ, Jiao H (2003) J Phys Chem A 107:6609–6612
12. Strout DL (2004) Chem Phys Lett 383:95–98
13. Wu HS, Jiao H (2004) Chem Phys Lett 386:369–372 (This relationship has been confirmed further: Zope RR, Baruah T, Pederson MR, Dunlap BI (2004) Chem Phys Lett 394:300–304)
14. Pokropivny VV, Skorokhod VV, Oleinik GS, Kurdyumov AV, Bartnitskaya TS, Pokropivny AV, Sisonyuk AG, Sceichenko DM (2000) J Solid State Chem 154:214–222
15. Becke AD (1993) J Chem Phys 98:5648–5652
16. Lee C, Yang W, Parr RG (1988) Phys Rev B 37:785–789
17. Dunning TH Jr (1989) J Chem Phys 90:1007–1023
18. Frisch MJ, Trucks GW, Schlegel HB, Scuseria GE, Robb MA, Cheeseman JR, Zakrzewski VG, Montgomery JA, Stratmann RE, Burant JC, Dapprich S, Millam JM, Daniels AD, Kudin KN, Strain MC, Farkas O, Tomasi J, Barone V, Cossi M, Cammi R, Mennucci B, Pomelli C, Adamo C, Clifford S, Ochterski J, Petersson GA, Ayala PY, Cui Q, Morokuma K, Malick DK, Rabuck AD, Raghavachari K, Foresman JB, Cioslowski J, Ortiz JV, Stefanov BB, Liu G, Liashenko A, Piskorz P, Komaromi I, Gomperts R, Martin R.L, Fox DJ, Keith T, Al-Laham MA, Peng CY, Nanayakkara A, Gonzalez C, Challacombe M, Gill PMW, Johnson BG, Chen W, Wong MW, Andres JL, Head-Gordon M, Replogle ES, Pople JA (1998) Gaussian 98. Gaussian Inc, Pittsburgh, PA
19. Wu HS, Cui XY, Qin XF, Jiao H (2005) J Mol Struct (Theochem) 714:153–155
20. Wu HS, Cui XY, Xu XH (2005) J Mol Struct (Theochem) 717:107–109

J Mol Model (2006) 12: 543–550
DOI 10.1007/s00894-005-0043-5

ORIGINAL PAPER

Mario J. F. Calvete · Danilo Dini · Michael Hanack
Juan Carlos Sancho-García · Weizhe Chen · Wei Ji

Synthesis, DFT calculations, linear and nonlinear optical properties of binuclear phthalocyanine gallium chloride

Received: 3 March 2005 / Accepted: 27 June 2005 / Published online: 8 November 2005

Abstract The axially substituted binuclear GaCl/GaCl phthalocyanine **1** with an unsymmetrical pattern of substitution has been prepared and its nonlinear optical (NLO) properties determined. The resulting binuclear complex retains approximately the same transition energies of monomeric $(RO)_8PcGaCl$ as far as the linear optical spectrum is concerned, although **1** has a double concentration of central atoms per molecule and an enlarged conjugated ligand. The lack of significant spectral shifts in passing from mononuclear to binuclear complexes has been rationalized theoretically by means of density functional theory calculations. The purpose of the present study is to determine whether binuclearity affects the optical limiting behavior of **1** with respect to monomeric $(RO)_8PcGaCl$ in the NLO regime determined by nanosecond laser pulses.

Keywords Phthalocyanine · Binuclear · Density functional theory · Aromaticity · Nonlinear optics

Introduction

Phthalocyanines (Pcs) are complexes with an extended conjugated network of π-electrons. They present in their structure a central cavity large enough to coordinate almost every positively charged atom of the periodic table [1]. This class of compounds is largely used as active material in various technologies [2–8] mostly because of fast blue-green color due to the HOMO/LUMO and HOMO-1/LUMO transitions [9–11], which have energy values in the broad ranges 1.5–2.0 and 2.5–4.0 eV, respectively [12, 13]. In the UV/vis absorption spectra of Pcs these transitions originate the typical Q- and B-bands within the approximate wavelength ranges 650–800 and 300–450 nm, respectively. The occurrence of broad absorption ranges is a consequence of the availability of a great variety of Pcs structures [14, 15] differing in the number, the nature and the position of substituents [16–20], the nature of the central atoms [12], the nature of the axial ligands [21,22], and the number and linkage between Pc-rings [23–25]. The modulation of the spectroscopic features of Pcs can be thus afforded since it is known that the number and the energy of allowed transitions in the UV/vis spectra of Pcs change with their structure through the modification of the molecular symmetry [26–33], the extent of electronic conjugation [23, 34–37] and the state of aggregation [38–41].

In the present paper, we explore the influence of the enlargement of the π-electron system on the linear and nonlinear optical (NLO) properties of a fused axially substituted binuclear Pc [42–44] with GaCl as central moieties (**1**, Fig. 1). The interest for this type of dimeric structure arises from the opportunity of dealing with condensed Pc-based systems that possess intermediate electronic properties in comparison to separated Pc molecules [36, 37, 42, 43, 45–50], and have double the concentration of central atoms and Pc rings per single molecular unit (Fig. 1). These structural features can be particularly appealing if GaCl is present as central unit when the NLO property of optical limiting (OL) is considered [51–54]. We have previously prepared and characterized a similar axially

Dedicated to Professor Dr. Paul von Ragué Schleyer on the occasion of his 75th birthday.

M. J. F. Calvete · D. Dini · M. Hanack (✉)
Institut für Organische Chemie, Universität Tübingen,
Auf der Morgenstelle 18, 72076 Tübingen, Germany
E-mail: hanack@uni-tuebingen.de
Tel.: +49-7071-2972432
Fax: +49-7071-295268

J. C. Sancho-García
Departamento de Química-Física, Universidad de Alicante,
03080 Alicante, Spain

W. Chen · W. Ji
Department of Physics, National University Singapore,
Science Drive 4, 117542 Singapore, Singapore

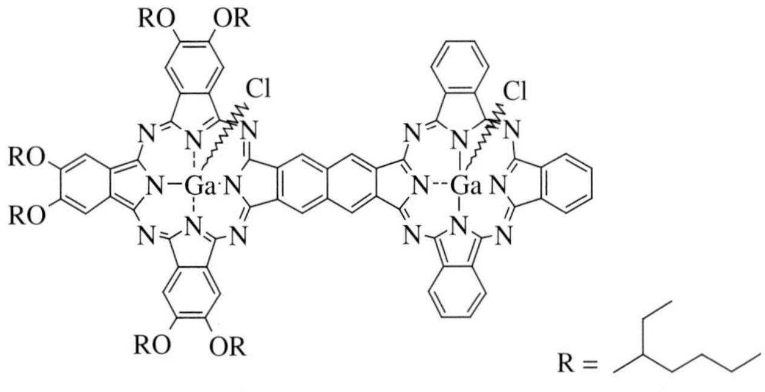

Fig. 1 Structure of binuclear fused GaCl/GaCl phthalocyanine **1**

substituted binuclear Pc with InCl as central moiety [55]. Due to the lack of a significant alteration of the electronic, linear and NLO properties in passing from dimeric side-by-side (InCl)Pc/Pc(InCl) to monomeric PcInCl [55], we decided to focus our attention towards an axially substituted binuclear complex coordinated by a smaller central unit, like GaCl. This is because the distortion of the ligand associated with the coordination of such a smaller central unit is reduced and, consequently, it is expected that the extension of the ligand in binuclear (GaCl)Pc/Pc(GaCl) **1** should lead to more relevant changes in the optical and electronic properties of **1** with respect to the analogous (InCl)Pc/Pc(InCl). The main reason for the choice of an unsymmetrical pattern of substitution with one unsubstituted ring in Pc **1** is due to the relative easiness of separation and purification in the process of synthesis (see Scheme 1).

In order to evaluate the electronic properties of the axially-substituted binuclear fused phthalocyaninato gallium chloride **1**, we have also characterized the extent of its π-delocalization by referring to the harmonic-oscillator model of aromaticity (HOMA) index [56, 57].

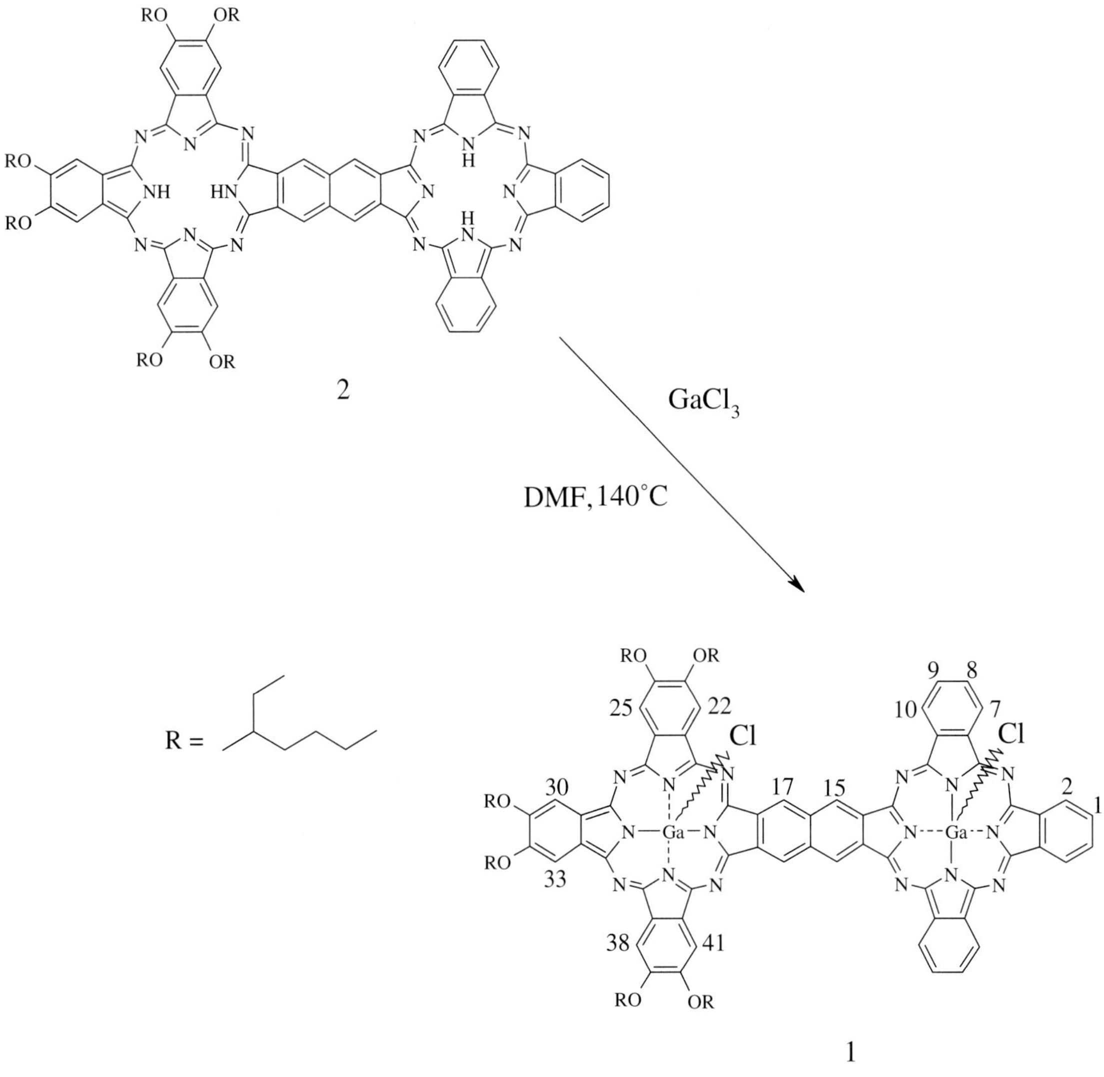

Scheme 1 Synthesis of the binuclear Pc **1**

Experimental part

Synthesis of binuclear GaCl/GaCl Pc 1 (Scheme 1)

Binuclear metal-free Pc **2** [58] (70 mg, 40 μmol) and an excess of $GaCl_3$ (90 mg, 500 μmol) were suspended in DMF. Quinoline (1 ml) was added, the mixture heated till 130°C, and then maintained at this temperature. After completion of the reaction, monitored by UV/vis spectroscopy and thin-layer chromatography for ~5 h, 20 ml of water was added dropwise to the mixture in order to precipitate the compound. After centrifugation, **1** was precipitated three times from hot methanol and dried in vacuum at 100°C. Yield: **1**, 63 mg, (79%), dark green powder. EA: Theory, C = 66.84%; H = 6.12%; N = 10.34%; Found: C = 66.02%; H = 5.51%; N = 9.19%. MS (FAB): 1975.7 [M^+], 1,478.2 [M^+-4OR + Na], 1,339.1 [M^+-5OR]. ^{1}H NMR (THF-d_8): δ = 0.90, 1.04, 1.17 1.30 [br, 36H, CH_3], 1.54, [br, 48H, CH_2], 2.04 [br, 6H, CH], 4.25, 4.52 [br, 12H, OCH_2], 6.55 [br, 2H, H-17], 6.68 [br, 2H, H-15], 7.13 [br, s, 6H, H-22, H-25, H-30, H-33, H-38, H-41], 7.92, 7.94 [br, d, 2H, H-2], 8.16, 8.19 [br, d, 2H, H-7], 8.42, 8.45 [br, d, 6H, H-9, H-10], 8.69, 8.71 [br, d, 2H, H-1, H-8]. ^{13}C NMR (THF-d_8): δ = 11.6, 11.9, 12.5, 14.4, 14.8 [CH_3], 23.4, 23.7, 24.1, 26.7, 30.3, 30.6, 31.5, 32.1, 32.9 [CH_2], 40.7, 41.0 [CH], 72.5, 72.7 [OCH_2], 104.2, 105.4, 105.8, 106.6 [C-22, C-25, C-30], 122.0, 123.4, 124.1, [C-1, C-8, C-9], 127.3 [C-15], 128.7 [C-17], 129.9, 130.2, 130.9, [C-2, C-7, C-10], 133.0–137.1 [C-3, C-6, C-11, C-14, C-16, C-18, C-21, C-26, C-29], 140.0–150.7 [C-4, C-5, C-12, C-13, C-19, C-20, C-27, C-28], 153.9, 154.6, 155.6 [C-23, C-24, C- 31]. IR (KBr): ν (cm^{-1}): 2,958, 2,927, 2,859, 1,602, 1,495, 1,457, 1,383, 1,276, 1,201, 1,098, 1,048, 892, 857, 740.7, 565. UV/vis (CH_2Cl_2): λ_{max} = 718.5, 687.5, 654.5, 344.0 nm.

Computational methods

The theoretical evaluation of aromaticity in the axially substituted binuclear Pc **1** used an approach that makes use of the HOMA index [56, 57]. The HOMA index approach is characterized by the evaluation of two main components, EN and GEO, which represent the energetic and geometric contributions of the extent of dearomatization, respectively [57]. The EN term describes changes in aromatic character due to deviation of the average bond length (R_{av}) from an optimal value (R_{opt}), while the GEO term reflects the consequences of bond length alternation. The index was naturally derived, by fixing the empirical value for the constant α (257.7), to give HOMA = 0 for the hypothetical Kekulé structures of aromatic systems and 1 for the system with all bond lengths equal to R_{opt}:

$$\text{HOMA} = 1 - \left[\alpha\left(R_{opt} - R_{av}\right)^2 + \frac{\alpha}{m}\sum_i \left(R_{av} - R_i\right)^2\right] = 1 - \text{EN} - \text{GEO}, \quad (1)$$

where R_i is the length of the ith bond between nonhydrogen atoms out of the m involved in the summation. Equation 1 has also been extended to hetero-π-electron systems, using the Pauling concept of bond number to average properly the parameters involving the heteroatoms [59]. The HOMA index has become one of the most effective ways to quantify the degree of π-delocalization [60–62] over a conjugated fragment. We calculate here the molecular geometries by means of density functional theory (DFT) methods [63–65]. The geometries of compound **1**, either with the chlorine atoms in *syn* or in *anti*, and its corresponding monomer were fully optimized with the B1PW91 exchange-correlation functional, which combines the Becke exchange [66] with the PW91 correlation functional [67] in a hybrid fashion [68]. This methodology is expected to describe the delocalization of the π-electronic cloud in a reliable way while keeping an affordable computational time [69]. A larger portion of exact Hartree-Fock exchange is excluded since the molecular geometries then progressively deteriorate [70]. The cost-effective 6-31G* basis set was accordingly used owing to its good balance between accuracy and computational resources. In the calculations, the oxoalkyl substituents were all replaced by hydrogen atoms since this exchange does not significantly affect the electronic and optical properties of the resulting unsubstituted dimer. On the other hand, this exchange significantly reduces the computational cost. All calculations were performed with the Gaussian 98 program package [71].

The differences between the HOMA values for dimeric metal-free Pc **2** and the dimeric metal complexes (GaCl)Pc/Pc(GaCl) (**1**) and (Mg)Pc/Pc(Mg) (**3**) (**3** has the same structure as **1**, but with Mg as central metal in both rings) [58] are expected to be rather small, as evinced from the small range in which the wavelength values of the Q_{0-0} band vary (c.f. Table 1) [72]. Hence, we confined ourselves to the sole theoretical evaluation of the differences in aromaticity between the topoisomers of compound **1** and its corresponding monomer $(RO)_8PcGaCl$, but not to the compared analysis of the aromaticity of **2** (**3**) with the respective metal-free (Mg) complex.

NLO transmission measurement

The light source used for the determination of NLO transmission of binuclear Pc **1** was a pulsed Nd:YAG laser whose second harmonics of emission were used. The incident beam had a Gaussian profile and the pulse duration was in the range 5–10 ns. NLO transmission curves were taken with f/number = 65 using a Gaussian beam as input beam when the sample is located in correspondence of the beam focus. The choice of such slow optics was made on purpose in order to minimize heating effects during the positioning of the sample around the focal plane. Beam waist is about 40 μm in the experiments of NLO transmission determination.

Table 1 Wavelengths of Q and B bands and molar extinction coefficients (ε) for binuclear Pcs **1**–**3**

Compound	Q_{0-1}/nm [log(ε/mol^{-1} cm^{-1} L)]	Q_{0-0}/nm [log(ε/mol^{-1} cm^{-1} L)]	Q_{1-1}/nm [log(ε/mol^{-1} cm^{-1} L)]	B_{0-0}/nm [log(ε/mol^{-1} cm^{-1} L)]	B_{0-1}/nm [log(ε/mol^{-1} cm^{-1} L)]
1	718.5(sh) [4.61]	687.5 [4.85]	654.5[4.65]	–	344.0 [4.85]
2	688.0 [4.28]	673.0 [4.35]	667.0 (sh) [4.27]	–	358.0 [4.45]
3	706.0 (sh) [4.78]	676.0 [4.89]	652.0 (sh) [4.79]	432.5 [4.73]	362.0 [4.90]
$(RO)_8$PcGaCl	–	694.5 [5.24]	625.0(sh) [4.62]	441.0 [4.38]	357.5 [4.96]
$(RO)_8$PcMg	–	679.0 [5.31]	613.0(sh) [4.69]	–	360.0 [5.01]

All the spectra were recorded in CH_2Cl_2 as solvent. (R = 2-ethylhexyl)

The collection of the transmitted beam was performed in the open-aperture configuration with the same optics as for the incident beam.

The time stability of the liquid samples for NLO experiments was checked through the following three procedures: once a day check per 1 week of the linear optical spectrum of the sample; comparison of the OL measurements obtained when pulse repetition rate was 10 Hz and when the sample was irradiated with a single shot; irradiation of the sample solution with laser pulses at a fixed energy with a repetition rate of 10 Hz for 30 min and successive check of the OL properties. Solvent effects on the OL of the sample solution were studied by measuring OL of the pure solvent. Solvents for OL experiments were used without further purification (commercial spectroscopic grade). Calibration of the optical apparatus was conducted using a reference sample of C_{60} as standard. The OL properties of the different samples were characterized under the same conditions of irradiation and compared with those of the standards.

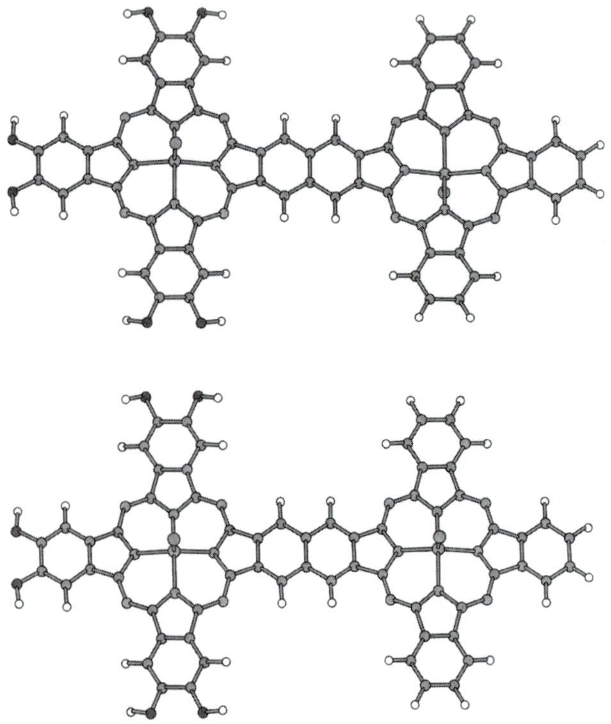

Fig. 2 Density functional theory-optimized structures of the *anti* (*top*) and *syn* (*bottom*) topoisomers of **1**

Results and discussion

The binuclear Pc **2** was prepared by a method recently developed by us [58]. From **2** we could obtain directly the binuclear Pc **1**, containing GaCl, by heating **2** and $GaCl_3$ (in excess) in DMF with a small amount of quinoline at a temperature just below the boiling point of DMF (Scheme 1). Such a procedure represents an extension of the method used for the preparation of dimeric (InCl)Pc/Pc(InCl) by replacing $InCl_3$ with $GaCl_3$ [55].

In the ^{1}H-NMR spectrum of **1**, the aromatic region (range: 6–9 ppm) shows asymmetrical peaks, which are ascribed to aggregation (see experimental). Protons H-25, H-30, H-33 and H-38 are present at 7.13 ppm (Scheme 1), while the signals at 6.55 and 6.68 ppm are attributed to H-17 and H-15, respectively. Protons H-10, H-9 and H-2 appear between 8.45 and 8.71 ppm, while the H-1 and H-8 protons are centered around 8.70 ppm.

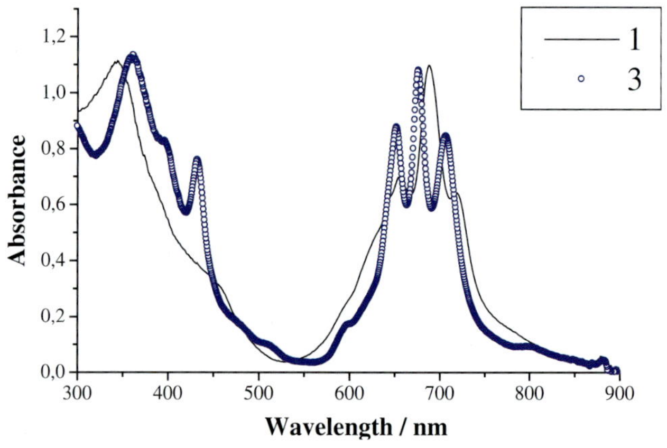

Fig. 3 UV/vis spectra of binuclear (InCl)Pc/Pc(InCl) **1** (*full line*) and (Mg)Pc/Pc(Mg) **3** (*open circles*) (cuvette thickness: 10 mm; concentrations < 1×10^{-4} M in CH_2Cl_2)

The ^{13}C-NMR spectrum of **1** shows the aromatic carbon peaks for C-22, C-25 and C-30 in the region around 105 ppm. The C-1, C-8, C-9 and C-2, C-7, C-10 carbon atoms of the unsubstituted part of the binuclear Pc are assigned around 123 and 130 ppm, respectively. Carbons C-15 and C-17 are located at 127.3 and 128.7, respectively. The alkyloxy substituted aromatic carbons C-23, C-24, C-31 are observed at about 155 ppm.

The IR spectrum of **1** resembles that of mononuclear Pcs, but the IR spectrum of **2** (metal-free binuclear Pc) shows the typical NH-stretching band at 3,298 cm^{-1} whereas **1** does not show such a band because of metalation.

From the analysis of the reported spectra we could not decide whether **1** is constituted by one isomer (with chlorines either in *syn* or *anti*), or a mixture of both forms [49]. The ordinary techniques would not allow the separation of the different topoisomers of **1**, as shown earlier with similar compounds [73].

From DFT-calculations, the *anti* conformation of **1** is predicted to be 6.35×10^{-23} J (=0.16 kcal mol^{-1}) more stable in the gas phase than the *syn* form. It must be noted that this tiny energy barrier is lower than K_BT at room temperature ($k_BT=4.11\times10^{-21}$ J at $T=298$ K being k_B the Boltzmann's constant). The steric hindrance of the *syn* isomer is the factor determining the relative stability of the two isomers, thus pointing to the existence of a subtle compromise between steric and possible π-delocalization effects in the binuclear complexes. The DFT-optimized structures are shown in Fig. 2. Note that the slight deviation from planarity for the rings of the macrocycle, driven by the presence of a coordinating central atom, might be largely reduced in the solid state due to packing effects.

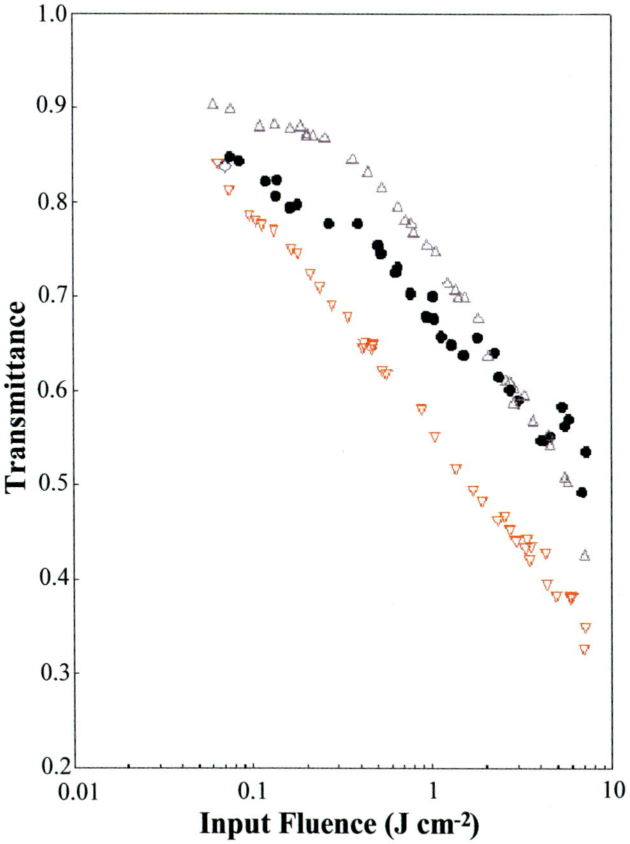

Fig. 4 Nonlinear transmission curves of binuclear Pc **1** (*up triangles*), C_{60} (*black circles*), and monomeric $(RO)_8PcGaCl$ [R=2-ethylhexyl] (*down triangles*) at 532 nm for 7 ns pulsed radiation. Cuvette thickness: 1 mm; concentrations (in g L^{-1}): 0.1, 0.80 in toluene for **1** and $(RO)_8PcGaCl$, respectively. Beam waist at the focus: 40 μm. Concentration of C_{60} solution was chosen in such a way that its linear transmission at 532 nm was $T_0=0.85$

UV/vis spectra of axially substituted binuclear Pc 1

The UV/vis maxima of binuclear Pcs **1** and **2** (in CH_2Cl_2) are given in Table 1. When comparing the optical spectra of the binuclear Pcs **1** and **3** [58] with the respective monomers $(RO)_8PcGaCl$ (R=2-ethylhexyl) and $(RO)_8PcMg$ (R=2-ethylhexyl) [58], respectively, a very weak blue-shift is observed for the more intense peak of the group of Q-bands of the binuclear systems. In other words, the two condensed Pc rings in **1** have an almost independent behavior in terms of π-electron delocalization. Actually, no mesomeric structures of compound **1** including both rings can be drawn. The shoulder of the Q-band in the UV/vis spectra of **1** (Fig. 3) shows a red-shift of the band of approximately 25 nm in the binuclear systems with respect to the mononuclear counterparts (Table 1). Similar to the trend observed with mononuclear complexes $(RO)_8PcGaCl$ and $(RO)_8PcMg$, the binuclear Pcs **1** and **3** (Fig. 1) also exhibit a red-shift in going from binuclear Mg/Mg **3** [58], to the binuclear (GaCl)Pc/Pc(GaCl) **1** (Fig. 3). The observed red-shifted shoulders are mainly due to the influence of the different central metals rather than a more extended delocalization of the conjugated electrons in the binuclear complexes. Similar results were also found by comparing the optical spectrum of monomeric $t\text{-}Bu_4PcInCl$ [74] with that of binuclear (InCl)Pc/Pc(InCl) [55].

Theoretical evaluation of aromaticity in binuclear Pc 1

The DFT-calculated EN, GEO, and HOMA values for the different isomers of **1** are shown in Table 2. The HOMA index for monomeric $(RO)_8PcGaCl$ was also computed and it results to be 0.794 at the B1PW91/6-31G* level. From the analysis of the HOMA indexes of the dimeric isomers and the monomer, it results that the *syn* isomer of **1** has the closest value (0.771) to that of the monomer (0.794) with respect to the *anti* isomer (0.715). Since there is no significant difference between the positions of the Q_{0-0} band of the monomer $(RO)_8Pc$-GaCl (694.5 nm) and the one of the dimeric compound **1** (687.5 nm) in the experimental UV/vis spectra (Table 1), a larger presence of the *syn* isomer in the actual mixture of isomers is predicted from HOMA index comparison. From calculations it is generally found a decrease in

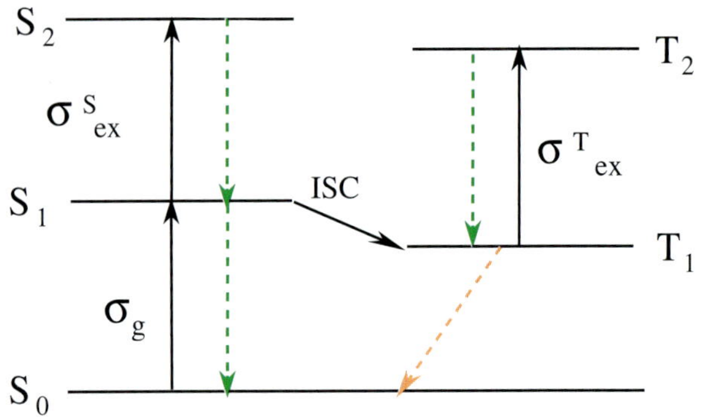

Fig. 5 Jablonski diagram for the description of the mechanism of reverse saturable absorption (*RSA*) upon short-pulse irradiation of cromophores like phthalocyanines, in the visible spectrum. S_0, $S_{1(2)}$ and $T_{1(2)}$ indicate the singlet ground state, the first (second) excited singlet state and the first (second) excited triplet state, respectively. The cromophore absorbs the first photon through the transition $[S_0 \rightarrow S_1]$, and, depending on the dynamics of the irradiated system, will absorb sequentially a second photon through either $[S_1 \rightarrow S_2]$ or $[T_1 \rightarrow T_2]$ transition. $\sigma_{ex}^{S(T)}$ and σ_g are the absorption cross sections from the excited singlet (triplet) state and the ground state, respectively. Verification of RSA implies that $\sigma_{ex}^{S(T)} > \sigma_g$. *Skew* and *straight downward arrows* indicate phosphorescence and fluorescence decays, respectively. ISC is the acronym for intersystem crossing $[S_1 \rightarrow T_1]$. For sake of clarity only the fundamental vibrational level of the various electronic levels is indicated

aromaticity in passing from the monomer to either of the dimeric isomers. This is determined by the relatively large values of the GEO terms, which are related with the bond length alternation. GEO terms result to be about twice the values of the EN terms (Table 2). Such findings imply that the dimer has rings with more localized double bonds and, therefore, is a less aromatic system with respect to the monomer. This is in accordance with the experimental observation of the red-shift of the $Q_{0\text{-}0}$ band in passing from monomeric $(RO)_8$Pc-GaCl to dimeric (GaCl)Pc/Pc(GaCl) **1** and it is theoretically confirmed by the lower HOMA values of the binuclear dimeric Pc with respect to the mononuclear system. Therefore, both the analyses of the position of the $Q_{0\text{-}0}$ absorption band and the values of HOMA index lead to the same conclusion: dimeric Pc **1** behaves almost like a monomeric Pc in terms of π-delocalization.

Nonlinear optical transmission of binuclear Pc 1

The variation of optical transmission of the axially substituted binuclear (InCl)Pc/Pc(InCl) **1** in toluene at 532 nm is shown in Fig. 4 when the incident radiation is produced by 7 ns laser pulses. For comparison, the profiles of NLO transmission under analogous conditions for mononuclear $(RO)_8$PcGaCl with R = 2-ethylhexyl [58] and C_{60} are also reported. At 532 nm the linear transmission of sample **1** is $T_0 = 0.88$. The limiting threshold T_{lim}, defined as the value of incident fluence at which the NLO transmission T of sample **1** corresponds to $T_0/2$, is 7.6 J cm^{-2}. From the trend in Fig. 4, it is found that **1** behaves as a reverse saturable absorber [75] since its solution in toluene reduces continuously its optical transmission upon increase of the incident fluence within the range 0–10 J cm^{-2}. The wavelength of analysis (532 nm) corresponds approximately to the minimum of linear optical absorption comprised between the enlarged Q and B absorption bands of **1** (Fig. 3). In correspondence of such a minimum of linear optical absorption (which is equivalent to a minimum of the ground-state absorption cross section), it is expected that the OL effect in the nanosecond pulse regime will have an optimum [74–77]. For complex **1**, the occurrence of such an NLO phenomenon at 532 nm is due to a mechanism of sequential two-photon absorption (Fig. 5). In this case an excited electronic state is formed upon absorption of the first photon, and it absorbs at the same wavelength more efficiently than the ground state [74, 76, 77]. For binuclear (GaCl)Pc/Pc(GaCl) phthalocyanine **1**, it has been observed that such an excited state absorbs light pulses with duration of several nanoseconds (pulse duration range: 3–10 ns). This implies that the lifetime of the absorbing excited state cannot be shorter than 10 ns for **1** under the experimental conditions used. If the absorbing excited state is not directly formed upon absorption of the first photon from the ground state, then the internal conversion time τ_{IC}, i.e. the time necessary for the system to be transformed into the most stable level of an excited state, must be shorter than the pulse duration, i.e. $\tau_{IC} < 3\text{–}10$ ns. Upon nanosecond pulses irradiation in Pcs, as intersystem crossing process occurs and the formation of the highly absorbing excited state is accompanied by the change of spin state after absorption of the first photon (Fig. 5) [78].

From the comparison of the OL curve of the binuclear system **1** and mononuclear $(RO)_8$PcGaCl [38] (Fig. 4) it is evident that the monomer has a stronger limiting effect when the linear transmission values are approximately the same. Such a difference is due to the higher tendency of aggregation for the dimer **1** with respect to the monomer $(RO)_8$PcGaCl, since the dimer presents an unsubstituted part, which has a somewhat stronger tendency to aggregate. As a consequence, the generation of the highly absorbing excited state is impeded by the presence of this unsubstituted half and brings about to a generally lower effectiveness of the OL effect for the dimeric complex **1**. As discussed before, an increase of electronic conjugation in dimer **1** with respect to the monomeric species due to the enlargement of the structure, is actually not verified as witnessed by the absence of red-shift in the linear absorption spectrum of **1** when compared to the monomeric counterpart.

Table 2 Density functional theory-calculated EN, GEO, and HOMA values for binuclear Pc **1**

	EN	GEO	HOMA
Syn	0.078	0.151	0.771
Anti	0.104	0.181	0.715

Therefore, it results that the OL effect from dimer **1** is a phenomenon that involves only the electronic excitation of a single Pc ring similarly to what we have previously found in the analysis of the NLO transmission of binuclear (InCl)Pc/Pc(InCl) [55].

Conclusions

The objective of the present work was the synthesis and full characterization of the binuclear axially substituted (GaCl)Pc/Pc(GaCl) **1**. Metal-free binuclear Pc **2** was the starting material for the preparation of binuclear Pc **1** containing GaCl (Scheme 1).

The theoretical evaluation of the extent of electronic delocalization and aromaticity in axially substituted binuclear Pc **1** has been accomplished by means of the harmonic oscillator model of aromaticity index. It is found that the two condensed Pc rings show almost independent behavior in terms of π-electrons delocalization, no mesomeric structures of compound **1** including both rings can be drawn. This is in agreement with the experimental spectral features of the binuclear complexes when compared to the spectrum of the mononuclear species $(RO)_8$PcGaCl. The absence of any electronic delocalization in **1** implies that this dimeric complex is equivalent to a species with a double concentration of Pc ring and central moieties per single molecular unit with no lateral interaction.

The OL effect produced by **1** was determined in the NLO regime determined by nanosecond laser pulses at 532 nm and it confirms the independent behavior of the two Pc rings. **1** behaves as a reverse saturable absorber via the mechanism of sequential two-photon (or excited state) absorption with a limiting threshold 7.6 J cm^{-2}. Binuclear Pc **1** is thus a less effective optical limiter than the monomeric system $(RO)_8$PcGaCl due to a higher tendency of aggregation for the dimer in solution.

Acknowledgements Financial support from EU project "NANO-CHANNEL" (Grant No. HPRN-CT-2002-00323) is gratefully acknowledged. The work in Alicante is supported by the "Ministerio de Educacion y Ciencia" of Spain and the "European Regional Development Fund"through project CTQ2004-06519/BQU; the regular grants of the "Vicerrectorado de Investigación" of the University of Alicante (VIGROB2004-36) are also acknowledged.

References

1. Leznoff CC, Lever ABP (eds) (1989–1996) Phthalocyanines: properties and applications, vol 1–4. VCH, Cambridge
2. Hanack M, Lang M (1994) Adv Mat 6:819–833
3. Erk P, Hengelsberg H (2003) In: Kadish KM, Smith KM, Guilard R (eds) The porphyrin handbook, vol 19. Elsevier, Amsterdam, pp 105–150
4. Simon JJ, Andre' HJ (1985) Molecular semiconductors. Springer, Berlin Heidelberg New York
5. Law KY (1993) Chem Rev 93:449–486
6. Dini D, Barthel M, Hanack M (2001) Eur J Org Chem 3759–3769
7. Hurditch R (2001) Adv Colour Sci Techn 4:33–40
8. Gregory P (1991) High technology applications of organic colorants. Plenum, New York
9. Goutermann M (1977) In: Dolphin R (ed) The porphyrins, vol 3. Academic, New York, pp 1–157
10. Stillman MJ, Nyokong T (1989) In: Leznoff CC, Lever AJP (eds) Phthalocyanines: properties and applications, vol 1. VCH, Weinheim, pp 133–257
11. Lever ABP, Pickens SR, Minor PC, Licoccia S, Ramaswamy BS, Magnell K (1981) J Am Chem Soc 103:6800–6806
12. Dini D, Hanack M (2003) In: Kadish KM, Smith KM, Guilard R (eds) The porphyrin handbook, vol 17. Elsevier, Amsterdam, pp 1–36
13. Dini D, Hanack M (2004) J Porphyrins Phthalocyanines 8:915–933
14. Hanack M, Heckmann H, Polley R (1997) In: Schaumann E (ed) Methods of organic chemistry (Houben-Weyle-Hetarenes IV: six-membered and larger hetero-rings with maximum insaturation), vol E9d. Georg Thieme Verlag, Stuttgart, pp 717–842
15. McKeown NB (2004) In: Weinreb SM (ed) Science of synthesis (Houben-Weyle, Methods of molecular transformations-hetarenes and related ring systems), vol 17. Georg Thieme Verlag, Stuttgart, pp 1237–1368
16. Cook MJ (2002) Chem Rec 2:225–236
17. Cook MJ (1994) J Mat Sci 5:117–128
18. Schmid G, Witke E, Schlick U, Knecht S, Hanack M (1995) J Mat Chem 5:855–859
19. Rager C, Schmid G, Hanack M (1999) Chem Eur J 5:280–288
20. Winter G, Heckmann H, Haisch P, Eberhardt W, Hanack M, Lueer L, Egelhaaf HJ, Oelkrug D (1998) J Am Chem Soc 120:11663–11673
21. Barthel M, Hanack M (2000) J Porphyrins Phthalocyanines 4:635–638
22. Chen Y, Subramanian LR, Barthel M, Hanack M (2002) Eur J Inorg Chem 1032–1034
23. Kobayashi N (2002) Coord Chem Rev 227:129–152
24. Cook MJ (1995) Adv Mat 7:877–880
25. Hanack M, Dini D (2003) In: Kadish KM, Smith KM, Guilard R (eds) The porphyrin handbook, vol 18. Elsevier, Amsterdam, pp 215–280
26. Kobayashi N, Nakajima SI, Ogata H, Fukuda T (2004) Chem Eur J 10:6294–6312
27. Kobayashi N, Ogata H (2004) Eur J Inorg Chem 906–914
28. Fukuda T, Ono K, Homma S, Kobayashi N (2003) Chem Lett 32:736–737
29. Kobayashi N, Mack J, Ishii K, Stillman MJ (2002) Inorg Chem 41:5350–5363
30. Kobayashi N, Miwa H, Nemykin V (2002) J Am Chem Soc 124:8007–8020
31. Kobayashi N, Konami H (2001) J Porphyrins Phthalocyanines 5:233–255
32. Kobayashi N, Muranaka A, Ishii K (2000) Inorg Chem 39:2256–2257
33. Nemykin VN, Kobayashi N, Nonomura T, Lukyanets EA (2000) Chem Lett 184–185
34. Orti E, Piqueras MC, Crespo R, Bredas JL (1990) Chem Mat 2:110–116
35. Orti E, Bredas JL, Clarisse C (1990) J Chem Phys 92:1228–1235
36. Kobayashi N, Higashi Y, Osa T (1994) J Chem Soc Chem Commun 1785–1786
37. Kobayashi N, Lam H, Nevin WA, Janda P, Leznoff CC, Koyama T, Monden A, Shirai H (1994) J Am Chem Soc 116:879–890
38. Snow AW (2003) In: Kadish KM, Smith KM, Guilard R (eds) The porphyrin handbook, vol 18. Elsevier, Amsterdam, pp 129–176
39. Monahan AR, Brado JA, De Luca AF (1972) J Phys Chem 76:1994–1996
40. George RD, Snow AW, Shirk JS, Barger WR (1998) J Porphyrins Phthalocyanines 2:1–7
41. Choi MTM, Li PPS, Ng DKP (2000) Tetrahedron 56:3881–3887

42. Calvete M, Hanack M (2003) Eur J Org Chem 2080–2083
43. Leznoff CC, Lam H, Marcuccio SM, Nevin WA, Janda P, Kobayashi N, Lever ABP (1987) J Chem Soc Chem Commun 699–701
44. Yang J, van de Mark MR (1993) Tetrahedron Lett 34:5223–5226
45. Kobayashi N, Higashi Y, Osa T (1991) J Chem Soc Chem Commun 1203–1205
46. Fukuda T, Stork JR, Potucek RJ, Olmstead MM, Noll BC, Kobayashi N, Durfee WS (2002) Angew Chem 114:2677–2680
47. De la Torre G, Martinez-Diaz MV, Torres T (1999) J Porphyrins Phthalocyanines 3:560–568
48. De la Torre G, Martinez-Diaz MV, Ashton PR, Torres T (1998) J Org Chem 63:8888–8893
49. Claessens CG, Torres T (2002) Angew Chem Int Ed 41:2561–2565
50. Ishii K, Kobayashi N, Higashi Y, Osa T, Lelievre D, Simon J, Yamauchi S (1999) Chem Commun 969–970
51. Bertagnolli H, Blau WJ, Chen Y, Dini D, Feth MP, O'Flaherty SM, Hanack M, Krishnan V (2005) J Math Chem 15:683–689
52. Chen Y, Fujitsuka M, O'Flaherty SM, Hanack M, Ito O, Blau WJ (2003) Adv Mat 15:899–902
53. Chen Y, O'Flaherty SM, Fujitsuka M, Hanack M, Subramanian LR, Ito O, Blau WJ (2002) Chem Mat 14:5163–5168
54. Chen Y, Subramanian LR, Fujitsuka M, Ito O, O'Flaherty SM, Blau WJ, Schneider T, Dini D, Hanack M (2002) Chem Eur J 8:4248–4254
55. Calvete MJF, Dini D, Flom SR, Hanack M, Pong RGS, Shirk JS (2005) Eur J Org Chem 3499–3509
56. Kruszewski J, Krygowski TM (1972) Tetrahedron Lett 13: 3839–3842
57. Krygowski TM, Cyrański M (1996) Tetrahedron 52:1713–1722
58. Calvete M (2003) PhD Thesis, University of Tuebingen
59. Krygowski TM, Cyrański M (1996) Tetrahedron 52:10255–10264
60. Krygowski TM, Cyrański M (2001) Chem Rev 101:1385–1419
61. Poater J, Fradera X, Durán M, Solá M (2003) Chem Eur J 9:1113–1122
62. Sancho-García JC, Brédas JL, Beljonne D, Cornil J, Martínez-Álvarez R, Hanack M, Poulsen L, Gierschner J, Mack HG, Egelhaaf HJ, Oelkrug D (2005) J Phys Chem B (to be published)
63. Kohn W, Sham LJ (1965) Phys Rev A 137:1697–1705
64. Lieb E (1983) Int J Quantum Chem 24:243–277
65. Parr RG, Yang W (1989) Density-functional theory of atoms and molecules. Oxford University Press, New York
66. Becke AD (1988) Phys Rev A 38:3098–3100
67. Perdew JP (1991) Electronic structure of solids. Akademie, Berlin
68. Perdew JP, Ernzenhof M, Burke K (1996) J Chem Phys 105:9982–9985
69. Sancho-García JC, Pérez-Jiménez AJ (2004) Recent Res Dev Chem Phys 5:157–171
70. Adamo C, Barone V (1997) Chem Phys Lett 274:242–250
71. Frisch MJ, Trucks GW, Schlegel HB, Scuseria GE, Robb MA, Cheeseman JR, Zakrzewski VG, Montgomery JA, Stratman RE, Burant JC, Dapprich S, Millam JM, Daniels AD, Kudin KN, Strain MC, Farkas O, Tomasi J, Barone V, Cossi M, Cammi R, Mennucci B, Pomelli C, Adamo C, Clifford S, Ochterski J, Petersson GA, Ayala PY, Cui Q, Morokuma K, Malick DK, Rabuck AD, Raghavachari K, Foresman JB, Cioslowski J, Ortiz JV, Baboul AG, Stefanov BB, Liu C, Liashenko A, Piskorz P, Komaromi I, Gomperts R, Martin RL, Fox DJ, Keith T, Al-Laham MA, Peng CY, Nanayakkara A, Gonzalez C, Challacombe M, Gill PMW, Johnson BG, Chen W, Wong MW, Andres JL, Gonzales C, Head-Gordon M, Replogle ES, Pople JA (1998) Gaussian 98. Gaussian Inc, Pittsburgh
72. Cyranski MK, Krygowski TM, Wisiorowski M, Van Eikema Hommes NJR, Von Ragué Schleyer P (1998) Angew Chem Int Ed 37:177–180
73. Hauschel B, Jung R, Hanack M (1999) Eur J Inorg Chem 693–703
74. Shirk JS, Pong RGS, Flom SR, Heckmann H, Hanack M (2000) J Phys Chem A 104:1438–1449
75. Van Stryland EW, Sheik-Bahae M, Said AA, Hagan DJ (1993) Progr Cryst Growth Charact 27:279–311
76. Perry JW, Mansour K, Lee IYS, Wu XL, Bedworth PV, Chen CT, Ng D, Marder SR, Miles P (1996) Science 273:1533–1536
77. Nalwa HS, Shirk JS (1996) In: Leznoff CC, Lever AJP (eds) Phthalocyanines: properties and applications, vol 4. VCH, Weinheim, pp 79–181
78. Turro NJ (1978) Modern molecular photochemistry. Benjamin/Cummings, Menlo Park

J Mol Model (2006) 12: 551–558
DOI 10.1007/s00894-006-0108-0

ORIGINAL PAPER

Fu-Qiang Zhang · Hai-Shun Wu · Yuan-Yuan Xu · Yong-Wang Li · Haijun Jiao

Structure and stability of neutral polyoxometalate cages: $(Mo_2O_6)_m$ (m=1–13)

Received: 4 March 2005 / Accepted: 2 December 2005 / Published online: 11 May 2006

Abstract The structure and stability of neutral polyoxometalate cages $(Mo_2O_6)_m$ (m=1–13) have been computed systematically. These neutral cages can be viewed topologically as polyhedra containing triangles (f_3) and squares (f_4). The relative stability of these polyhedra is associated with the location and separation of the f_3. The initial stable isomers were preselected by the number of shared triangle edges (N_{33}), and the predicted stability was validated further at the GGA-PW91/DND level of density function theory with the fine quality of mesh size. For large clusters, the square neighbor signature (P_{4444}), which is similar to the hexagon neighbor rule for fullerene, becomes more applicable. The calculated disproportionation energies indicate that Mo_6O_{18} (O_h, Lindqvist), $Mo_{12}O_{36}$ (O_h, α Keggin), $Mo_{18}O_{54}$ (D_{3h}, Wells–Dawson) and $Mo_{24}O_{72}$ (O_h) cages have enhanced stability.

Keywords POM · Neutral cage · Keggin · Stability · DFT

Introduction

Molybdenum and tungsten polyoxometalates (POM) constitute an extreme variety of compounds in number and diversity [1]. Their interesting and potentially valuable physical and chemical properties have made impact on catalysis, biology, medicine, and materials science [2–4]. Besides the typical architectures, such as the Lindqvist [5] anion $[M_6O_{19}]^{2-}$ (M=Mo, W), Tungstate Y [6] anion $[M_{10}O_{32}]^{x-}$, the Keggin [7] anion $[XM_{12}O_{40}]^{y-}$, and the Wells–Dawson [8] anion $[X_2M_{18}O_{62}]^{y-}$ (X=Al, P, S, As, etc), as well as a huge number of their derivatives, many other isopoly and heteropoly structures have resulted in an unmatched structural variety.

Many POM can be considered as supramolecular species with identifiable host cages and encapsulated guests [9]. One important hypothesis referred to as the *clathrate* model suggests that a Keggin anion can be viewed as an $[XO_4]^{y-}$ anion encapsulated in a neutral $M_{12}O_{36}$ cage [10]. Poblet et al. [11] have verified this model for a set of Keggin anions with density functional theory calculations. Recently, this model was also found to be applicable to Wells–Dawson anions [12]. Thus, the clathrate model enables us to study a wider range of properties of POM, despite their intrinsic multiformality, complexity, and the arbitrary encapsulated guest anions.

The shape of the framework and the composition of the metal define the POM properties. Most of the typical POM structures exhibit interpenetrating closed loops, formed by the metal center and the bridging oxygen atoms. Based on this idea, a structure stability index was developed by Nomiya and Miwa [13], and they proposed the analogy of closed loops of this type to macrocyclic π-bonding systems. However, this stability index cannot be applied to $[PM_{12}O_{40}]^{3-}$, as indicated by Bridgeman et al. [14]. With graph theory and topology as bases, King [15] studied the aromaticity of POM and pointed out that the closed loops might play a significant role in electron transfer processes. In particular, upon analyzing a set of α/β Keggin anions, Poblet et al. [11] found that the intrinsic stability of the neutral α-$M_{12}O_{36}$ cage over the β one controls stability.

Knowledge of the structure and stability of metal-oxide clusters offers valuable insight into the physical and chemical properties of condensed matter [16]. Because of the properties of similar discrete fragments of metal oxide

Dedicated to Professor Dr. Paul von Ragué Schleyer on the occasion of his 75th birthday.

F.-Q. Zhang · Y.-W. Li · H. Jiao
The State Key Laboratory of Coal Conversion,
Institute of Coal Chemistry, Chinese Academy of Sciences,
Taiyuan, People's Republic of China

F.-Q. Zhang · H.-S. Wu (✉) · Y.-Y. Xu
Department of Chemistry, Shanxi Normal University,
Linfen, People's Republic of China
e-mail: wuhs@dns.sxtu.edu.cn

H. Jiao
Leibniz-Institut für Organische Katalyse
an der Universität Rostock eV,
Albert-Einstein-Strasse 29a,
18059 Rostock, Germany

structures of definite sizes and shapes, these species can be viewed as soluble metal-oxide analogues and are therefore of special interest as models for the reactions and properties of oxides. [10, 17] Despite the abundant experimental data and considerable activity in POM research in the last 30 years [18], the basic structural properties still need to be rationalized systematically.

On the basis of the clathrate model and density functional theory computations, the stability of all neutral Keggin ($M_{12}O_{36}$) and Wells–Dawson ($M_{18}O_{54}$) (M=Mo, W) cages has been investigated systematically [19]. These cages can be viewed topologically as polyhedra, and their stabilities obey the isolated triangle rule (ITR), e.g., the more the isolated triangles, the more stable the isomer. As an extension, we report our results of all neutral POM cages $(Mo_2O_6)_m$ (m=1–13). It is found that these cages obey the ITR rule, which is similar to the isolated square rule (ISR) [20] for boron nitride (BN) inorganic fullerenes and isolated pentagon rule (IPR) [21] for carbon fullerenes. Therefore, neutral POM cages can be considered as inorganic fullerenes.

Computational methods

All calculations were performed within density functional theory using the DMol^3 program available from the Materials Studio package of Accelrys [22]. The initial structures obtained were relaxed and adjusted to their highest symmetry using the Visualizer build-in. For selecting more stable isomers, the initial structures were sorted by the number of the shared triangle edges (N_{33}) [19] and then optimized fully using the local density approximation with the Vosko–Wilk–Nusair [23] parameters, double numerical basis set (DN) and the *coarse*-quality mesh size for numerical integration. The five most stable structures obtained for each $(Mo_2O_6)_m$ were further refined using the generalized gradient corrected (GGA) functional by Perdew and Wang (PW91) [24], the DND basis set (the DN set augmented by a set of *d* polarization functions), and the *fine*-quality mesh size for numerical integration. Effect core potentials were used for metallic elements throughout the calculations. The tolerances of the final energy convergence, gradient, and the displacement convergence were 2×10^{-5} a.u., 4×10^{-4} a.u./Å, and 5×10^{-4} Å, respectively. The real space cutoff of atomic orbitals was set at 5.5 Å. Increasing the convergence criteria to 2×10^{-7} a.u., 4×10^{-6} a.u./Å, and 5×10^{-6} Å, changes the total electronic energies of the two most stable $Mo_{26}O_{78}$ isomers in D_{2d} and D_2 symmetries by no more than 0.004 kcal mol^{-1}, and their relative energies by less than 0.007 kcal mol^{-1}.

The quality of the method used (GGA-PW91/DND) has been validated on the basis of the nearly perfect agreement between the computed and estimated α/β stability, as well as the structural parameters of $[AlW_{12}O_{40}]^{5-}$ [25]. For example, the computed α/β stability of $[AlW_{12}O_{40}]^{5-}$ of 2.02 kcal mol^{-1} agrees very well with the estimated 2.1± 0.5 kcal mol^{-1} by experiment [26], and the available theoretical data (2.37 kcal mol^{-1}) [27]. The computed structural parameters also agree very well with the available X-ray data.

Results and discussion

Design principle

Due to the structural complexity of POM, it is difficult to take all conditions into consideration. In this paper, only the type I structure of Pope [28], with tractable M/O ratio (1:3), closed shell configuration, and *ideal* four-connected metal structures, was considered [29].

The Euler equation (Eq. 1) [30] holds for every convex polyhedron, where v is the number of vertices, f the number of faces, and e the number of edges. For a polyhedron of a POM cage in which each metal center and the connecting oxygen atoms can be considered topologically as a vertex, each vertex is connected with four edges and there are many possible kinds of coexisting faces, such as triangle (f_3), square (f_4), pentagon (f_5), and hexagon (f_6) faces. Their relationships are expressed in Eqs. 2, 3, and 4. Because of the absence of larger face types such as f_5 and f_6, Eq. 6 can be simplified to Eqs. 7 and 8 for small clusters of Mo_v ($v \le 26$). All topological structures can be generated easily using the plantri [31] program (which realizes the sphere quadrangulation algorithms) or the modified fullerene spiral algorithms [32].

$$v + f = e + 2 \tag{1}$$

$$4v = 2e \tag{2}$$

$$f = \sum f_i \tag{3}$$

$$\sum (i \times f_i) = 2e \tag{4}$$

Considering Eqs. 1, 2, 3, and 4, we obtained Eqs. 5 and 6

$$f = v + 2 \tag{5}$$

$$f_3 - \sum (i - 4) f_i = 8 \quad (i \ge 4) \tag{6}$$

$$f_3 = 8 \tag{7}$$

$$f_4 = v - 6 \tag{8}$$

Structure and stability

Figure 1 shows the most stable $(Mo_2O_6)_m$ structures obtained, and some selected bond distances are given in Table 1. Free from the influence of the encapsulated anions, the structural parameters obtained are very close to the available X-ray data. For $Mo_{12}O_{36}/O_h$ and $Mo_{18}O_{54}/O_h$, the calculated bond length of the metal-bridging oxygen atom (Mo–O_b) and Mo–Mo distances are in the scope of the experimental data. Similar to previous theoretical studies [11, 12, 33], the bond length of the metal terminal oxygen (M–O_t) is overestimated by about 0.03 Å on average, and Guo et al. [34] have attributed this to the absence of crystal environment, and the largest deviation of the nonbonded Mo–Mo distances in Mo_6O_{18}/O_h can be attributed to the absence of the central O^{2-}.

The stability of the most stable $(Mo_2O_6)_m$ clusters was tested using the binding energy (ΔE_m) on the basis of the dimer structure [$(Mo_2O_6)_m$, m=1], as defined in Eq. 9. The same method has been used to estimate the most stable isomers and the effect of cage sizes of $(AlN)_x$ clusters [35], $(BN)_x$ clusters [36], $(MeAlO)_x$ cages [37], and Al_x clusters [38]. The computed data are given in Table 2 and plotted in Fig. 2. One can see clearly that the binding energy increases strongly for $m \leq 6$, and this change is mainly attributed to the reduced strain with increased cage size. [36] However, the binding energy hardly changes for $m \geq 6$, indicating that the strain contribution is nearly constant. However, detailed analysis shows that cluster with m=6 has enhanced stability, and this is also found for $(AlN)_{12}$ [35], $(BN)_{12}$ [36], and $(MeAlO)_{12}$ [37].

$$\Delta E_m = E(Mo_2O_6) - E_m/m \qquad (9)$$

Table 1 Selected distances (in Ångstrom, interval values) of the most stable $(Mo_2O_6)_m$ (m=1–13) cages

$(Mo_2O_6)_m$/sym	Mo–O_t	Mo–O_b	Mo–Mo
Mo_2O_6/C_{2v}	1.719–1.720	1.935	2.857
Mo_4O_{12}/C_{4v}	1.719	1.911	3.619
Mo_6O_{18}/O_h	1.703	1.933	3.458
Expt[a]	1.67–1.69	1.90–1.94	3.28–3.29
Mo_8O_{24}/D_{4d}	1.702	1.931–1.938	3.397–3.690
$Mo_{10}O_{30}/D_2$	1.703	1.922–1.956	33.462–3.838
$Mo_{12}O_{36}/O_h$	1.701	1.923	3.693
Expt[b]	1.67–1.69	1.85–1.97	3.42–3.73
$Mo_{14}O_{42}/D_{2d}$	1.700–1.703	1.914–1.933	3.503–3.745
$Mo_{16}O_{48}/D_{4d}$	1.702	1.917–1.924	3.690–3.692
$Mo_{18}O_{54}/D_{3h}$	1.701–1.703	1.917–1.926	3.634–3.750
Expt[c]	1.672–1.689	1.902–2.024	3.432–3.764
$Mo_{20}O_{60}/D_2$	1.701–1.704	1.907–1.937	3.668–3.826
$Mo_{22}O_{66}/D_{2d}$	1.701–1.703	1.903–1.942	1.684–1.765
$Mo_{24}O_{72}/O_h$	1.702	1.915–1.923	3.762–3.763
$Mo_{26}O_{78}/D_4$	1.700–1.702	1.886–1.942	3.666–3.776

[a]X-ray data from [51]
[b]X-ray data from [52]
[c]X-ray data from [53]

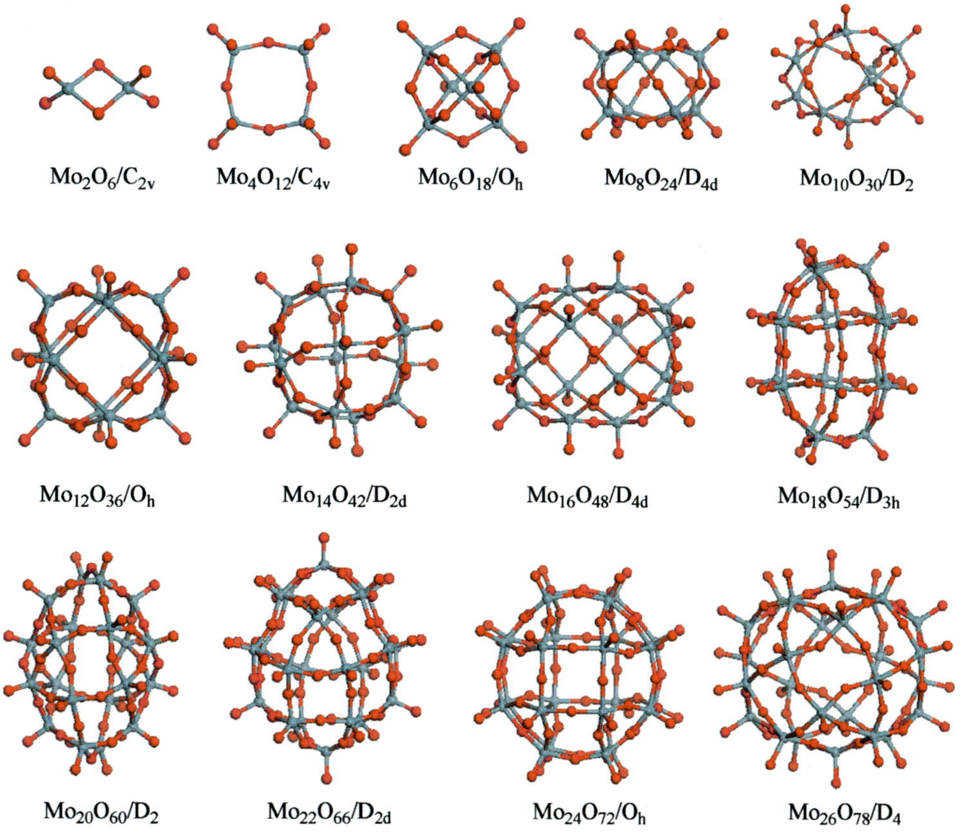

Fig. 1 The most stable $(Mo_2O_6)_m$ clusters (m=1–13)

Table 2 Total electronic energies (E_{tot}, a.u.) and relative energies (kcal mol^{-1}) of the most stable $(Mo_2O_6)_m$ (m=1–13) cages

$(Mo_2O_6)_m$/sym	E_{tot} [a]	ΔE_m [b]	$\Delta^2 E_m$ [c]	Isomers[d]	Isomers[e]
Mo_2O_6/C_{2v}	−588.4325	0.0		0	0
Mo_4O_{12}/C_{4v}	−1,176.9502	27.2	−11.0	0	0
Mo_6O_{18}/O_h	−1,765.5030	43.7	10.9	0	0
Mo_8O_{24}/D_{4d}	−2,354.0208	46.3	2.2	0	0
$Mo_{10}O_{30}/D_2$	−2,942.5316	47.1	−8.5	2	0
$Mo_{12}O_{36}/O_h$	−3,531.0696	50.4	13.1	5	1
$Mo_{14}O_{42}/D_{2d}$	−4,119.5660	49.0	−6.9	8	0
$Mo_{16}O_{48}/D_{4d}$	−4,708.0844	49.8	0.6	12	2
$Mo_{18}O_{54}/D_{3h}$	−5,296.6008	50.2	2.7	25	3
$Mo_{20}O_{60}/D_2$	−5,885.1086	49.9	−3.4	30	3
$Mo_{22}O_{66}/D_{2d}$	−6,473.6271	50.4	−1.9	51	5
$Mo_{24}O_{72}/O_h$	−7,062.1516	51.1	8.1	76	12
$Mo_{26}O_{78}/D_4$	−7,650.6502	50.4		109	17

[a]At GGA-PW91/DND/Fine
[b]Binding energy ΔE_m per Mo_2O_6 unit (Eq. 8)
[c]Disproportionation energy (Eq. 9)
[d]Total number of all the possible topological isomers (not including ringlike isomers)
[e]Number of isolated-triangle rule isomers in parenthesis

Insight into the enhanced stability can be gained from the second difference in energy $\Delta^2 E_m$ in Eq. 10, where E_{m+1}, E_{m-1}, and E_m are the energies of $(Mo_2O_6)_{m+1}$, $(Mo_2O_6)_{m-1}$, and $(Mo_2O_6)_m$. Because Eq. 10 can be considered as a disproportionation reaction of a molecule into two fragments, the calculated reaction energy ($\Delta^2 E_m$) can reveal the relative stability of the cluster, i.e., a positive value (endothermic) for enhanced stability toward fragmentation and a negative value (exothermic) for instability. The same method has been used to estimate the stability of $(AlN)_x$ clusters, [35] $(BH)_x^{2-}$ cages, [39] and Al_x clusters [38]. As given in Table 2 and shown in Fig. 3, the relationship between $\Delta^2 E_m$ and m alternates with increasing cage sizes, where Mo_6O_{18} (Lindqvist), $Mo_{12}O_{36}$ (Keggin), and $Mo_{24}O_{72}$ stand out as *stable* cages. To a lesser extent, $Mo_{18}O_{54}$ (Wells–Dawson) also appears to be

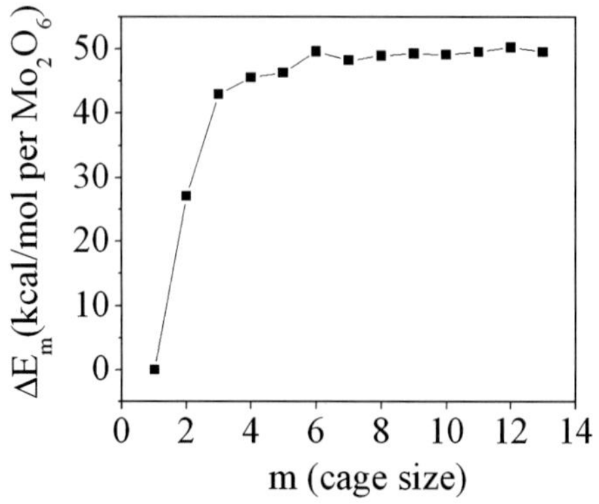

Fig. 2 Binding energy ΔE_m per Mo_2O_6 unit as a function of the cage size (m)

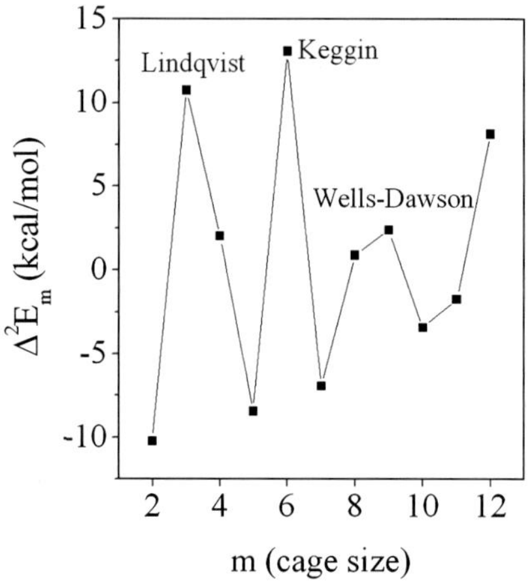

Fig. 3 Disproportionation energy $\Delta^2 E_m$ as a function of the cage size (m)

stable. The largest disproportionation energy of 13.1 kcal mol^{-1} identifies the neutral α-Keggin cage $Mo_{12}O_{36}$ as the most stable cluster. It is noteworthy that the anions of Mo_6O_{12}, $Mo_{12}O_{36}$, and $Mo_{18}O_{54}$ are known experimentally, but that the anions of $Mo_{24}O_{72}$ are unknown.

$$\Delta^2 E_m = (E_{m+1} + E_{m-1})/2 - E_m \quad (10)$$

The stability of a given POM cage depends on the structure and bonding environment. Assigning certain coefficients to the {O_4Mo=O} unit in different bonding environments is an effective means for predicting energies. To understand the relationship between the stability and structure of POM, a least-squares fit is done using the bonding environment index, as suggested by Ziegler [37]. For simplification, each pyramidal {O_4Mo=O} unit[1] (corresponding to the octahedron {O_5Mo=O} in the POM anions) is extracted as a point, and there are six possible points (P_{4444}, P_{4443}, P_{4343}, P_{4433}, P_{4333}, and P_{3333}) according to their surrounding number of f_4 and f_3, respectively, e.g., P_{4443} for the point with three f_4 and one f_3 around, and P_{4343} for point surrounded by two f_4 and two f_3 alternatively.

The fit gives the energy expression in Eq. 11 for $(MoO_3)_n$ cages with {O_4Mo=O} units at the GGA-PW91/DND/FINE level. $E(v)$ is the total electronic energy, and n (equal to m/2) is the number of the {MoO_3} units. The root-mean-square deviation of this fit for the total energy is 21.2 kcal mol^{-1}. The fit has been checked on all stable cages in Fig. 3 and the largest differences are found for

[1] Because each bridged-oxygen linked by two metal atoms, so in fact each pyramid reduced to one {MO_3} unit in average.

Mo_6O_{18} and $Mo_{24}O_{78}$ with about 33.9 kcal mol^{-1} (<0.03%). The significant energy deviation can be attributed to the large difference of local bonding environment, as discussed later.

$$E(v, \text{a.u.}) = -294.2550v - [0.9331P_{4443} + 0.2854P_{4343} + (-1.1118)P_{4444} + (-1.3305)P_{4433} + (-2.6286)P_{4333} + (-4.5031)P_{3333}] \times 10^{-3}. \quad (11)$$

Moreover, it must be noticed that the coefficients pertaining to each specific point provide a means to gauge the stability of a particular environment [37]. The more positive the coefficient, the more stable the environment. Hence, the stability is, in decreasing order, P_{4443}>P_{4343}>P_{4444}>P_{4433}>P_{4333}>P_{3333}. Thus, P_{4443} and P_{4343} are more stable types, while P_{3333} and P_{4333} are less stable types. It is clearly seen that P_{3333} and P_{4333} exhibit a large amount of ring strain and destabilize the structure. Along with the IPR [21] for carbon fullerenes or ISR [20] for BN inorganic fullerenes, these neutral POM cages obey the ITR.

However, there exists a large difference between the predicted and calculated energies, and this can be attributed to the fact that in this simple model the local bonding environment is neglected. For example, the peripheral MMM angles of the P_{4444} in small cages, such as the most stable $Mo_{14}O_{42}$ (which shares the similar skeleton with $([PMo_{12}O_{40}(VO)_2]^{5-})$ [40] reaches 57.2°, while in large clusters, such as $Mo_{26}O_{78}$, the MMM angle of P_{4444} is 87.3°. This difference in structural parameters will result in the large deviation of the energy fit.

Beside the isolated arrangement of triangles, the squares of the polyhedron should have environments as similar as possible for spreading the steric strain evenly. This is similar to the hexagon neighbor rule (HNR) [41] of fullerenes. Here, we use the number of P_{4444} to depict simply the square neighbor signatures. Taking the first ten most stable isomers of $Mo_{22}O_{66}$ and $Mo_{24}O_{72}$ as examples (Table 3), besides the general trend on the stability obeying the ITR, the number of P_{4444} also correlates obviously with the stability. For example, the most stable isomer **1** of $Mo_{22}O_{66}$ is the exclusive one, not only with the largest triangles isolation (N_{33}=0) but also with the most isolated squares (P_{4444}=0); while isomers (**2**–**10**), with P_{4444}=2–3, are all higher in energy. For $Mo_{24}O_{72}$, the most stable isomers **11** and **12** are the only ones with N_{33}=0 and P_{4444}=0, the third most stable isomer **13**, with P_{4444}=2 and N_{33}=0, is higher in energy, while other isomers with P_{4444}>2 are much higher in energy. Thus, the square neighbor rule (SNR) is applicable in selecting the first few stable isomers, and for big clusters, it shows obvious superiority than N_{33}. Thus, the more reliable method is combining P_{4444} with N_{33}. For the Keggin cages ($Mo_{12}O_{36}$), the most stable cage is the α isomer with N_{33}=0, while the β isomer with N_{33}=3 is higher in energy by 8.0 kcal mol^{-1} [19]. For the Wells–Dawson cage ($Mo_{18}O_{54}$), the most stable cage is the α isomer with N_{33}=0, while second most stable cage in C_{2v} symmetry with N_{33}=2 is higher in energy by 1.8 kcal mol^{-1}. It is also noteworthy that with increased cage sizes, these isomers become very close in energy and their relative stability and cannot be estimated simply by such rules.

The analogy with fullerenes can be perused further. For example, the α-Keggin cage ($Mo_{12}O_{36}$) with f_3=8 and f_4=6 is the first structure composed only of perfectly isolated triangles, which corresponds to C_{60} without pentagon adjacency (f_5=12 and f_6=20) and $(BN)_{12}$ without square adjacency (f_4=6 and f_6=8). The second example is the Lindqvist cage (Mo_6O_{18}) composed only of triangles (f_3=8), and the analogous fullerene counterpart is C_{20} with only pentagons (f_5=12) [42] and the $(BN)_n$ counterpart is $(BN)_4$ with only squares (f_4=6) [20]. The next analogue is $Mo_{16}O_{48}$, which obeys the ITR (f_3=8 and f_4=10), and the counterpart is C_{70} following the IPR (f_5=12 and f_6=25), and $(BN)_{15}$ is following the ISR (f_4=8 and f_6=11) [20]. In addition, no

Table 3 Total electronic energies (E_{tot}, a.u.) and relative energies (E_{rel}, kcal mol^{-1}) of the first ten stable of $Mo_{22}O_{66}$ and $Mo_{24}O_{72}$ isomers

	P_{4444}	N_{33}	E_{tot} [a]	E_{rel} [a]
$Mo_{22}O_{66}$				
1/D_{2d}	0	0	−6,473.6271	0.0
2/D_4	2	0	−6,473.6144	8.0
3/C_2	2	0	−6,473.6125	9.2
4/C_{2v}	2	0	−6,473.6116	9.7
5/C_1	2	1	−6,473.6100	10.7
6/C_s	2	0	−6,473.6095	11.0
7/C_2	2	2	−6,473.6081	11.9
9/C_2	2	1	−6,473.6018	15.9
8/C_2	2	2	−6,473.6006	16.6
10/C_2	3	2	−6,473.5978	18.4
$Mo_{24}O_{72}$				
11/D_{4d}	0	0	−7,062.1516	0.0
12/D_{4h}	0	0	−7,062.1516	0.0
13/C_2	2	0	−7,062.1247	16.9
14/C_2	4	2	−7,062.1225	18.3
15/C_1	3	1	−7,062.1198	20.0
16/D_2	4	0	−7,062.1184	20.8
17/C_1	3	1	−7,062.1170	21.7
18/C_1	4	1	−7,062.1167	21.9
19/C_2	4	2	−7,062.1136	23.8
20/C_2	4	1	−7,062.1128	24.4

[a] At GGA-PW91/DND/Fine

Table 4 Total electronic energies (E_{tot}, a.u.) and relative energies (E_{rel} to the most stable one, kcal mol^{-1}) of the second most stable $(Mo_2O_6)_m$ (m=2–13) clusters

$(Mo_2O_6)_m$/sym	E_{tot}	E_{rel} [a,b]
Mo_4O_{12}/C_{2v}	−1,176.9486	1.0
Mo_6O_{18}/D_{3d}	−1,765.4347	42.9
Mo_8O_{24}/C_2	−2,353.9357	53.4
$Mo_{10}O_{30}/D_{4h}$	−2,942.5211	6.6
$Mo_{12}O_{36}/C_{3v}$	−3,531.0568	8.0
$Mo_{14}O_{42}/C_2$	−4,119.5534	7.9
$Mo_{16}O_{48}/C_2$	−4,708.0795	3.1
$Mo_{18}O_{54}/C_{2v}$	−5,296.5972	1.8
$Mo_{20}O_{60}/D_{2d}$	−5,885.1081	0.3
$Mo_{22}O_{66}/D_4$	−6,473.6144	8.0
$Mo_{24}O_{72}/D_{4d}$	−7,062.1516	0.0
$Mo_{26}O_{78}/C_s$	−7,650.6353	9.3

[a] At GGA-PW91/DND/Fine
[b] Relative energies to the most stable isomers in Table 1

perfect ITR POM structures can exist for 12<v<16, while no IPR fullerene structures can exist for 60<m<70, and no ISR $(BN)_x$ structures can exist for 12<x<15 [20]. On the basis of the similarity in stability and structure between POMs and fullerene as well as BN fullerene, this kind of neutral shell can be considered as inorganic fullerene.

It is noteworthy that, similar to fullerene [43] and BN fullerene [35, 44], the energy differences between these isomers are small. For simplicity, only the second most stable isomers are discussed in this paper; the optimized structures are shown in Fig. 4 and the energies are given in Table 4. Apart from Mo_4O_{12}/D_{2d}, $Mo_{18}O_{54}/C_{2v}$, $Mo_{20}O_{60}/D_{2d}$, and $Mo_{24}O_{72}/D_{4d}$, which are very close in energy to the most stable isomers (1.0, 1.8, 0.3, and 0.0 kcal mol^{-1}, respectively), all the other isomers are higher in energy than the most stable isomers (Table 1).

As pointed out previously, this topological analysis can also be used for clusters with large sizes (big or giant) with f_5, and the relationship between f_5 and f_3 is $f_3 = f_5 + 8$ (Eq. 6 is changed to $f_3 - f_5 = 8$). For example, Presyssler's anion ($[NaP_5W_{30}O_{110}]^{14-}$) [45] has a neutral cage of $M_{30}O_{90}$, and it can be considered as a polyhedron with f_3=10, f_4=20, and f_5=2. The giant keplerates $\{Mo_{102}\}$ [46], $\{Mo_{72}Fe_{30}\}$ [47], and $\{Mo_{132}\}$ [48], in striking similarity with C_{60}, can be viewed as derivatives of $M_{60}O_{180}$ (I_h, f_3=20, f_4=30, and f_5=12). They were described as $\{\mathbf{Mo_{60}}\}(pentagon)_{12}(linker)_{30}$ for $\{Mo_{102}\}$ and $\{Mo_{70}Fe_{30}\}$, as well as $\{\mathbf{Mo_{60}}\}(pentagon)_{12}(linker)_{60}$ for $\{Mo_{132}\}$ by Müller [46] recently. Accordingly, $\{Mo_{75}V_{20}\}$ [49] as derivatives of $M_{30}O_{90}$ (I_h, f_3=20, f_4=0, and f_5=12) can be viewed as $\{\mathbf{Mo_{30}}\}(Mo_6)(Mo_5)$ [46]. Within the topological analysis, all these clusters have perfectly arranged triangles and, therefore, obey the ITR.

Apart from the discussion above, note that our topological analysis and computations are on the basis of the clathrate model. However, POM cages are seldom empty, and the interaction between the host and guest exists. It has often been noted that the assembly of the flexible and versatile metal oxide compounds is probably

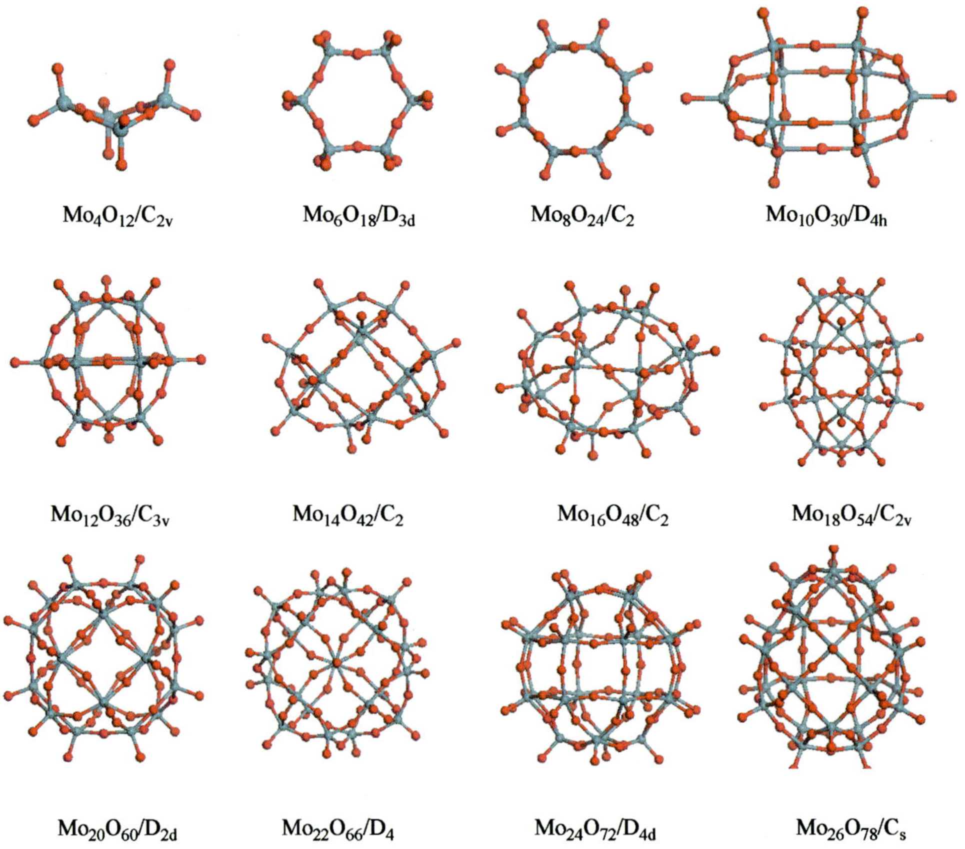

Fig. 4 The second most stable $(Mo_2O_6)_m$ (m=2–13) clusters

determined by the shape and symmetry of small guest molecule [50]. Although there are several factors competing in the control of the stability of POMs [11], the sizes of host and guest must be compatible for a chemical interaction with the formation of pseudooctahedral metal coordination. For example, considered as a stable cage, the POM of the $Mo_{24}O_{72}$ (O_h) cage is still unknown; one of the problems is the appropriate guest to be encapsulated and to be stabilized in solution, as Pope et al. rationalized [45]. Combining chemical rationality and topological stability, the WD structures are more similar to C_{70} than the $Mo_{16}O_{48}$ as the second ITR, as shown in Fig. 2. Because of the unavailability of appropriate guests, the anions of some stable neutral cages cannot be formed experimentally.

Conclusions

Using sphere quadrangulation algorithms or modified fullerene spiral algorithms, all possible cage isomers of $(Mo_2O_6)_m$ (m=1–13) structures have been generated. Their relative stabilities were estimated qualitatively by using the number of the shared triangle cages (N_{33}) and calculated at the GGA-W91/DND level of density functional theory with the fine quality of mesh size. Like the HNR of fullerene, the SNR was also found useful for selecting the most stable isomers. The most stable neutral cages of $(Mo_2O_6)_m$ (m=3–13) that mimic the IPR of fullerene and ISR of inorganic fullerene $(BN)_x$ are found to obey the ITR. Based on the disproportionation energy, the Mo_6O_{12}, $Mo_{12}O_{36}$, and $Mo_{18}O_{54}$ cages have enhanced stability, and they correspond to the neutral cages of the Lindqvist, Keggin, and Wells–Dawson anions, respectively. Among them, the neutral the α-Keggin $Mo_{12}O_{36}$ is the first ITR isomer and represents the most stable cage of these clusters. The Wells–Dawson ($Mo_{18}O_{54}$) cage represents the second ITR structure, which corresponds to C_{70} as the second IPR cage of fullerene. Therefore, neutral cages of $(Mo_2O_6)_m$ (m=3–13) can be viewed as inorganic fullerenes.

Acknowledgements The work was supported by Chinese Academy of Sciences and the National Natural Science Foundation China (20471034).

References

1. (a) Pope MT (1983) Heteropoly and isopoly oxometalates. Springer, Berlin Heidelberg New York; (b) Pope MT, Müller A (1991) Angew Chem Int Ed Engl 30:34–48
2. Pope MT, Müller A (1994) Polyoxometalates: from platonic solids to anti-retroviral activity. Kluwer, Dordrecht, The Netherlands
3. Hill CL, Weeks MS, Schinazi RF (1990) J Med Chem 22:2767–2772
4. (a) Hill CL (1998) Chem Rev 98:1–390 (Special issue on polyoxometalates); (b) Chen YG, Gong J, Qu LY (2004) Coord Chem Rev 248:245–260
5. (a) Gouzerh P, Jeannin Y, Proust A, Robert F (1989) Angew Chem Int Ed Engl 28:1363–1364; (b) Lindqvist I (1952) Ark Kemi 5:247–250
6. (a) Fuchs J, Hartl H, Schiller W, Gerlach U (1976) Acta Crystallogr B32:248; (b) Chemseddine A, Sanchez C, Livage J, Launay JP, Fournier M (1984) Inorg Chem 23:2609–2613; (c) Duncan DC, Hill CL (1996) Inorg Chem 35:5828–5835
7. Keggin JF (1933) Nature 131:908–909
8. (a) Dawson B (1953) Acta Crystallogr 6:113–126; (b) D'Amour H (1976) Acta Crystallogr C 32:729–740
9. Müller A (1991) Nature 352:115–115
10. Day VW, Klemperer WG (1985) Science 228:533–541
11. (a) López X, Maestre JM, Bo C, Poblet JM (2001) J Am Chem Soc 123:9571–9576; (b) Maestre JM, López X, Bo C, Poblet JM, Casaò-Pastor N (2001) J Am Chem Soc 123:3749–3758; (c) Maestre JM, López X, Poblet JM (2002) Inorg Chem 41:1883–1888; (d) Poblet JM, López X, Bo C (2003) Chem Soc Rev 32:297–308; (f) Rohmer MM, Bénard M, Blaudeau JP, Maestre JM, Poblet JM (1998) Coord Chem Rev 178: 1019–1049
12. (a) López X, Bo C, Poblet JM (2002) J Am Chem Soc 124:12574–12582; (b) López X, Bo C, Poblet JM (2003) Inorg Chem 42:2634–2638
13. (a) Nomiya K, Miwa M (1984) Polyhedron 3:341–346; (b) Nomiya K, Miwa M (1984) Polyhedron 4:89–95; (c) Nomiya K, Miwa M (1985) Polyhedron 4:675–679; (d) Nomiya K, Miwa M (1985) Polyhedron 4:1407–1412; (e) Nomiya K (1987) Polyhedron 6:309–314
14. Bridgeman AJ, Cavigliasso G (2003) J Phys Chem A 107:6613–6621
15. (a) King RB (2001) J Chem Inf Comput Sci 41:517–526; (b) King RB (1992) J Chem Inf Comput Sci 32:42–47; (c) King RB (1991) Inorg Chem 30:4437–4440
16. Sambrano JR, Andres J, Beltran A, Sensato F, Longo E (1998) Chem Phys Lett 287:620–626
17. Baker LCW (1961) Advances in the chemistry of coordination compounds. Macmillan, New York, p 604
18. Dividson ER (2000) Chem Rev 100:351–818 (special issue on computational transition metal chemistry)
19. Zhang FQ, Wu HS, Cao DB, Zhang XM, Li YW, Jiao H (2005) J Mol Struct Theochem 755:119–126
20. Fowler PW, Heine T, Mitchell D, Schmidt R, Seifert G (1996) J Chem Soc Faraday Trans 92:2197–2201
21. (a) Kroto HW (1987) Nature (Lond) 329:529–531; (b) Schmalz TG, Seitz WA, Klein DJ, Hite GE (1988) J Am Chem Soc 110:1113–1127
22. (a) Delley B (1990) J Chem Phys 92:508–517; (b) Delley B (2000) J Chem Phys 113:7756–7764 (DMol[3] is available as part of Materials Studio.)
23. Vosko SH, Wilk L, Nusair M (1980) Can J Phys 58:1200–1211
24. Wang Y, Perdew JP (1991) Phys Rev B 44:13298–13307
25. Zhang FQ, Wu HS, Qin XF, Li YW, Jiao, H (2005) J Mol Struct Theochem 755:113–117
26. Weinstock IA, Cowan JJ, Barbuzzi EMG, Zeng H, Hill CL (1999) J Am Chem Soc 121:4608–4617
27. López X, Poblet JM (2004) Inorg Chem 43:6863–6865
28. Pope MT (1972) Inorg Chem 11:1973–1974
29. (a) Friedrichs OD, Dress AWM, Hudson DH, Klinowski J, Mackay AL (1999) Nature 400:644–647; (b) Friedrichs OD, Dress AWM, Müller A, Pope MT (1993) Mol Eng 3:9–28
30. (a) Coxeter HSM (1961) Introduction to geometry. Wiley, New York; (b) Cerari M, Cucinella S (1987) The chemistry of inorganic homo and heterocycles. In: Haiduc I, Sowerby DB (eds) Academic, New York, pp 167–190
31. Brinkman G, McKay B (2001) Plantri and fullgen. http://www.mathematik.uni-bielefeld.de/~senkel/CAGE/contents.html
32. Manolopoulos DE, May JC, Down SE (1991) Chem Phys Lett 181:105–111

33. (a) Bridgeman AJ (2004) Chem Eur J 10:2935–2941; (b) Bridgeman AJ, Cavigliasso G (2003) J Phys Chem A 107:6613–6621; (c) Bridgeman AJ (2003) Chem Phys 287:55–69; (d) Bridgeman AJ, Cavigliasso G (2002) Inorg Chem 41:1761–1770; (e) Bridgeman AJ, Cavigliasso G (2002) Chem Phys 279:143–159; (f) Bridgeman AJ, Cavigliasso G (2002) Inorg Chem 41:3500–3507; (g) Bridgeman AJ Cavigliasso G (2002) J Chem Soc Dalton Trans 2244–2249
34. Guo YR, Pan QJ, Wei YD, Li ZH, Li X (2004) J Mol Struct Theochem 676:55–64
35. Wu HS, Zhang FQ, Xu XH, Zhang CJ, Jiao H (2003) J Phys Chem A 107:204–209
36. Seifert G, Fowler PW, Mitchell D, Porezag D, Frauenheim T (1997) Chem Phys Lett 268:352–358
37. Zurek E, Woo TK, Ziegler T (2001) Inorg Chem 40:361–370
38. Lloyd LD, Johnston RL (1998) Chem Phys 236:107–121
39. Schleyer PvR, Najafian K, Mebel AM (1998) Inorg Chem 37:6765–6772; see also Schleyer PvR, Najafian K (1998) Inorg Chem 37:3454–3470
40. Chen Q, Hill CL (1996) Inorg Chem 35:2403–2405
41. (a) Raghavachari K (1992) Chem Phys Lett 190:397–400; (b) Achiba Y, Fowler PW, Mitchell D, Zerbetto F (1998) J Phys Chem A 102:6835–6841
42. Prinzbach H, Weller A, Landenberger P, Wahl F, Worth J, Scott LT, Gelmont M, Olevano D, Issendorff BV (2000) Nature 407:60–63
43. (a) Cioslowski J, Rao N, Moncrieff D (2000) J Am Chem Soc 122:8265–8270; (b) Zhao X, Slanina Z, Goto H (2004) J Phys Chem A 108:4479–4484; (c) Wu HS, Xu XH, Jiao H (2004) J Phys Chem A 108:3813–3816
44. (a) Strout DL (2004) Chem Phys Lett 383:95–98; (b) Strout DL (2001) J Phys Chem A 105:261–263
45. (a) Alizadeh MH, Harmalker SP, Jeannin Y, Martin-Frère J, Pope MT (1985) J Am Chem Soc 107:2662–2669; (b) Dickman MH, Gama GJ, Kim KC, Pope MT (1996) J Clust Sci 7: 567–583
46. Müller A, Kögerler P, Dress AWM (2001) Coord Chem Rev 222:193–218
47. Müller A, Das SK, Kögerler P, Bögge H, Schmidtmann M, Trautwein AX, Schünemann V, Krickemeyer E, Preetz W (2000) Angew Chem Int Ed Engl 39:3414–3417
48. Müller A, Krickemeyer E, Bögge H, Schmidtmann M, Preetz W, Peters F (1998) Angew Chem Int Ed Engl 37:3359–3363
49. Müller A, Koop M, Bögge H, Schmidtmann M, Peters F (1999) Chem Commun 1885–1886
50. (a) Müller A, Reuter H, Dillinger S (1995) Angew Chem Int Ed Engl 34:2328–2361; (b) Müller A, Rohlfing R, Krickemeyer E, Bögge H (1993) Angew Chem Int Ed Engl 32:909–912; (c) Müller A, Krickemeyer E, Penk M, Rohlfing R, Armatage A, Bögge H (1991) Angew Chem Int Ed Engl 30:1674–1677
51. (a) Dahlstrom P, Zubie J, Neaves B, Dilworth JR (1982) Cryst Struct Commun 11:463–469; (b) Garner CD, Giwkaderm NC, Mabbs FE, McPhail AT, Miller RW, Onan KD (1978) J Chem Soc Dalton Trans 1582–1589
52. Bi LH, Wang EB, Xu L, Huang RD (2000) Inorg Chim Acta 305:163–171
53. Neier R, Trojanoski C, Mattes R (1995) J Chem Soc Dalton Trans 2521–2528

J Mol Model (2006) 12: 559–562
DOI 10.1007/s00894-005-0094-7

ORIGINAL PAPER

Golam Rasul · G. K. Surya Prakash · George A. Olah

Theoretical study of AlH_n^{2+} (n=1–7) dications

Received: 8 March 2005 / Accepted: 30 November 2005 / Published online: 8 March 2006

Abstract The structures and stability of $AlH_n^{2+}(n = 1 - 7)$ **1–7** dications were calculated at the ab initio MP2/aug-cc-pVTZ level of theory. The dications AlH^{2+} **1** and AlH_2^{2+} **2** were characterized to be unstable thermodynamically. However, these and the stable dications, AlH_n^{2+} $(n = 3 - 7)$ **3–7** have considerable kinetic barriers for deprotonation. Each of the structures **3–7** contains one or more two-electron three-center (2e–3c) bonds. Aluminum atoms of these dications carry most of the positive charges, as indicated by NBO charge calculations.

Keywords ab initio calculations · Alane dications · Main group hydrido dications

Introduction

The main-group hydrido dications have been a topic of numerous theoretical and experimental investigations [1]. Recently, we reported a study of structures and stabilities of $BH_n^{2+}(n = 1 - 7)$ dications using ab initio and density functional theory (DFT) methods at the MP2/aug-cc-pVTZ and B3LYP/ aug-cc-pVTZ levels, respectively [2]. The dications BH^{2+} and BH_2^{2+} were found to be highly unstable both kinetically and thermodynamically. However, dications $BH_n^{2+}(n = 3 - 7)$ have considerable kinetic barriers for deprotonation. Each of these structures contains at least one two-electron three-center (2e–3c) bond. We have previously reported calculated structures and energetics of the CH_3^{2+}, CH_5^{2+} and CH_6^{2+} carbodications [3–5]. Wong and Radom [6] have reported the calculated structure of the CH_4^{2+} dication. We have also recently reported structures and energetics of onium dications XH_4^{2+} and XH_5^{2+} (X=N, P, and As) [7, 8], boronium BH_5^{2+} and BH_7^{2+} dications [4, 9] as well as their corresponding aluminum based analogs, AlH_5^{2+} and AlH_7^{2+} dications [10]. Koch, Frenking and Schwarz reported [11] the ab initio HF/6-31G(d) calculated structures of $SiH_n^{2+}(n = 1 - 5)$ dications [12]. Structures and bonding of such highly electron-deficient dications are of substantial theoretical significance. In continuation of our study of the main-group hydrido dications, we have now extended our investigations to the structures and stabilities of $AlH_n^{2+}(n = 1 - 7)$ dications by the ab initio method at the MP2/aug-cc-pVTZ level. The thermodynamic and kinetic barriers for deprotonation processes of dications were also estimated.

Calculations

All calculations were carried out using the Gaussian 03 program system [13]. Initial geometry optimizations were performed at the DFT B3LYP/aug-cc-pVTZ level [14]. Vibrational frequencies at the B3LYP/aug-cc-pVTZ// B3LYP/aug-cc-pVTZ level were used to characterize stationary points as minima (number of imaginary frequencies (NIMAG)=0) or transition state structures (NIMAG=1) and to evaluate zero-point vibrational energies (ZPE), which were scaled by a factor of 0.98 [15]. Further geometry optimizations were performed at the MP2/aug-cc-pVTZ level. Final energies were calculated at the MP4(SDTQ)/aug-cc-pVTZ//MP2/aug-cc-pVTZ + ZPE level (Table 1). Atomic charges and Wiberg bond indices [16] (bond orders) were obtained at the MP2/aug-cc-pVTZ// MP2/aug-cc-pVTZ level using the Natural Bond Orbital analysis (NBO) method [17].

Dedicated to Professor Dr. Paul von Ragué Schleyer on the occasion of his 75th birthday

G. Rasul (✉) · G. K. S. Prakash · G. A. Olah
Loker Hydrocarbon Research Institute and Department of Chemistry, University of Southern California, University Park, Los Angeles, CA 90089-1661, USA
e-mail: rasul@usc.edu
e-mail: olah@usc.edu

Table 1 Total energies (-au), ZPE[a] and relative energies (kcal mol^{-1})[b]

	B3LYP/ aug-cc-pvtz	ZPE	MP2/ aug-cc-pvtz	MP4(SDTQ)/ aug-cc-pvtz// MP2/aug-cc-pvtz	Relative energy (kcal mol^{-1})
1	242.08106	2.1	241.61877	241.62898	0.0
1t	242.04353	0.0	241.56657	241.57523	31.6
2	242.64366	3.3	242.15951	242.16987	42.7
2a	242.70304	6.9	242.23529	242.24372	0.0
2t	242.62847	2.6	242.12291	242.13294	65.2
2at	242.63855	3.0	242.12546	242.13599	63.7
3	243.31906	10.6	242.84048	242.85704	0.0
3t	243.22642	7.1	242.73659	242.75233	62.2
4	243.91925	14.6	243.43305	243.44950	
5	244.54012	19.5	244.04686	244.07046	0.0
5t	244.42832	15.8	243.92239	243.94541	74.8
6	245.12901	22.8	244.62752	244.65187	
7	245.75552	29.3	245.24866	245.27953	

[a]Zero-point vibrational energies (ZPE) at B3LYP/aug-cc-pvtz// B3LYP/aug-cc-pvtz scaled by a factor of 0.98
[b]At MP4(SDTQ)/aug-cc-pvtz// MP2/aug-cc-pvtz + ZPE level

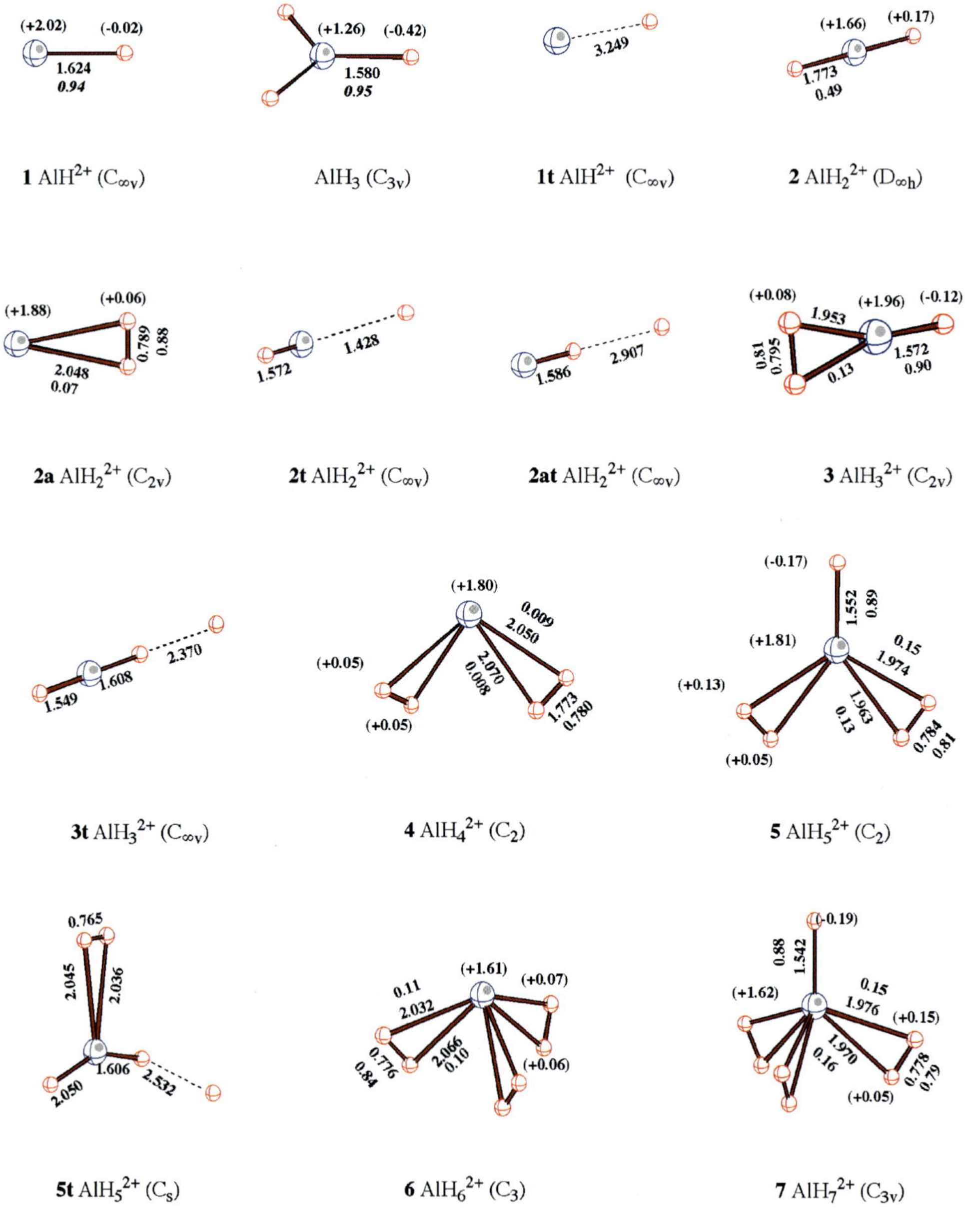

Fig. 1 MP2/aug-cc-pVTZ structures, NBO charges (in parentheses) and Wiberg bond index (in italics) of **1**–**7**

Results and discussion

Structure **1** was found to be a minimum (Fig. 1) on the potential energy surface (PES) of singlet AlH^{2+} at the MP2/aug-cc-pVTZ level. The computed bond length of Al–H bond was found to be 1.624 Å. For comparison, we also calculated the neutral monomeric AlH_3 at the same MP2/aug-cc-pVTZ level (Fig. 1). The Al–H bond of **1** is only 0.044 Å longer than the Al–H bond of AlH_3. The relative bond strength of the Al–H bond of **1** was estimated using the Wiberg bond index [16] analysis (Fig. 1). The computed bond index of Al–H bond was found to be 0.94. This is practically same as the Al–H bond index of AlH_3 (0.95). NBO charge calculations show that in dication **1** the aluminum carries (+2.02) all of the charges (Fig. 1). Dissociation of **1** into Al^+ and H^+ was computed to be exothermic by 53.7 kcal mol^{-1} (Table 2). The transition structure **1**t for the dissociation lies 31.6 kcal mol^{-1} higher in energy than structure **1** at the MP4(SDTQ)/aug-cc-pVTZ//MP2/aug-cc-pVTZ + ZPE level. These results indicate that the dication **1** is quite stable kinetically. In contrast, the boron dication [2] BH^{2+} was found to be very unstable kinetically, computed at the same level.

The linear $C_{\infty v}$ symmetric structure **2** and the trigonal C_{2v} symmetric structure **2a** were located as minima on the PES of doublet AlH_2^{2+} (Fig. 1). The structure **2** is characterized by the long Al–H bond length of 1.773. The electron deficient longer Al–H bond is expectedly weak as indicated by the calculated bond index of 0.49 (Fig. 1). On the other hand, the structure **2a** is characterized as a complex of a doubly charged aluminum atom and a hydrogen molecule. Previously Koch et al. reported [18] a detailed theoretical investigation of CH_2XH^{2+} and CH_3X^{2+} dications (X=NH_2, OH, F). It was shown that the geometries and stabilities of these types of dications are best explained by the donor-acceptor interaction of (neutral) donor and (doubly charged) acceptor molecules. The model was also used to explain the structures, stabilities and bonding of the compounds of noble gases [19]. However, the structure **2a** is 42.7 kcal mol^{-1} more stable than **2**. In the dication **2**, aluminum carries a charge of 1.66 and the two hydrogens carry 0.34 (Fig. 1). On the other hand, in the dication **2a**, aluminum carries a charge of 1.88 and the two hydrogens carry only 0.12. Deprotonation of both **2** and **2a** are exothermic by 45.6 and 2.9 kcal mol^{-1} (Table 2), respectively. However, these processes have high kinetic barriers of 63.7 and 62.2 kcal mol^{-1}, respectively, through transition states **2**t and **2at** (Fig. 1). Dissociation of **2a** into Al^{2+} and H_2 is, however, quite endothermic by 25.8 kcal mol^{-1} (Table 2).

The C_{2v} symmetric structure **3** is the only minimum that was located on the PES of singlet dication AlH_3^{2+} (Fig. 1). The structure **3** is characterized with a 2e–3c bond resembling a complex between AlH^{2+}**1** and a hydrogen molecule. The Al–H and H–H bond distances of the 2e–3c interaction are 1.953 and 0.795 Å, respectively. Deprotonation of **3** is exothermic by only 9.2 kcal mol^{-1}, but the process has a kinetic barrier of 62.2 kcal mol^{-1} through the transition state **3**t (Fig. 1). However, dissociation of **3** into **1** and H_2 is quite endothermic by 33.0 kcal mol^{-1} (Table 2).

Table 2 Dissociation energies (ΔE_0) and kinetic barriers at 298 K for the selected processes[a]

Process	ΔE_0 (kcal mol^{-1})[a]	Barrier (kcal mol^{-1})
AlH^{2+}**1**$\rightarrow Al^+ + H^+$	−53.7	31.6
AlH_2^{2+}**2** $\rightarrow AlH^+ + H^+$	−45.6	65.2
AlH_2^{2+}**2a** $\rightarrow AlH^+ + H^+$	−2.9	63.7
AlH_3^{2+}**3** $\rightarrow AlH_2^+ + H^+$	−9.2	62.2
AlH_4^{2+}**4** $\rightarrow AlH_3^+ + H^+$	+14.1	
AlH_5^{2+}**5** $\rightarrow AlH_4^+ + H^+$	+10.2	74.8
AlH_6^{2+}**6** $\rightarrow AlH_5^+ + H^+$	+27.7	
AlH_7^{2+}**7** $\rightarrow AlH_6^+ + H^+$	+25.7	
AlH_2^{2+}**2a** $\rightarrow Al^{2+} + H^2$	+25.8	
AlH_3^{2+}**3** $\rightarrow AlH^{2+}$**1** $+ H_2$	+33.0	
AlH_4^{2+}**4** $\rightarrow AlH_2^{2+}$**2a**$+H_2$	+19.8	
AlH_5^{2+}**5** $\rightarrow AlH_3^{2+}$**3** $+ H_2$	+23.4	
AlH_6^{2+}**6** $\rightarrow AlH_4^{2+}$**4** $+ H_2$	+17.2	
AlH_7^{2+}**7** $\rightarrow AlH_5^{2+}$**5** $+ H_2$	+19.8	

[a]at MP4(SDTQ)/ aug-cc-pVTZ //MP2/ aug-cc-pVTZ + ZPE level

The C_2 symmetric form **4** was located as a minimum for doublet AlH_4^{2+} (Fig. 1). The structure **4** resembles a complex between Al^{2+} and two hydrogen molecules resulting in the formation of two 2e–3c bonds. The calculated charges of aluminum and each of the hydrogen of **4** are +1.80 and +0.05, respectively (Fig. 1). The deprotonation of **4** is endothermic by 14.1 kcal mol^{-1} (Table 2). However, no transition state could be located for the deprotonation of **4**.

Olah and Rasul have previously reported [10] the structure and energetics of AlH_5^{2+} at the MP2/6-31G(d,p) level of theory. At the MP2/6-31G(d,p) level, the planar C_{2v}-symmetric form was found to be the minimum. However, at the MP2/aug-cc-pVTZ level, a slightly different C_2-symmetric form **5** with two 2e–3c bonds was located as the only minima for singlet AlH_5^{2+} (Fig. 1). The optimized structure of AlH_5^{2+} (**5**) is shown in Fig. 1. Structure **5** resembles a complex between AlH^{2+}**1** with two hydrogen molecules with an empty *p*-orbital orthogonal to the plane of the molecule. Deprotonation of **5** is endothermic by 10.2 kcal mol^{-1}, but the process has a kinetic barrier of 74.8 kcal mol^{-1} through the transition state **5**t (Fig. 1).

The six coordinate propeller-shaped C_3-symmetrical structure **6** with three 2e–3c bonds is the only minimum on the PES of doublet AlH_6^{2+} (Fig. 1). The Al–H bond distances of **6** are 2.032 and 2.066 Å, slightly shorter than those in AlH_4^{2+} **4** (2.050 and 2.070 Å). The six hydrogen atoms bear charges of only +0.39. The deprotonation process in **6** is endothermic by 27.7 kcal mol^{-1}. Even the trication AlH_6^{3+} was found to be a minimum, as shown in earlier calculations [10].

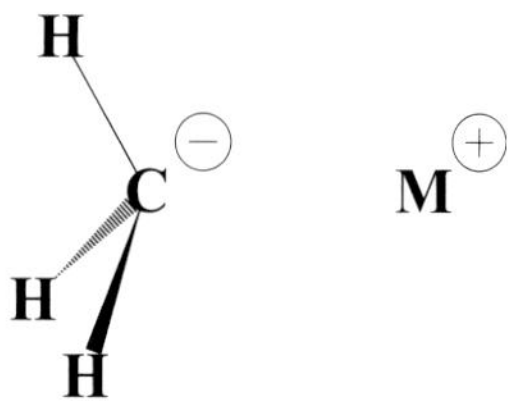

2

teracting predominantly electrostatically with only marginal covalent character (see Structure **2**).

The purpose of the present study is twofold. In the first place, we wish to augment the scarce structural and thermodynamic information on organoalkali metal compounds (vide infra) with a set of data that is obtained consistently at the same level of theory and, thus, to enable a systematic analysis of trends. Thus, we compute at the BP86/TZ2P level of density functional theory (DFT) [26] how the structure and thermochemistry (e.g., C–*M* bond lengths and strengths) of methyl alkali metal compounds CH_3M depend on the metal atom, along *M*=Li, Na, K and Rb. A second purpose is to understand the emerging trends in terms of quantitative Kohn–Sham molecular orbital (KS-MO) theory [27]. In particular, we show that a description of the bonding mechanism in terms of a polar C–*M* electron-pair bond between the methyl radical and alkali metal atom is just as natural as an ionic description (i.e., in terms of $CH_3^-+M^+$) and that it provides a straightforward way of understanding all observed trends. In the course of this work, we present a new implementation into our Amsterdam Density Functional (ADF) program [26] for computing Kohn–Sham Fock matrix elements in terms of fragment molecular orbitals (MOs).

Materials and methods

Computational details

All computations were carried out with the ADF program [26] at the BP86/TZ2P level of the generalized gradient approximation (GGA) of Kohn–Sham density functional theory (DFT) [27]. The TZ2P basis set consists of uncontracted set of Slater-type orbitals (STOs), including diffuse functions. It is of triple-ζ quality for all atoms and has been augmented with two sets of polarization functions: 3d and 4f on C, Li, Na; 4d and 4f on K, Rb; and 2p and 3d on H. In addition, an extra set of *p*-functions was added to the basis sets of Li (2p), Na (3p), K (4p) and Rb (5p). The 1s core shell of carbon and lithium, the 1s2s2p core shells of sodium and potassium, and the 1s2s3s2p3p3d core shells of rubidium were treated by the frozen-core (FC) approximation. An auxiliary set of s-, p-, d-, f- and g-STOs was used to fit the molecular density and to represent the Coulomb and exchange-correlation potentials accurately in each SCF cycle. For a more detailed description of our methods and the ADF program, see [26] and [27].

Fock matrix elements in terms of fragment MOs

The ADF program has the option to analyze the interaction energy between two molecular fragments by building up the molecule from these either mono- or polyatomic fragments. The molecular orbitals $\left\{\psi_j^\Gamma\right\}$ that are computed are then expressed as linear combinations of the symmetry-adapted fragment orbitals $\left\{\varphi_j^\Gamma\right\}$, i.e., symmetry-adapted linear combinations of the molecular orbitals *of the fragments under consideration*, which in turn can be written as linear combinations of STO basis functions $\{\chi_k\}$:

$$\psi_{\mathrm{i}}^\Gamma = \sum_{\mathrm{j}} a_{\mathrm{ij}}^\Gamma \varphi_{\mathrm{j}}^\Gamma = \sum_{\mathrm{k}} b_{\mathrm{ik}}^\Gamma \chi_k \quad (1)$$

For the Kohn–Sham Fock matrix $\mathbf{F}_{\mathrm{KS}}$ in the basis of the symmetry-adapted fragment orbitals, the generalized Eigenvalue problem holds, i.e.,

$$\mathbf{F}_{\mathrm{KS}}\mathbf{C} = \mathbf{S}\,\mathbf{C}\,\mathbf{E} \quad (2)$$

with the Kohn–Sham Fock matrix elements given by

$$(F_{KS})_{ij} = \left\langle \varphi_i^\Gamma \middle| \hat{F}_{KS} \middle| \varphi_J^\Gamma \right\rangle \quad (3)$$

The Kohn–Sham Fock matrix elements correspond the coupling or interaction strength between two orbitals φ_i^Γ and φ_j^Γ. They can be obtained either with the potential of the promolecule density (i.e., the initial sum of fragment densities) or with the final molecular density. The former case corresponds to the field that the orbitals are exposed to before they start interacting but after the fragments have adopted their positions in the overall molecule. In the latter case, solution of the generalized Eigenvalue problem of Eq. (2) yields the molecular Kohn–Sham orbitals ψ_i^Γ and orbital energies $\varepsilon_i = \left\langle \psi_i^\Gamma \middle| \hat{F}_{KS} \middle| \psi_i^\Gamma \right\rangle$. In the present study, we use the latter interaction matrix elements. We wish to point out, however, that the two choices yield similar values that show exactly the same trends.

Results and discussion

Structures and energies

The results of our BP86/TZ2P computations on methyl lithium, sodium, potassium and rubidium are collected in Table 1. Our computed homolytic bond dissociation enthalpies (BDE for 298 K) and bond lengths (see Table 1) agree within a few kcal mol^{-1} and a few hundredths of an angstrom (Å), respectively, with the available experimental [28] and/or with other DFT and ab initio results [10, 11, 16–25]. The computed strength (i.e., BDE) of the C–*M* bond in

Table 1 Analysis of the C–M bond between $CH_3^{\bullet}$ and $M^{\bullet}$ in CH_3M[a]

Property	CH_3Li	CH_3Na	CH_3K	CH_3Rb
Geometry (in Å, degree)				
C–M	2.010	2.376	2.747	2.821
C–H	1.105	1.098	1.100	1.096
H–C–H	106.48	109.84	109.14	109.63
Bond energy decomposition (in kcal mol^{-1})[b]				
ΔE_{oi}	−63.1	−42.2	−38.4	−42.2
ΔV_{elstat}	−30.3	−23.3	−19.0	−20.4
ΔE_{Pauli}	38.4	27.8	23.4	30.7
ΔE_{int}	−55.0	−37.7	−34.0	−31.9
ΔE_{prep}	10.2	6.7	7.4	6.9
ΔE	−44.8	−31.0	−26.6	−25.0
ΔH^{298}=−BDE	−44.0	−30.3	−26.2	−24.6
Miscellaneous bonding parameters[c]				
S_{bond}	0.31	0.28	0.21	0.19
F_{bond}	−70.3	−61.5	−40.2	d
$\varepsilon(ns)$−$\varepsilon(2a_1)$	3.2	3.4	3.8	4.0
Q(M)	0.386	0.351	0.428	0.466

[a]Computed at BP86/TZ2P
[b]$\Delta E=\Delta E_{prep}+\Delta E_{int}$; $\Delta E_{int}=\Delta V_{elstat}+\Delta E_{Pauli}+\Delta E_{oi}$ (see text)
[c]Interaction matrix elements between SOMOs F_{bond} in kcal mol^{-1}; bond overlap S_{bond}; orbital energy difference $\varepsilon(ns)$−$\varepsilon(2a_1)$ in eV; VDD atomic charges Q(M) in a.u
[d]Cannot yet be computed for Rb, for technical reasons

methyl alkali metal monomers decreases if one descends the periodic table, from 44.0 (Li) to 30.3 (Na) to 26.2 (K) to 24.6 kcal mol^{-1} (Rb). The computed C–M bond distances increase monotonically from 2.010 (Li) to 2.376 (Na) to 2.747 (K) to 2.821 Å (Rb), in agreement with mm-wave spectroscopy experiments of Grotjahn et al. for M=Li, Na and K [28, 29]. Note that our H–C–H angles are systematically smaller for CH_3Li (106.48°) than for the heavier homologous (109–110°), in line with CCSD(T) calculations [18] for CH_3Li and CH_3Na. The mm-wave experiments [28, 29], on the other hand, yield essentially constant M–C–H angles of 107.2, 107.3 and 107.0°, along the series Li, Na and K. Thus, for methyl rubidium, for which an experimental structure has not yet been published, we predict a C–M bond length that is about 0.07 Å longer than that in methyl potassium, whereas the methyl has essentially the same structure in both molecules.

Bonding mechanism

Our analyses of the C–M bond comprize three complementary approaches: (1) decomposition of the electronic bond energy ΔE; (2) quantitative analysis of the Kohn–Sham orbital electronic structure; (3) analysis of the electron-density distribution using various atomic charge schemes. The bond energy ΔE corresponds to the reaction $CH_3^{\cdot}+M^{\cdot}\rightarrow CH_3M$ and is first divided into the preparation energy ΔE_{prep} associated with deforming the methyl radical (from its planar equilibrium structure to the pyramidal geometry it adopts in CH_3M) and the actual interaction energy ΔE_{int} between pyramidal $CH_3^{\cdot}$ and $M^{\cdot}$ radicals. ΔE_{prep} is somewhat more destabilizing in methyl lithium than in the heavier homologues owing to the stronger C–Li interaction, which causes the above-mentioned higher extent of methyl pyramidalization in CH_3Li. The interaction energy ΔE_{int} shows the same trend as ΔE and ΔH^{298}: it weakens strongly from Li to Na and then more moderately along Na, K and Rb. We have further decomposed ΔE_{int} into the electrostatic attraction ΔV_{elstat} between the unperturbed charge distributions of the $CH_3^{\cdot}$ and $M^{\cdot}$ fragments, the Pauli repulsive orbital interaction ΔE_{Pauli} (between occupied orbitals) and the bonding orbital interactions ΔE_{oi} that are provided by electron-pair bonding and donor-acceptor interactions [27].

Interestingly, the more dominant of the two bonding terms is the orbital interactions ΔE_{oi} (which is between −63 and −38 kcal mol^{-1}) and not the electrostatic attraction ΔV_{elstat} (which is between −30 and −19 kcal mol^{-1}). This may be surprizing at first if one proceeds from the ionic model, in which the cohesion in the C–M bond is ascribed to electrostatic attraction between CH_3^- and M^+ ions in a nearly salt-like, ionic CH_3M. Note, however, that, electrostatically, the charge separation across the C–M bond is energetically unfavorable in the gas-phase compared to a more balanced (less polar) charge distribution.

Of course, this changes in solution as the solvent molecules interact strongly with, and thus stabilize, the ionic fragments of the heterolytic dissociation pathway. Note however that in solution, a physically correct description of a methyl alkali metal consequently involves (at least) a few solvent molecules of the first solvation shell and that the system is rather to be conceived as a methyl alkali metal–solvent complex. Here, we focus our analysis to the intrinsic nature of carbon–metal bonding as it exists, unaffected by solvent molecules, in the gas phase.

However, also in the gas phase, charge separation still occurs because of the electronegativity difference between $CH_3^{\cdot}$ and $M^{\cdot}$, that is, the fact that the energy of the $2a_1$ singly occupied molecular orbital (SOMO) on the $CH_3^{\cdot}$ radical is significantly lower than that of the ns SOMO on the $M^{\cdot}$ atom (see orbital energy differences $\varepsilon(ns)$-$\varepsilon(2a_1)$ in Table 1). This is nicely illustrated by examining the heterolytic dissociation of CH_3M into $CH_3^-+M^+$ (values not shown in Table 1). Here, the electrostatic interaction dominates by far: ΔV_{elstat} is between −197 (Li) and −158 kcal mol^{-1} (K) whereas ΔE_{oi} is only between −21 (Li) and −18 kcal mol^{-1} (K). In CH_3Li, for example, the heterolytic bond energy ΔE_{hetero} between $CH_3^-+M^+$ amounts to −174.2 kcal mol^{-1}, which arises from −197.4 kcal mol^{-1} of electrostatic attraction ΔV_{elstat}, −21.1 kcal mol^{-1} of orbital interactions ΔE_{oi} and 44.0 kcal mol^{-1} of Pauli repulsion. The weakening of the electrostatic interaction in the heterolytic dissociation on going from Li to Rb is due to the increase in the M–C bond length, which is not compensated by the smaller charge transfer from CH_3^- to M^+ along the Li–Rb series. However, the heterolytic process is much more endothermic (−ΔE-

hetero=175–124 kcal mol^{-1}) than homolytic dissociation ($-\Delta E$=45–25 kcal mol^{-1}) because of the ionization energy that must be invested to remove an electron from $M^{\cdot}$ and transfer it to $CH_3^{\cdot}$. It is of course perfectly valid to compute the homolytic bond energy in two steps by adding the reaction energies of $CH_3^{\cdot}+M^{\cdot} \rightarrow CH_3^{-}+M^{+}$ (charge separation) and $CH_3^{-}+M^{+} \rightarrow CH_3M$ (heterolytic bond formation) [30]. However, ΔE_{hetero} alone provides only incomplete information about the quantity we are actually interested in, i.e., the homolytic bond strength ΔE in the gas phase. Moreover, we feel that analyzing the homolytic bond energy ΔE, i.e., the reaction energy for $CH_3^{\cdot}+M^{\cdot} \rightarrow CH_3M$, directly in terms of the bonding between the corresponding neutral molecular fragments provides a more straightforward description and understanding of its origin and of trends therein, than first analyzing the bonding between the ionic fragments CH_3^{-} and M^{+}, associated with a different, energetically (ΔE_{hetero}) much disfavored ionic dissociation, and adding the charge-transfer energy also to obtain ΔE. On the other hand, the origin in the heterolytic bond energy ΔE_{hetero}, i.e. the reaction energy for $CH_3^{-}+M^{+} \rightarrow CH_3M$, is obviously most directly described in terms of the interaction between the methyl anion and alkali metal cation, which indeed is predominantly electrostatic (vide supra).

Both, electrostatic attraction ΔV_{elstat} and Pauli repulsion ΔE_{Pauli} between $CH_3^{\cdot}$ and $M^{\cdot}$ become weaker along Li–K. The common origin of these trends, which partially cancel each other, is the decreasing overlap between occupied orbitals of $CH_3^{\cdot}$ and $M^{\cdot}$ as the metal AOs become more extended and diffuse. The dominant contributor to the trend in ΔE is the orbital interaction ΔE_{oi}, which (just as ΔV_{elstat}) weakens along Li–K from −63.1 to −42.2 to −38.4 kcal mol^{-1}. The trend in ΔE_{oi} and, thus, in the thermodynamic stability ΔE (or ΔH^{298}=−BDE) can be traced directly to covalent features in the bonding mechanism, i.e., the bond overlap between and mixing of the SOMOs that yield the electron-pair bond. Our quantitative analyses of the orbital electronic structure show that this electron-pair bonding $2a_1+ns$ combination between the methyl and alkali metal SOMOs is polarized toward methyl. However, they also reveal that the alkali metal *ns* contribution is not at all marginal: in terms of Gross Mulliken contributions [31] the composition is approximately 70% $2a_1$+25% *ns*. In case of methyl lithium, the situation is 70% $2a_1$+24% 2s with, in addition, a sizeable contribution of 7% from the lithium $2p_\sigma$ AO. In terms of mixing coefficients [31], this is $0.72\,2a_1 + 0.53\,2s$ $(+0.32\,2p_\sigma)$. We recall that the stabilization of the bonding $2a_1+ns$ combination relative to $\varepsilon(2a_1)$, i.e., the covalent contribution to the orbital interactions ΔE_{oi}, is to second order equal to $F_{bond}^2/\varepsilon(2a_1)-\varepsilon(ns)$, that is, the interaction matrix element $F_{bond} = \langle 2a_1 | \hat{F} | ns \rangle$ squared and divided by the difference in orbital energies and that F_{bond} is in good approximation proportional to the bond overlap $S_{bond} = \langle 2a_1 | ns \rangle$ [32], as illustrated by **3**.

The bond overlap S_{bond} is sizeable and decreases from 0.31 to 0.28 to 0.21 to 0.19 along *M*=Li, Na, K and Rb. This is because the metal *ns* AOs become more diffuse and

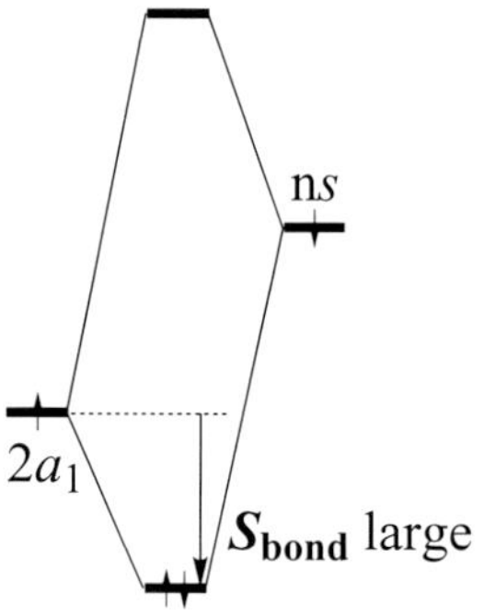

3

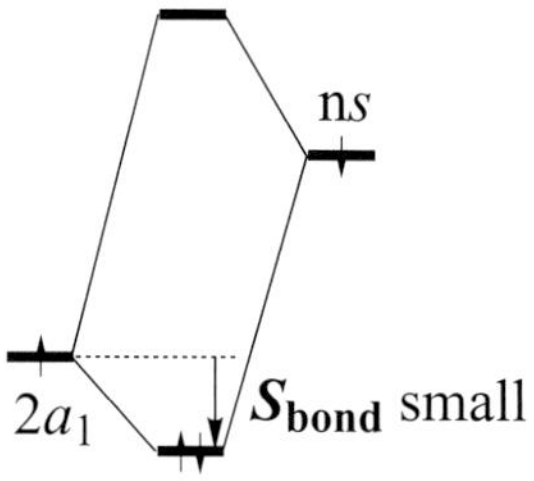

4

extended along this series, which causes the optimum overlap to become smaller and to be achieved at longer bond distances [33]. The corresponding interaction-matrix elements F_{bond} show the same trend as S_{bond}; they are sizeable and weaken from −70.3 to −61.5 to −40.2 kcal mol^{-1} along Li, Na and K (see Table 1). This leads to the observed weakening of ΔE_{oi} along Li–K [34], as illustrated by **3**→**4**, and contributes significantly to the same trend in the overall C-M bond strength. The loss of ΔE_{oi} is largest from Li to Na because, in addition to the decreasing bond overlap, the stabilizing admixture of the low-energy lithium $2p_\sigma$ AO into the electron-pair bonding combination is lost.

Note that ionic bonding would have led to the opposite trend in ΔE_{oi}, that is, an increase of the orbital interactions along Li–Rb. To understand this, we must first realize that ionic bonding between $CH_3^{\cdot}$ and $M^{\cdot}$ radicals corresponds, in terms of orbital interactions, to the complete transfer of an electron from the metal *ns* to the methyl $2a_1$ orbital and the absence of mixing between these SOMOs. This is illustrated schematically below in **5**.

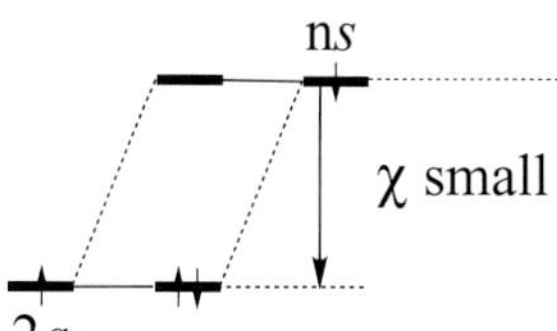

5

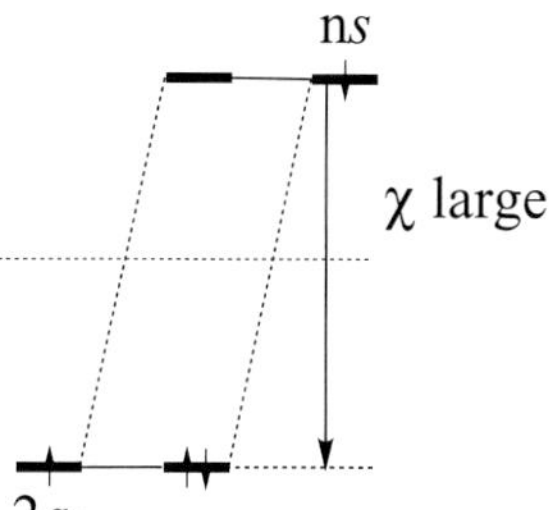

6

The electron of the metal drops, so to say, into the methyl $2a_1$ (which becomes a lone-pair like MO) and the orbital interactions ΔE_{oi} now originate from the associated orbital energy difference $\varepsilon(2a_1)-\varepsilon(ns)$. The latter, which can also be associated with the difference in electronegativity $\Delta\chi$ [35], increases along Li–Rb and so should ΔE_{oi} (see **5**→**6**) if the ionic aspect were to dominate the trend in the $CH_3^{\cdot}+M^{\cdot} \rightarrow CH_3M$ process. This is apparently not what happens.

Alkali metal atomic charges

The above description of the C–M bond (see **3** and **4**) as an electron-pair bond with significant covalent character (i.e., stabilization through bond overlap) is not at all in contradiction with a high polarity. Indeed, we find a wave function (electron-pair bonding MO) that is polarized toward carbon, and our Li atomic charge is +0.89, +0.50 and +0.39 a.u. according to the atoms in molecules (AIM), Hirshfeld and Voronoi deformation density (VDD) methods (see Table 1 for VDD charges *Q*(M) along Li–Rb) [10, 36]. The VDD atomic charge appears to be very close to the GAPT (generalized atomic polar tensors) charge of +0.4178 a.u. computed by Cioslowski at the HF/6-31G** level [37]. The polarity across the C–*M* bond arises of course from the difference in energy of the SOMOs of the methyl radical and the lithium atom (i.e., the electronegativity difference $\Delta\chi$). As pointed out in the introduction, the current ionic picture of the C–Li bond is based to an important extent on this polarization, in particular, on the atomic charge of Li in methyl lithium that in earlier studies was mostly found being close to +1 a.u. (i.e., +0.8, +0.85 and +0.9 a.u. according to IPP [9], NPA [10], and AIM [12] methods). This was interpreted as indicating that the C–Li bond is 80–90% ionic. The present study confirms that the C–Li bond is highly polar. However, as has been pointed out recently [36], atomic charges are not absolute bond-polarity indicators. Different atomic charge methods have different scales: the atomic charge of exactly the same atom in exactly the same chemical environment may differ dramatically from one method to another (e.g. Li in CH_3Li is +0.89 a.u. and +0.39 a.u. according to AIM and VDD). Atomic charges become meaningful only through the comparison of trends in values computed with the same method. Thus, one can conclude that all atomic-charge schemes indicate that the C–*M* bond is significantly more polar than most other bonds [36]. Furthermore, a high polarity does not rule out a prominent role for a stabilization of the C–*M* electron-pair bonding MO due to bond overlap between the SOMOs of the $CH_3^{\bullet}$ and $M^{\bullet}$ radicals. In fact, as illustrated in **3** and **4**, "covalent" and "ionic" stabilization (i.e., stabilization caused by bond overlap and electronegativity difference, respectively) occur in general simultaneously in polar bonds. In the case of the C–*M* bond in CH_3M, the trend of the *decreasing* bond overlap S_{bond} along *M*=Li, Na, K and Rb overrules the trend of the *increasing* electronegativity difference $\Delta\chi$, which derives from the difference in SOMO energies $\varepsilon(ns)-\varepsilon(2a_1)$ (see table 1). This is not because the electronegativity difference is small but because it changes relatively little compared to changes in the stabilization due to S_{bond}.

Conclusions

The C–*M* bond in methyl alkali metal systems becomes longer and *weaker* if one goes from Li to the more electropositive Rb as follows from our BP86/TZ2P computations. For methyl rubidium, the structure of which has not yet been characterized experimentally, we predict a C–*M* bond length of 2.821 Å, that is, about 0.07 Å longer than that in methyl potassium, whereas the methyl has essentially the same structure in both molecules. Also, the polarity of the C–*M* bond increases along Li–Rb but it preserves a strong intrinsic preference to homolytic over ionic dissociation.

We have shown that a description of the bonding mechanism in terms of a polar C–*M* electron-pair bond between the methyl radical and alkali metal atom is just as natural as an ionic description (i.e., in terms of $CH_3^-+M^+$). In this picture, the weakening of the C–M bond along Li–Rb is ascribed to the decreasing bond overlap between the SOMOs of the $CH_3^{\bullet}+M^{\bullet}$ radicals into which this bond preferentially dissociates. This is not in contradiction with the fact that the C–M bond is highly polar.

Acknowledgements We thank the following organizations for financial support: HPC-Europa, the Deutsche Akademische Austauschdienst (DAAD), the National Research School Combination-Catalysis (NRSC-C), the Netherlands Organization for Scientific Research (NWO), the Ministerio de Educación y Cultura (MEC), the Training and Mobility of Researchers (TMR) program of the European Union, the Dirección General de Enseñanza Superior e Investigación Científica y Técnica (MEC-Spain) and the DURSI (Generalitat de Catalunya). Excellent service by the Stichting Academisch Rekencentrum Amsterdam (SARA) and the Centre de Supercomputació de Catalunya (CESCA) is gratefully acknowledged.

References

1. Elschenbroich C, Salzer A (1992) Organometallics. A concise indroduction, 2nd edn. VCH, Weinheim, Germany
2. Dill JD, Schleyer PR, Binckley JS, Pople JA (1977) J Am Chem Soc 99:6159–6173
3. Graham GD, Marynick DS, Lipscomb WN (1980) J Am Chem Soc 102:4572–4578
4. Schiffer H, Ahlrichs R (1986) Chem Phys Lett 124:172–176
5. Günther H, Moskau D, Bast P, Schmalz D (1987) Angew Chem, Int Ed Engl 26:1212–1220
6. Bauer W, Schleyer PR (1992) Adv Carbanion Chem 1:89–175
7. Bauer W (1995) Lithium Chemistry. Wiley-Interscience, New York
8. Fraenkel G, Martin KV (1995) J Am Chem Soc 117:10336–10344
9. Streitwieser A, Williams JE, Alexandratos S, McKelvey JM (1976) J Am Chem Soc 98:4778–4783
10. Bickelhaupt FM, van Eikema Hommes NJR, Fonseca Guerra C, Baerends EJ (1996) Organometallics 15:2923–2931
11. Lambert C, Kaupp M, Schleyer PvR (1993) Organometallics 12:853–859
12. Wiberg KB, Rablen PR (1993) J Comput Chem 14:1504–1518
13. Streitwieser Jr A (1978) J Organometal Chem 156:1–3
14. Bushby RJ, Steel HL (1987) J Organometal Chem 336:C25–C32
15. Bushby RJ, Steel HL (1990) J Chem Soc, Perkin Trans 2:1143–1153
16. Kwon O, Sevin F, McKee ML (2001) J Phys Chem A 105:913–922
17. Breidung J, Thiel W (2001) J Molec Struct 599:239–254
18. Scalmani G, Brédas JL (2000) J Chem Phys 112:1178–1191
19. Fressigné C, Maddaluno J, Giessner-Prettre C (1999) J Chem Soc, Perkin Trans 2:2197–2201
20. Kremer T, Harder S, Junge M, Schleyer PvR (1996) Organometallics 15:585–595

21. Tyerman SC, Corlett GK, Ellis AM, Claxton TA (1996) J Molec Struct (Theochem) 364:107–119
22. Lambert C, Schleyer PvR (1994) Angew Chem 106:1187–1199; Angew Chem Int Ed Engl 33:1129–1140
23. Wiberg K, Breneman CM (1990) J Am Chem Soc 112:8765–8775
24. Kaufmann E, Raghavachari K, Reed AE, Schleyer PvR (1988) Organometallics 7:1597–1607
25. Bauer W, Winchester WR, Schleyer PvR (1987) Organometallics 6:2371–2379
26. te Velde G, Bickelhaupt FM, Baerends EJ, van Gisbergen SJA, Fonseca Guerra C, Snijders JG, Ziegler T (2001) J Comput Chem 22:931–967
27. Bickelhaupt FM, Baerends EJ (2000) In: Lipkowitz KB, Boyd DB (eds) Reviews in computational chemistry, vol. 15. Wiley-VCH, New York p 1–86
28. Grotjahn DB, Pesch TC, Brewster MA, Ziurys LM (2000) J Am Chem Soc 122:4735–4741
29. Grotjahn DB, Pesch TC, Xin J, Ziurys LM (1997) J Am Chem Soc 119:12368–12369
30. Schmid R (2003) J Chem Ed 80:931–937
31. The description of the MO in terms of fragment MO *coefficients* in stead of Gross Mulliken contributions yields the same picture but it has the disadvantage of not being normalized, that is, the figures do not add up to 1 (or to 100%)
32. Albright TA, Burdet JK, Whangbo M-H (1985) Orbital interactions in chemistry. Wiley-Interscience, New York
33. The C-M bond distance also increases along Li–Rb because of the increasing number of metal core shells that enter into Pauli repulsion with closed shells on the methyl fragment. For a detailed discussion of how Pauli repulsion affects bond distances, see: Bickelhaupt FM, DeKock RL, Baerends EJ (2002) J Am Chem Soc 124:1500–1505
34. More subtle effects begin to play a role from K to Rb as the covalent character of the bond, and especially changes therein, become smaller
35. Mann JB, Meek TL, Knight ET, Capitani JF, Allen LC (2000) J Am Chem Soc 122:5132–5137
36. Fonseca Guerra C, Handgraaf J-W, Baerends EJ, Bickelhaupt FM (2004) J Comput Chem 25:189–210
37. Cioslowski J (1989) J Am Chem Soc 111:8333–8336

J Mol Model (2006) 12: 569–576
DOI 10.1007/s00894-005-0035-5

ORIGINAL PAPER

Anastassia N. Alexandrova · Eldon Koyle
Alexander I. Boldyrev

Theoretical study of hydrogenation of the doubly aromatic B_7^- cluster

Received: 8 March 2005 / Accepted: 29 July 2005 / Published online: 29 October 2005

Abstract We have studied the influence of hydrogenation on the relative stability of the low-lying isomers of the anionic B_7^- cluster, computationally. It is known that the pure-boron B_7^- cluster has a doubly (σ- and π-) aromatic C_{6v} ($^3A_1$) quasi-planar wheel-type triplet global minimum (structure 1), a low-lying σ-aromatic and π-antiaromatic quasi-planar singlet C_{2v} ($^1A_1$) isomer **2** (0.7 kcal mol^{-1} above the global minimum), and a planar doubly (σ- and π-) antiaromatic C_{2v} ($^1A_1$) isomer **3** (7.8 kcal mol^{-1} above the global minimum). However, upon hydrogenation, an inversion in the stability of the species occurs. The planar $B_7H_2^-$ (C_{2v}, $^1A_1$) isomer **4**, originated from the addition of two hydrogen atoms to the doubly antiaromatic B_7^- isomer **3**, becomes the global minimum structure. The second most stable $B_7H_2^-$ isomer **5**, originated from the quasi-planar triplet wheel isomer **1** of B_7^-, was found to be 27 kcal mol^{-1} higher in energy. The inversion in stability occurs due to the loss of the doubly aromatic character in the wheel-type global minimum isomer (C_{6v}, $^3A_1$) of B_7^- upon H_2–addition. In contrast, the planar isomer of B_7^- (C_{2v}, $^1A_1$) gains aromatic character upon addition of two hydrogen atoms, which makes it more stable.

Keywords Multiple aromaticity · Boron hydrides · Boron clusters

Dedicated to Professor Dr. Paul von Ragué Schleyer on the occasion of his 75th birthday.

A. N. Alexandrova · E. Koyle · A. I. Boldyrev (✉)
Department of Chemistry and Biochemistry,
Utah State University, 0300 Old Main Hill,
Logan, UT 84322-0300, USA
E-mail: Boldyrev@cc.usu.edu
Tel.: +1-435-7971630
Fax: +1-435-7973390

Introduction

Pure-boron clusters have received much attention in the literature over the past couple of decades [1–12]. The geometries and electronic properties of these species were addressed, while sometimes the results obtained by different groups were not in agreement. Recently, we published a series of articles on the chemical bonding of neutral and anionic planar or quasi-planar pure-boron clusters from B_3 to B_{15} [13–18]. Both theoretical and photoelectron spectroscopic characterizations were performed for these species. We have explained the planar or quasi-planar structure of small boron clusters on the basis of the presence of double (σ- and π-) aromaticity, or σ- and π-antiaromaticity, or a mixture of them. Double aromaticity (the simultaneous presence of σ- and π-aromaticity) was introduced in chemistry by Schleyer and coworkers [19] in the late 1970 s for explaining properties of the 3,5-dehydrophenyl cation. Double aromaticity and antiaromaticity were first used by Martin-Santamaria and Rzepa [20] for explaining chemical bonding in small carbon rings. Berndt and coworkers have shown that small carborane molecules containing three- and four-membered rings also exhibit both σ- and π-aromaticity [21–25]. Double aromaticity in boron clusters forces them to adopt the wheel-type structures, such as D_{7h} ($^1A_1'$) B_8^{2-} or D_{8h} ($^1A_{1g}$) B_9^- [17], while the doubly antiaromatic boron clusters, such as B_6^{2-} [15], adopt highly distorted polygonal topology. Switching from an aromatic form to an antiaromatic form and vice versa can be accomplished by adding or removing a pair of electrons, or by forming two B-H bonds during the partial hydrogenation of boron clusters.

We thought that it would be important to understand how hydrogenation may affect the structure of boron clusters, because when 1:1 hydrogenation is reached, a transformation from planar or quasi-planar structures of pure-boron clusters into 3D structures of hydrogenated species is expected. The $B_xH_x^{2-}$ boranes are known to have deltahedral structures [26–28].

Hydrogenated small boron clusters have been studied before. Ricca and Bauschlicher characterized cationic B_nH^+ clusters theoretically [29]. Small systems B_2H^+, $B_2H_2^+$, and $B_3H_2^+$ were studied by Curtiss and Pople [30]. Larger B_nH_n species ($n = 3–6$) were calculated by many groups using ab initio methods (see reference [31] and references therein). A series of theoretical and experimental studies on other similar systems have been reported by various authors [32–34]. Experimental and theoretical studies of the pure B_7 cluster in neutral and various ionic forms were also previously reported [8, 35–39]. The B_7^- anion has a triplet doubly (σ- and π-) aromatic C_{6v} ($^3A_1$) quasi-planar wheel-type global minimum, a low-lying singlet σ-aromatic and π-antiaromatic C_{2v} ($^1A_1$) quasi-planar isomer (0.7 kcal mol^{-1} above the global minimum), and a planar doubly (σ- and π-) antiaromatic C_{2v} ($^1A_1$) isomer [16].

In the current article we present results of our theoretical study of the hydrogenation of a doubly aromatic boron cluster B_7^-. This includes our computational results on the relative stability of low-lying B_7^- isomers upon the influence of addition of two hydrogen atoms.

Computational methods

While the structure of the B_7^- anion was firmly established in our previous work [16], we decided to perform the search for its global minimum again in order to test our Gradient Embedded Genetic Algorithm (GEGA) program, written by Alexandrova et al. [40]. We then used the GEGA program to find low-lying isomers for the $B_7H_2^-$ anion also. The GEGA procedure [40] is based upon the geometry interplay proposed by Hartke [41] and Deaven and Ho [42], and involves some unique mutation and optimization specifics. The hybrid method, known as B3LYP [43–45], with relatively small basis set 3-21G was employed throughout the execution of the GEGA. Briefly, within the GEGA procedure, the initial geometries of individuals in the population are randomly generated and further optimized to the nearest local minima on the potential energy surface, using the Gaussian 03 package [46]. If a saddle point is encountered, the normal mode of the first imaginary frequency is followed until a local minimum is found. Then the population, composed of the good individuals thus selected, undergoes breeding and mutations. Probabilities to be bred are assigned according to the best-fit (lowest-energy) criterion. Based on these probabilities, pairs of parents are selected randomly. The geometries of the parents are cut by a random cutting plane, and the halves thus obtained (genes) are then recombined either in a simple or in a head-to-tail manner forming a child. The number of atoms in the newly generated geometry is checked, and the child is optimized to the nearest local minimum. After the number of individuals in the population has doubled within the breeding process, the best-fit group is selected and convergence of the algorithm is checked. The GEGA is considered converged if the current lowest energy species (global minimum or at least very stable local minimum) remains leading for 20 iterations. If convergence is not yet met, the highest energy species in the population undergo mutations. The mutation rate is set to 33.33%. Mutations are shifts of random atoms of a species in random directions, with the purpose of changing the initial geometry so as to push the structure out of the current local minimum to another well on the potential energy surface. Mutants are optimized to the nearest local minima. After that, the algorithm proceeds with the new cycle of breeding. All low-lying isomers are detected and stored throughout the execution and they are reported to the user at the end of the run. A few runs of GEGA are done on the system in order to confirm the global minimum structure found.

The geometry and vibrational frequencies of the global minimum thus identified, as well as low-lying isomers, were refined further at higher levels of theory. The B3LYP and coupled-cluster CCSD(T) [47–49] methods with polarized split-valence basis sets (6-311++G**) [50, 51] were used for this purpose. We also optimized geometry and calculated frequencies for the $B_7H_2^-$ global minimum structure at the CASSCF(4,4)/6-311++G** [52–57] level of theory. Single-point calculations were performed at the CASSCF(8,8)/6-311++G**//CASSCF(4,4)/6-311++G** level of theory in order to test the applicability of the B3LYP and CCSD(T) methods. The Natural Bond Order (NBO) analysis [58–62] was used for detailed chemical bonding examination. Molecular orbitals (MOs) were calculated at the HF/6-311++G** level of theory. B3LYP, HF, and CCSD(T) calculations were performed using the Gaussian 03 program. MOs were made using the MOLDEN 3.4 program [63].

Computational results and discussion

The GEGA search revealed the same three low-lying isomers **1**–**3** (Fig. 1) for B_7^- as we reported in our previous work [16]. Thus, our GEGA program performed well. The search for the most stable structure of the $B_7H_2^-$ anion, performed by GEGA and accompanied by the higher level calculations, revealed that the planar **4** (C_{2v}, $^1A_1$) (Fig. 2) structure is the global minimum. The order of the lowest structures for $B_7H_2^-$ was the same at the B3LYP/3-21G, B3LYP/6-311+G*, and CCSD(T)/6-311+G* levels of theory. The global minimum $B_7H_2^-$ cluster has the following electronic configuration: $1a_1^21b_2^22a_1^22b_2^23a_1^24a_1^23b_2^21b_1^24b_2^25a_1^26a_1^21a_2^2$. The molecular properties of the global minimum structure at three levels of theory are summarized in Table 1. The geometric parameters and harmonic frequencies calculated at these levels of theory were in reasonable agreement with each other. The Hartree–Fock configuration was found to be dominant ($C_{HF} = 0.969$) in the CASSCF(8,8)/

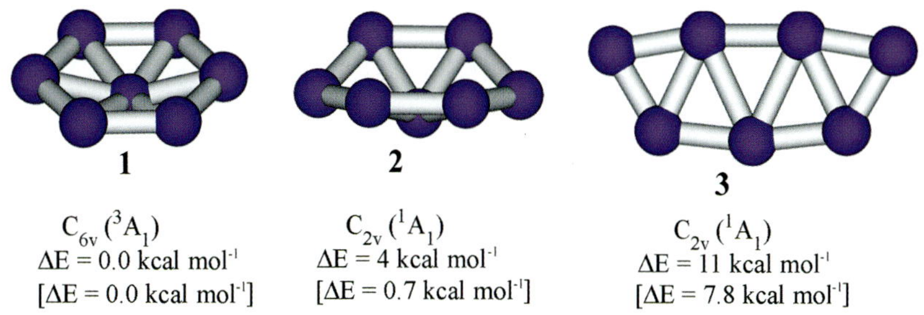

Fig. 1 Low-energy isomers of the B_7^- cluster at the B3LYP/6-311+G* level of theory (relative energies computed at the CCSD(T)/6-311+G(2df) level are shown in square brackets)

6-311++G** expansion and thus our B3LYP and CCSD(T) calculations should be reliable.

The second lowest isomer (structure 5 in Fig. 2) is 27 kcal mol^{-1} higher in energy at the B3LYP/6-311++G** level of theory. Molecular properties of the second isomer at B3LYP/6-311++G** and CCSD(T)/6-311++G** levels of theory are summarized in Table 2. All low-energy structures of $B_7H_2^-$ identified are shown in Fig. 2. The global minimum of $B_7H_2^-$ is a planar species; it originates from the planar isomer **3** of the bare B_7^- cluster (Fig. 1) with two hydrogen atoms bonded to the end boron atoms of the cluster. Other alternative isomers **8**, **10**, and **14**, having hydrogen atoms bound in different positions, also derive from the structure 3 of B_7^-. Among the isomers depicted in Fig. 2, the singlet and triplet species **5**–**7**, **9**, **11**, **12**, and **15** arise from either the isomer **1** or isomer **2** of B_7^- with different bonding positions of hydrogen atoms.

Thus, the addition of two hydrogen atoms completely rearranges the relative stability of the isomers of B_7^-. In the following section, we present our explanation of why that is happening.

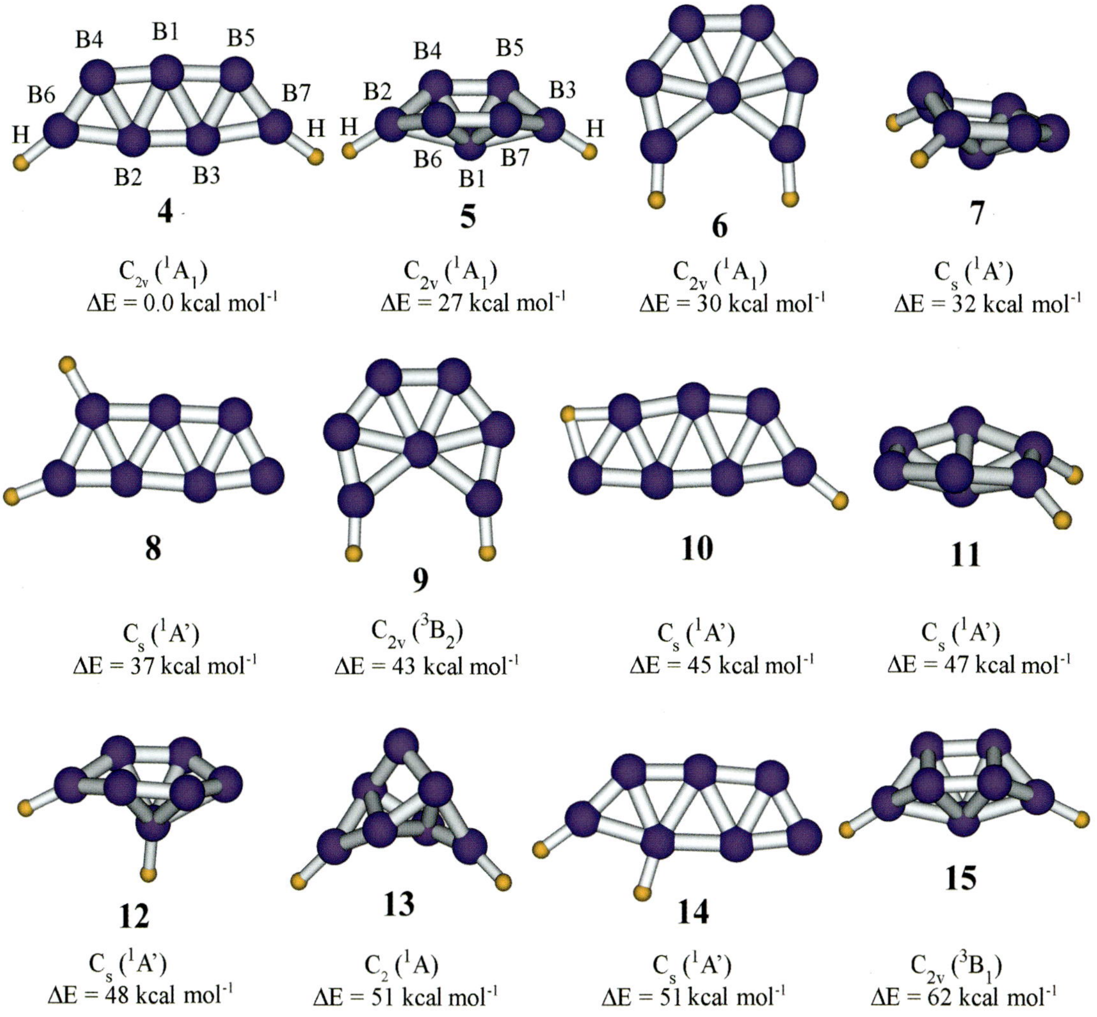

Fig. 2 Alternative structures of the $B_7H_2^-$ hydride (B3LYP/6-311++G** level of theory)

Table 1 Molecular properties of the $B_7H_2^-$ (C_{2v}, $^1A_1$) global minimum species

	B3LYP/6-311++G**		CASSCF(4,4)/6-311++G**		CCSD(T)/6-311++G**
E_{total}, a.u.	−175.143862		−173.860826[b]		−174.554521
ZPE, kcal mol^{-1}	27.452		29.468		[c]
Geometry	R(B1-B2,3) = 1.7496 Å		R(B1-B2) = 1.7859 Å		R(B1-B2) = 1.7735 Å
	R(B1-B4,5) = 1.5954 Å		R(B1-B4) = 1.5969 Å		R(B1-B4) = 1.6179 Å
	R(B4-B6,7) = 1.5166 Å		R(B4-B6) = 1.5087 Å		R(B4-B6) = 1.5428 Å
	R(B2-B3) = 1.5975 Å		R(B2-B3) = 1.6150 Å		R(B2-B3) = 1.6196 Å
	R(B2-B6) = 1.6730 Å		R(B2-B6) = 1.6938 Å		R(B2-B6) = 1.6884 Å
	R(B6-H) = 1.1833 Å		R(B6-H) = 1.1815 Å		R(B6-H) = 1.1902 Å
	R(B4-B2) = 1.6763 Å		R(B4-B2) = 1.6910 Å		R(B4-B2) = 1.7071 Å
	∠(B4-B6-H) = 162.895°		∠(B4-B6-H) = 163.756°		∠(B4-B6-H) = 162.245°
Frequencies, cm^{-1}	$\omega_1(a_1)$ 2686 (84)[a]	$\omega_{12}(b_1)$ 705 (1)	$\omega_1(a_1)$ 2766	$\omega_{12}(b_1)$ 764	[c]
	$\omega_2(a_1)$ 1312 (8)	$\omega_{13}(b_1)$ 469 (2)	$\omega_2(a_1)$ 2042	$\omega_{13}(b_1)$ 485	
	$\omega_3(a_1)$ 1210 (0)	$\omega_{14}(b_1)$ 168 (1)	$\omega_3(a_1)$ 932	$\omega_{14}(b_1)$ 166	
	$\omega_4(a_1)$ 858 (8)	$\omega_{15}(b_2)$ 2684 (284)	$\omega_4(a_1)$ 794	$\omega_{15}(b_2)$ 2762	
	$\omega_5(a_1)$ 754 (13)	$\omega_{16}(b_2)$ 1314 (14)	$\omega_5(a_1)$ 726	$\omega_{16}(b_2)$ 1381	
	$\omega_6(a_1)$ 683 (1)	$\omega_{17}(b_2)$ 1128 (2)	$\omega_6(a_1)$ 642	$\omega_{17}(b_2)$ 1362	
	$\omega_7(a_1)$ 566 (12)	$\omega_{18}(b_2)$ 897 (10)	$\omega_7(a_1)$ 567	$\omega_{18}(b_2)$ 1151	
	$\omega_8(a_1)$ 381 (0)	$\omega_{19}(b_2)$ 767 (0)	$\omega_8(a_1)$ 398	$\omega_{19}(b_2)$ 882	
	$\omega_9(a_2)$ 731 (0)	$\omega_{20}(b_2)$ 641 (17)	$\omega_9(a_2)$ 809	$\omega_{20}(b_2)$ 809	
	$\omega_{10}(a_2)$ 471 (0)	$\omega_{21}(b_2)$ 521 (0)	$\omega_{10}(a_2)$ 480	$\omega_{21}(b_2)$ 523	
	$\omega_{11}(a_2)$ 258 (0)		$\omega_{11}(a_2)$ 174		

[a]Infrared intensities (km mol^{-1}) are shown in parentheses
[b]At the CASSCF(8,8)/6-311++G**//CASSCF/6-311++G** level E_{total} = −173.891026 a. u., C_{HF} = 0.969
[c]Vibrational frequencies were not calculated at this level of theory

Chemical bonding in $B_7H_2^-$ isomers

First, we performed an NBO analysis of the global minimum structure of $B_7H_2^-$. We found nine classical 2c–2e bonds, seven of which are localized on the peripheral edges of the B_7^- unit (bonds B1–B4, B4–B6, B6–B2, B2–B3, B3–B7, B7–B5, B5–B1, see Fig. 2), and the remaining two are simply covalent B–H bonds. The rest of the electrons in the system participate in delocalized bonding over the entire cluster. A similar bonding picture was found in structure 3 of the bare B_7^- species. The NBO analysis showed the presence of seven peripheral 2c–2e B–B bonds in the B_7^- cluster. However, the difference between the dihydride and the original B_7^- cluster occurs in the structure of delocalized bonding patterns.

The MOs of the global minimum isomer of $B_7H_2^-$ are shown in Fig. 3a. Figure 3b represents the MOs of the structure 3 of the bare B_7^- cluster. In order to elucidate the MOs responsible for the localized bonding, described

Table 2 Molecular properties of the second isomer of $B_7H_2^-$ (C_{2v}, $^1A_1$) – species **2**, Figure 1

	B3LYP/6-311++G**		CCSD(T)/6-311++G**
E_{total}, a.u.	−175.100763		−174.53084463
ZPE, kcal mol^{-1}	27.592		[b]
Geometry	R(B1-B2,3) = 1.9267 Å		R(B1-B2,3) = 1.9338 Å
	R(B1-B4,5,6,7) = 1.6640 Å		R(B1-B4,5,6,7) = 1.6901 Å
	R(B4-B5) = 1.5905 Å		R(B4-B5) = 1.6175 Å
	R(B2-B3) = 3.7470 Å		R(B2-B3) = 3.7730 Å
	R(B2-B4) = 1.6232 Å		R(B2-B4) = 1.6462 Å
	R(B2-H) = 1.1956 Å		R(B2-H) = 1.1981 Å
	∠(B1-B2-H) = 133.181°		∠(B1-B2-H) = 133.283°
Frequencies, cm^{-1}	$\omega_1(a_1)$ 2572 (83)[a]	$\omega_{12}(b_1)$ 1109 (9)	[b]
	$\omega_2(a_1)$ 1277 (1)	$\omega_{13}(b_1)$ 823 (11)	
	$\omega_3(a_1)$ 967 (6)	$\omega_{14}(b_1)$ 566 (16)	
	$\omega_4(a_1)$ 757 (7)	$\omega_{15}(b_2)$ 2570 (409)	
	$\omega_5(a_1)$ 695 (0)	$\omega_{16}(b_2)$ 1032 (1)	
	$\omega_6(a_1)$ 438 (0)	$\omega_{17}(b_2)$ 1015 (25)	
	$\omega_7(a_1)$ 272 (0)	$\omega_{18}(b_2)$ 901 (31)	
	$\omega_8(a_2)$ 875 (0)	$\omega_{19}(b_2)$ 792 (12)	
	$\omega_9(a_2)$ 818 (0)	$\omega_{20}(b_2)$ 484 (1)	
	$\omega_{10}(a_2)$ 614 (0)	$\omega_{21}(b_2)$ 317 (62)	
	$\omega_{11}(a_2)$ 359 (0)		

[a]Infrared intensities are shown in parentheses
[b]Vibrational frequencies were not calculated at this level of theory

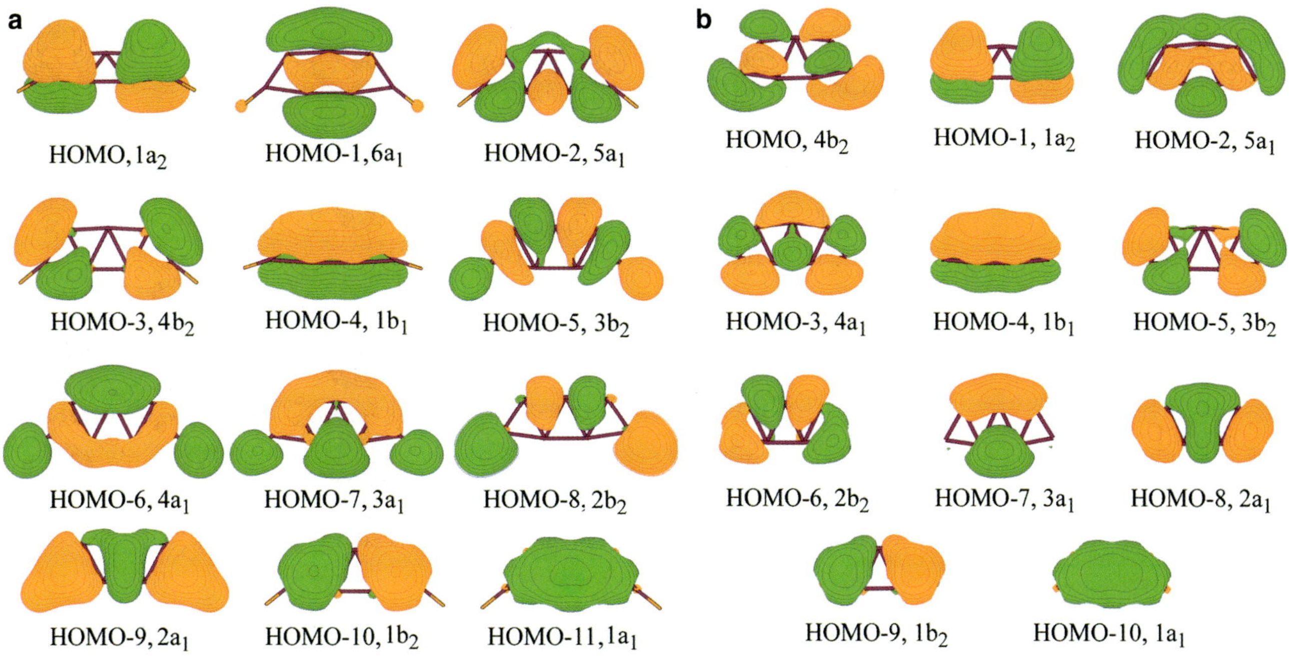

Fig. 3 Molecular orbital picture of **a** the global minimum of $B_7H_2^-$ (C_{2v}, $^1A_1$) and **b** of the bare B_7^- cluster (C_{2v}, $^1A_1$)

above, we performed an NBO calculation on the $B_7H_2^{5+}$ cation. We found that if the HOMO-2, HOMO-3, HOMO-5, HOMO-6, HOMO-7, HOMO-8, HOMO-9, HOMO-10, and HOM0-11 are occupied in $B_7H_2^{5+}$, the MOs can be localized into seven 2c–2e bonds as explained earlier. Thus, the localized bonding described is due to the presence of these low-lying MOs. The remaining HOMO, HOMO-1, and HOMO-4 are responsible for the bonding over the whole system. A similar analysis performed on the bare B_7^- cluster revealed that HOMO-3, HOMO-5, HOMO-6, HOMO-7, HOMO-8, HOMO-9, and HOMO-10 (Fig. 3b) form the peripheral σ-framework, while the remaining MOs represent delocalized bonding in the cluster. One may notice that, upon the introduction of two electrons to the system (addition of two H-atoms), the localized bonding region is enriched with two classical 2c–2e B–H bonds. Correspondingly, the delocalized region sacrifices one molecular orbital. Indeed, the HOMO of the B_7^- cluster is missing in $B_7H_2^-$, while the HOMO-1 in B_7^- corresponds to the HOMO in $B_7H_2^-$, the HOMO-2 corresponds to the HOMO-1 in $B_7H_2^-$, and the shapes of the HOMO-4 are identical in the two species (see Fig. 3).

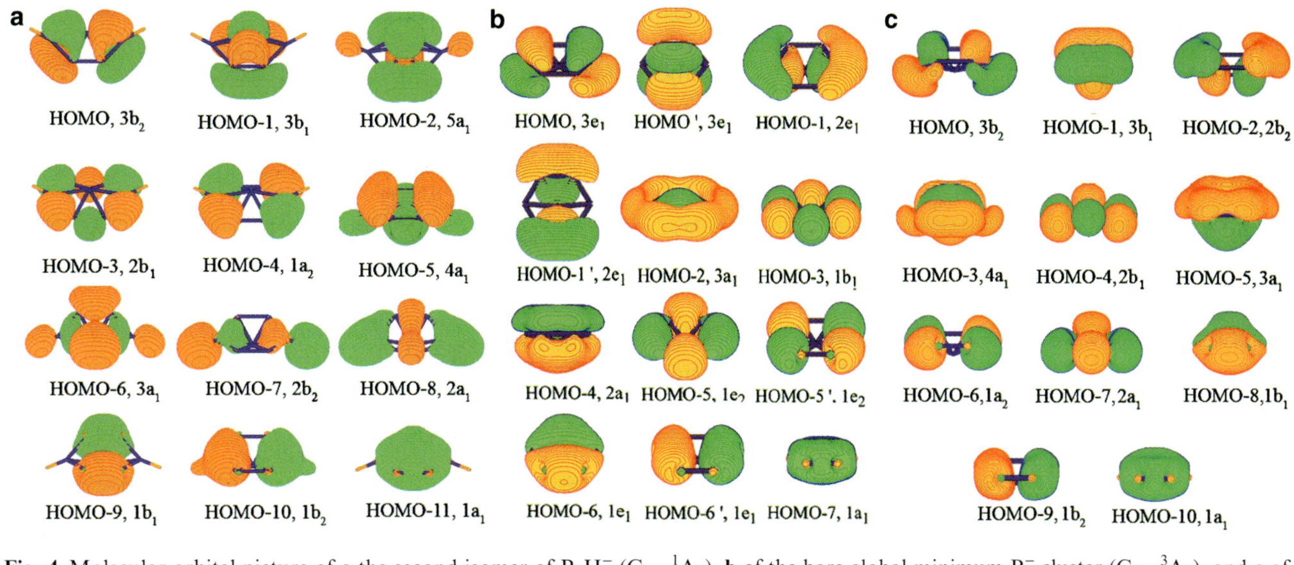

Fig. 4 Molecular orbital picture of **a** the second isomer of $B_7H_2^-$ (C_{2v}, $^1A_1$), **b** of the bare global minimum B_7^- cluster (C_{6v}, $^3A_1$), and **c** of the C_{2v}, $^1A_1$ isomer 2 of B_7^-

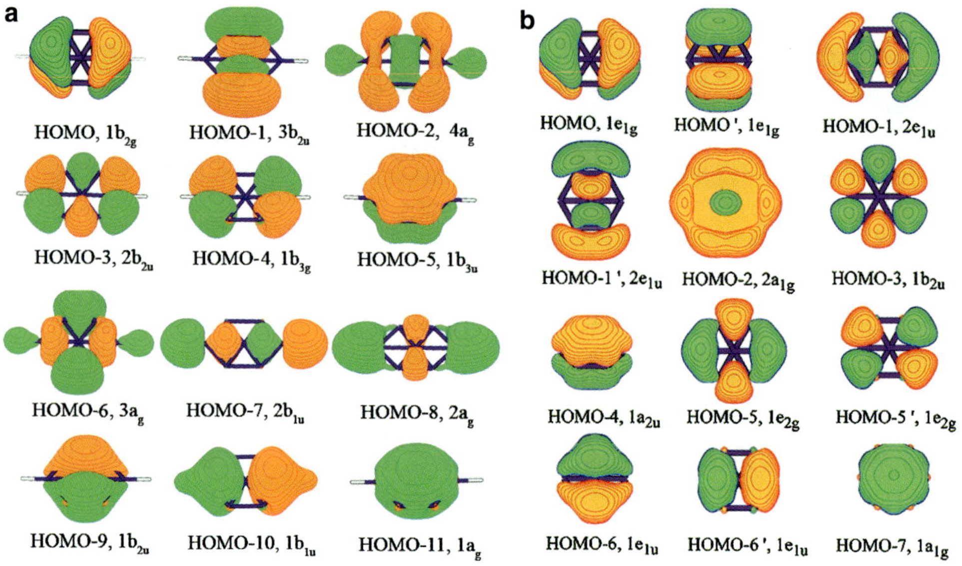

Fig. 5 Adjusted to the planarity molecular orbital picture of **a** the second isomer of $B_7H_2^-$ and **b** of the bare B_7^- cluster with corresponding geometry

As has been discussed in our previous work [16], the planar isomer **3** of the B_7^- cluster possesses doubly antiaromatic character. The HOMO-4 ($1b_1$) is a completely bonding π-molecular orbital, and the HOMO-1 ($1a_2$) is partially bonding π-MO with one nodal plane. Four π-electrons on these MOs make structure 3 of B_7^- π-antiaromatic. The HOMO-2 ($5a_1$) and the HOMO ($4b_2$) are MOs of σ-type formed by the radial overlap of the 2p-AOs on boron atoms, where the former MO is completely bonding, and the latter one is partially bonding. Hence, the four electrons occupying these MOs make the system also σ-antiaromatic.

In the hydride, the partially bonding σ-molecular orbital (HOMO in B_7^-) mentioned earlier is now a virtual orbital. The only delocalized σ-MO in the species is the HOMO-1 ($6a_1$) (Fig. 3a). Since the cluster obeys the ($4n+2$) Hückel rule for aromatic compounds, it possesses σ-aromaticity. The π-MOs in the $B_7H_2^-$ cluster are analogous to those in the B_7^- species. Four electrons in these MOs make the global minimum of $B_7H_2^-$ π-antiaromatic. Consequently, upon the molecular hydrogen addition, the planar B_7^- cluster gains σ-aromaticity and loses the antibonding character in the σ-set of MOs.

We also performed a molecular orbital analysis for the wheel-type B_7^- and $B_7H_2^-$ moieties. The MOs of these species are shown in Fig. 4. Figure 5 depicts the MOs of the clusters in a simplified manner, that is, when geometries of both clusters are adjusted to planarity. This adjustment is done in order to make the bonding picture clearer (the sets of valence MOs shown are analogous to the ones in the original species). Removal of the electrons from the HOMO, HOMO-1, HOMO-2, and HOMO-5 (Fig. 5a) in $B_7H_2^-$ and subsequent localization demonstrates that the remaining MOs are responsible for the formation of six 2c–2e B–B bonds: B2–B4, B4–B5, B5–B3, B3–B7, B7–B6, B6–B2, and two B–H bonds. Similarly, in the B_7^- (C_{6v}, $^3A_1$) cluster, removal of electrons from the HOMO, HOMO′, HOMO-1, HOMO-1′, HOMO-2, and HOMO-4 (Fig. 5b) leaves the cluster with six peripheral B–B σ-bonds forming the six-membered ring. Thus, HOMO, HOMO-1, HOMO-2, and

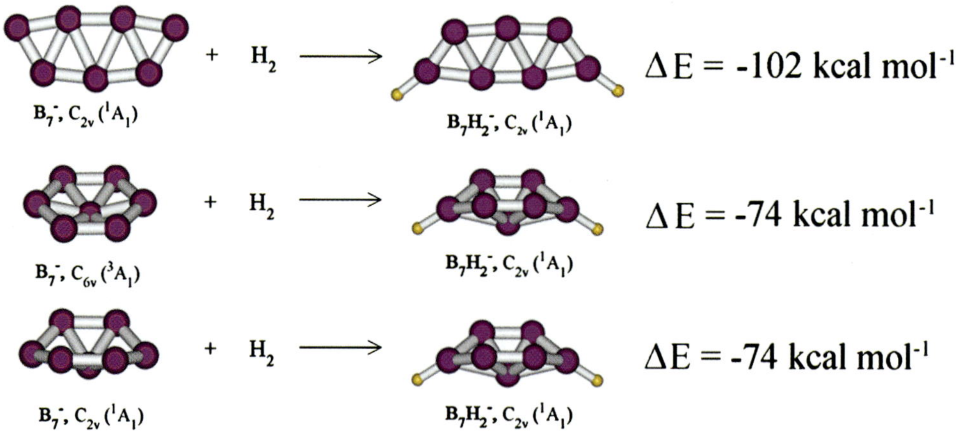

Fig. 6 Calculated hydrogenation energies (at the CCSD(T)/6-311+G(2df) level of theory) for the lowest B_7^- isomers

HOMO-5 in $B_7H_2^-$, and HOMO, HOMO′, HOMO-1, HOMO-1′, HOMO-2, and HOMO-4 in B_7^- are responsible for the bonding over the whole system. As has been reported earlier [16], the triplet species **1** is doubly aromatic. The HOMO-4 ($1a_{2u}$) in Fig. 5b is a completely bonding π-MO and the HOMO-2 ($4a_{1g}$) is a completely bonding σ-MO. Degenerate pairs HOMO ($1e_{1g}$) and HOMO-1 ($2e_{1u}$) are partially bonding MOs of π- and σ-radial types, respectively. Thus the structure 1 is σ-aromatic (six σ-electrons) and π-aromatic (four π-electrons with single occupancy of the every degenerate HOMO). In the second isomer of B_7^- (singlet structure 2, Fig. 1), the HOMO (Fig. 5b) is doubly occupied, leaving the second degenerate MO unoccupied. Thus, the regular 4*n* Hückel's rule governs the shape of the cluster, making it π-antiaromatic. However, the second lowest isomer of $B_7H_2^-$ (structure 5, Fig. 2) possesses double antiaromaticity. The HOMO-5 ($1b_{3u}$) and HOMO ($1b_{2g}$) are π-MOs populated by four electrons, and thus the system is π-antiaromatic. The HOMO-2 ($4a_g$) and HOMO-1 ($3b_{2u}$) are σ-radial MOs populated by four σ-electrons, and thus the system is also σ-antiaromatic. Hence, upon hydration the most stable triplet B_7^- cluster **1** loses both π- and σ-aromatic character. The singlet isomer **2** of B_7^-, being σ-aromatic and π-antiaromatic, becomes doubly antiaromatic during the hydration process.

Energy of the hydrogenation

The effect of hydrogenation on the relative stability of the B_7^- isomers can also be illustrated by the energies of reactions of hydrogenation (at the CCSD(T)/6-311++G** level of theory) (Fig. 6). One can see that the energy of the hydrogenation is significantly higher for the isomer **3** of B_7^-.

Conclusions

Using the GEGA program, we found low-energy structures of B_7^- and $B_7H_2^-$ at the B3LYP/3-21G level of theory. The low-energy species were then re-optimized at the B3LYP/6-311++G** and CCSD(T)/6-311++G** levels of theory and the results were found to be in good agreement. CASSCF(8,8)/6-311++G** calculations showed that the Hartree–Fock configuration was dominant in the CASSCF expansion, and thus our B3LYP and CCSD(T) calculations must be reliable.

The B_7^- cluster has a wheel-type quasi-planar doubly (σ- and π-) aromatic triplet C_{6v} ($^3A_1'$) global minimum, a low-lying σ-aromatic and π-antiaromatic singlet C_{2v} ($^1A_1$) isomer with a quasi-planar shape (0.7 kcal mol^{-1} above the global minimum), and a planar doubly (σ- and π-) antiaromatic C_{2v} ($^1A_1$) isomer (7.8 kcal mol^{-1} above the global minimum). Upon hydrogenation, an inversion in stability of the species occurs. The planar $B_7H_2^-$ (C_{2v}, $^1A_1$) isomer **4**, formed by the addition of two hydrogen atoms to the doubly antiaromatic C_{2v} ($^1A_1$) B_7^- isomer **3**, becomes the global minimum structure. It is 27 kcal mol^{-1} more stable than isomer **5** of $B_7H_2^-$, originated from the quasi-planar triplet wheel isomer of B_7^-. We have shown that the reason for the inversion of stability of adducts is in the loss of the doubly aromatic character in the wheel-type global minimum isomer (C_{6v}, $^3A_1$) of B_7^- upon the H_2 addition. In contrast, the planar isomer of B_7^- (C_{2v}, $^1A_1$) gains aromatic character upon addition of two hydrogen atoms, which makes it more stable.

Acknowledgments This work was supported partially by the donors of The Petroleum Research Fund (ACS-PRF# 38242-AC6), administered by the American Chemical Society, partially by the National Science Foundation (CHE-0404937), and partially by the Summer Research Institute at the Pacific Northwest National Laboratory operated by Battelle, Richland, Washington, USA.

References

1. Hanley L, Whitten JL, Anderson SL (1988) J Phys Chem 92:5803–5812
2. Hanley L, Anderson SL (1987) J Phys Chem 91:5161–5163
3. Hanley L, Anderson SL (1988) J Chem Phys 89:2848–2860
4. Hintz PA, Ruatta SA, Anderson SL (1990) J Chem Phys 92:292–303
5. Ruatta SA, Hintz PA, Anderson SL (1991) J Chem Phys 94:2833–2847
6. Hintz PA, Sowa MB, Ruatta SA, Anderson SL (1991) J Chem Phys 94:6446–6458
7. Placa SJL, Roland PA, Wynne JJ (1992) Chem Phys Lett 190:163–168
8. Bonacic-Koutecky V, Fantucci P, Koutecky J (1991) Chem Rev 91:1035–1108
9. Kato H, Yamashita K, Morokuma K (1992) Chem Phys Lett 190:361–366
10. Ray AK, Howard IA, Kanal KM (1992) Phys Rev B 45:14247–14255
11. Boustani I (1997) Phys Rev B 55:16426–16438
12. Wang ZX, Schleyer PvR (2001) Science 292:2465–2469
13. Zhai H-J, Wang LS, Alexandrova AN, Boldyrev AI, Zakrzewski VG (2003) J Phys Chem A 107:9319–9328
14. Zhai H-J, Wang LS, Alexandrova AN, Boldyrev AI (2002) J Chem Phys 117:7917–7924
15. Alexandrova AN, Boldyrev AI, Zhai HJ, Wang LS, Sheiner E, Fowler PW (2003) J Phys Chem A 107:1359–1369
16. Alexandrova AN, Boldyrev AI, Zhai H-J, Wang LS (2004) J Phys Chem A 108:3509–3517
17. Zhai H-J, Alexandrova AN, Birch KA, Boldyrev AI, Wang LS (2003) Angew Chem Int Ed 42:6004–6008
18. Zhai H-J, Kiran B, Li J, Wang LS (2003) Nat Mater 2:827–833
19. Chandrasekhar J, Jemmis ED, Schleyer PvR (1979) Tetrahedron Lett 39:3707–3710
20. Martin-Santamaria S, Rzepa HS (2000) Chem Commun 16:1503–1504
21. Präsang C, Mlodzianowska A, Sahin Y, Hofmann M, Geiseler G, Massa W, Berndt A (2002) Angew Chem Int Ed 41:3380–3382
22. Präsang C, Hofmann M, Geiseler G, Massa W, Berndt A (2002) Angew Chem Int Ed 41:1526–1529
23. Präsang C, Mlodzianowska A, Geiseler G, Massa W, Hofmann M, Berndt A (2003) Pure Appl Chem 75:1175–1182
24. Amseis P, Mesbah W, Präsang C, Hofmann M, Geiseler G, Massa W, Berndt A (2003) Organometallics 22:1594–1596
25. Mesbah W, Präsang C, Hofmann M, Geiseler G, Massa W, Berndt A (2003) Angew Chem Int Ed 42:1717–1719

26. Lipscomb WN (1963) Boron hydrides. Benjamin, New York
27. Muetterties EL (1975) Boron hydride chemistry. Academic, New York
28. Cotton FA, Wilkinson G, Murillo CA, Bochmann M (1999) Advanced inorganic chemistry, 6th edn. Wiley, New York
29. Ricca A, Bauschlicher CW Jr (1997) J Chem Phys 106:2317–2322
30. Curtiss LA, Pople JA (1989) J Chem Phys 91:4809–4812
31. Schleyer PvR, Najafian K, Mebel A (1998) Inorg Chem 37:6765–6772
32. Goss JP, Briddon PR, Jones R, Teukam Z, Ballutaud D, Jomard F, Chevallier J, Bernard M, Deneuville A (2003) Phys Rev B 68:235209–235218
33. DiLabio GA, Matusek DR (2000) Chem Phys Lett 317:597–602
34. Wang P, Orimo S, Tanabe K, Fujii H (2003) J Alloys Comp 350:218–221
35. Boustani I (1995) Chem Phys Lett 240:135–140
36. Boustani I, Quandt A, Hernandez E, Rubio A (1999) J Chem Phys 110:3176–3185
37. Boustani I (1994) Int J Quantum Chem 52:1081–1111
38. Ricca A, Bauschlicher CW Jr (1996) Chem Phys 208:233–242
39. Kato H, Tanaka E (1991) J Comput Chem 12:1097–1107
40. Alexandrova AN, Boldyrev AI, Fu Y-J, Wang X-B, Wang L-S (2004) J Chem Phys 121:5709–5719
41. Hartke B (1993) J Phys Chem 97:9973–9976
42. Deaven DM, Ho KM (1995) Phys Rev Lett 75:288–291
43. Parr RG, Yang W (1989) Density-functional theory of atoms and molecules. Oxford University Press, Oxford
44. Becke AD (1993) J Chem Phys 98:5648–5652
45. Perdew JP, Chevary JA, Vosko SH, Jackson KA, Pederson MR, Singh DJ, Fiolhais C (1992) Phys Rev B 46:6671–6687
46. Frisch MJ, Trucks GM, Schlegel HB, Scuseria GE, Robb MA, Cheeseman JR, Montgomery JA, Vreven T, Kudin KN, Burant JC, Millam JM, Iyengar SS, Tomasi J, Barone V, Mennucci B, Cossi M, Scalmani G, Rega N, Petersson GA, Nakatsuji H, Kitao O, Nakai H, Klene M, Li X, Knox JE, Hratchian HP, Cross JB, Adamo C, Jaramillo J, Gomperts R, Stratmann RE, Yazyev O, Austin AJ, Cammi R, Pomelli C, Ochterski JW, Ayala PY, Morokuma K, Voth GA, Salvador P, Dannenberg JJ, Zakrzewski VG, Dapprich S, Daniels AD, Strain MC, Farkas O, Malick DK, Rabuck AD, Raghavachari K, Foresman JB, Ortiz JV, Cui Q, Baboul AG, Clifford S, Cioslowski J, Stefanov BB, Liu A, Liashenko A, Piskorz P, Komaromi I, Martin RL, Fox DJ, Keith T, Al-Laham MA, Peng CY, Nanayakkara A, Challacombe M, Gill PMW, Johnson BG, Chen W, Wang MW, Gonzales C, Pople JA (2003) Gaussian 03, Revision A.1. Gaussian Inc., Pittsburgh
47. Krishnan R, Binkley JS, Seeger R, Pople JA (1980) J Chem Phys 72:650–654
48. Cizek J (1969) Adv Chem Phys 14:35–89
49. Knowles PJ, Hampel C, Werner H-J (1993) J Chem Phys 99:5219–5227
50. Clark T, Chandrasekhar J, Spitznagel GW, Schleyer PvR (1983) J Comput Chem 4:294–301
51. Frisch MJ, Pople JA, Binkley JS (1984) J Chem Phys 80:3265–3269
52. Hegarty D, Robb MA (1979) Mol Phys 38:1795–1812
53. Eade RHE, Robb MA (1981) Chem Phys Lett 83:362–368
54. Schlegel HB, Robb MA (1982) Chem Phys Lett 93:43–46
55. Bernardi F, Bottini A, McDouglas JJW, Robb MA, Schlegel HB (1984) Far Symp Chem Soc 19:137–147
56. Yamamoto N, Vreven T, Robb MA, Frisch MJ, Schlegel MA (1996) Chem Phys Lett 250:373–378
57. Frisch MJ, Ragazos IN, Robb MA, Schlegel HB (1992) Chem Phys Lett 189:524–528
58. Carpenter JE, Weinhold F (1988) J Mol Struct (Theochem) 169:41–62
59. Carpenter JE (1987) PhD Thesis, University of Wisconsin, Madison
60. Foster JP, Weinhold F (1980) J Am Chem Soc 102:7211–7218
61. Reed AE, Weinhold F (1983) J Chem Phys 78:4066–4073
62. Reed AE, Curtiss LA, Weinhold F (1988) Chem Rev 88:899–926
63. Schaftenaar G (1998) MOLDEN 3.4. CAOS/CAMM Center, The Netherlands

J Mol Model (2006) 12: 577–589
DOI 10.1007/s00894-005-0065-z

ORIGINAL PAPER

Govindan Subramanian · Douglas B. Kitchen

Computational approaches for modeling human intestinal absorption and permeability

Received: 4 April 2005 / Accepted: 28 September 2005 / Published online: 1 April 2006

Abstract Human intestinal absorption (HIA) is an important roadblock in the formulation of new drug substances. Computational models are needed for the rapid estimation of this property. The measurements are determined via in vivo experiments or in vitro permeability studies. We present several computational models that are able to predict the absorption of drugs by the human intestine and the permeability through human Caco-2 cells. The training and prediction sets were derived from literature sources and carefully examined to eliminate compounds that are actively transported. We compare our results to models derived by other methods and find that the statistical quality is similar. We believe that models derived from both sources of experimental data would provide greater consistency in predictions. The performance of several QSPR models that we investigated to predict outside the training set for either experimental property clearly indicates that caution should be exercised while applying any of the models for quantitative predictions. However, we are able to show that the qualitative predictions can be obtained with close to a 70% success rate.

Dedicated to Professor Dr. Paul von Ragué Schleyer on the occasion of his 75th birthday.

Electronic Supplementary Material Supplementary material is available for this article at http://dx.doi.org/10.1007/s00894-005-0065-z.

G. Subramanian · D. B. Kitchen (✉)
Computer-Aided Drug Discovery Department,
Albany Molecular Research, Inc.,
21 Corporate Circle,
P.O. Box 15098 Albany, NY 12212-5098, USA
e-mail: douglask@albmolecular.com
Tel.: +1-518-4640279
Fax: +1-518-4640289
e-mail: gsubramanian@ttpharma.com

Present address:
G. Subramanian
Transtech Pharma.,
4170 Mendenhall Oaks Parkway,
High Point, NC 27265, USA

Keywords Human intestinal absorption · Caco-2 permeability · QSPR · QSAR · ADME

Introduction

Understanding the physicochemical and pharmacokinetic properties of known drugs and potential drug candidates is a major bottleneck for the low success rate of compounds in clinical development [1]. Traditional structure–activity relationship (SAR) studies optimize the potency and efficacy of a congeneric compound series on a protein target. A rapid understanding of the absorption, distribution, metabolism, and excretion (ADME) characteristics of compounds still impedes significant progress in this area [2–4]. For instance, oral dosing is usually the most desirable way to administer drugs and therefore the therapeutic efficacy of a compound often involves the efficient transport or absorption of a drug to the blood stream. Therefore, the plasma solubility, membrane permeability, protein binding, transport properties, and the diffusion kinetics are some of the components influencing the overall bioavailability of the drug. Metabolism or elimination of a drug may also decrease the efficacy of a compound or increase toxicity or unwanted side effects.

The primary barrier towards good bioavailability is human intestinal absorption (HIA). Therefore, there is an increasing need to understand and measure the effect of physicochemical properties of the drug on the intestinal absorption process [5, 6]. It requires dissolution, passage through the gut and finally diffusion or transport into the blood stream. For example, the P-glycoprotein (P-gp) activity in the apical cell membrane may limit the bioavailability, while absorption through transcellular (membrane diffusion, carrier-mediated) or paracellular routes alter the pharmacokinetic profile of the compounds [7, 8]. The cost and time factors involved in in vivo and in situ experiments [9–11], make it difficult to collect sufficient data to analyze structural contributions to the rate of intestinal absorption. Recent advances in high-throughput screening and combinatorial chemistry synthesis have elevated the need to gain a priori information regarding the intestinal absorption on a variety of new chemical entities. Determining the amount of HIA requires expensive experiments and hence simpler in vitro models have been developed.

In order to obtain pharmacokinetic information earlier in drug discovery projects, in vitro Caco-2 (immortalized human colon adenocarcinoma cell line) monolayers are now used because they exhibit remarkable morphological and functional similarity to the small intestinal columnar epithelium [12–15]. The permeability measured from these experiments agrees, in general, with HIA and hence is useful for experimental modeling of in vivo absorption. The measured apparent permeabilities (P_{app}) depend on the cell culture, growth factors, and other experimental conditions, resulting in slightly different values reported by various groups (see supporting information) [16–23]. There is a need to develop quantitative and predictive mathematical models that relate various physicochemical properties to intestinal permeability and absorption so that poorly absorbed compounds are eliminated in the initial stages of drug discovery. The primary goal of this work is to develop computational models to describe the initial absorption of compounds from the gut into the blood stream.

Cellular permeability can be understood as a series of partitioning and associated diffusion of a molecule from one region to another in a lipid bilayer that surrounds the cell. Therefore, early physical and computational studies intended to explain intestinal absorption of drugs focused mainly on a single physicochemical property like octanol–water partition coefficient or the distribution coefficient [24–29]. Reasonable correlations were obtained for a homologous series of compounds, although structural diversity impeded the model's predictivity. Other properties such as the molecular weight, size, and shape, polar van der Waals surface area (PSA), and H-bonding capability are believed to be important for modeling intestinal permeability [30]. Hence, Lipinski proposed a scheme to classify the intestinal absorption using simpler yet powerful 1D-descriptors such as the count of polar atoms, log*P*, and the molecular weight [31]. Similarly, Clark used the computed PSA to categorize (good, medium, poor) the extent of intestinal absorption for structurally diverse compounds [32].

Table 1 Statistical results for $\log P_{app}$ (Eqs. 3, 8, 9, 10 and 11) and logHIA[a] (Eqs. 4, 5, 6, 7 and 12) for training and prediction (sub)set data and comparisons with QSPR models reported in literature

Models	Type[b]	N_{tr}[c]	r^2 (tr)	r^2 (cv)	N_{pc}[d]	rmse[e]	N_{pr}[f]	r^2 (pr)[g]	rmse[h]	N_{pr}[i]	r^2 (ss)[j]	rmse[k]
logP $\boldsymbol{\log P_{app}}$												
Reference [23]	PSA	9	0.907			0.36						
Equation (8)	PSA+MW	17	0.693									
Reference [54]	PLS (9)	17	0.909	0.852	2	0.305						
Equation (9)	PSA+MW	17	0.771	0.654	2	0.458	50	0.455	0.734	42	0.672	0.533
Equation (10)	G/PLS (6)	17	0.972	0.549	1	0.161	50	0.010	1.526	35	0.632	0.493
Equation (11)	PSA+MW	22	0.610	0.437	2	0.561	37	0.622	0.570	35	0.691	0.520
Equation (3)	G/PLS (6)	22	0.892	0.535	1	0.295	37	0.629	0.840	31	0.780	0.530
logHIA &	***%HIA***											
Reference [40]	PLS (10)	20	0.916	0.798	3	8.1						
Reference [41]	Hashkeys	20	0.691			21.7						
Reference [37]	PSA	20	0.94			9.2						
Reference [32]	PSA	20	0.94			9.1						
Equation (12)[l]	PSA	20	0.886			12.9	62	0.463	29.1			
Equation (12)[m]	PSA	30	0.549			21.3	46	0.655	19.3			
Equation (4)	G/PLS (6)	30	0.907	0.808	3	9.0	46	0.166	31.7	37	0.452	21.0
Equation (5)	G/PLS (6)	30	0.908	0.814	3	7.6	46	0.191	31.2	38	0.462	19.6
Equation (6)	G/PLS (6)	30	0.814	0.617	1	13.4	46	0.304	28.5	37	0.539	17.9
Equation (7)	G/PLS (6)	30	0.668	0.343	3	17.5	46	0.367	26.4	39	0.559	17.9

[a]The rms error for Eqs. (4, 5, 6, 7) and that reported by Norinder are given in %HIA after the computed logHIA values are transformed using Eq. (2)
[b]Variables and methods used in fitting the experimental data. The values in parenthesis indicate the number of independent variables used in the multivariate analysis
[c]Number of molecules in the training set
[d]Number of PLS components
[e]Root mean square error for the training set
[f]Number of molecules in the prediction set
[g]square of the correlation coefficient for the prediction set
[h]Root mean square error for the prediction set
[i]Number of molecules in the prediction subset
[j]square of the correlation coefficient for the prediction subset (the outliers that were not included are underlined in Table 3)
[k]Root mean square error for the prediction subset
[l]PSA_{50}=83.987 and the slope is −15.246
[m]PSA_{50}=104.673 and the slope is −28.97

A number of quantitative structure–property relationship (QSPR) models have also been proposed in the literature for predicting $\log P_{app}$ and %HIA [33–35]. Most of the reported QSPR models utilize PSA as an indicator of intestinal permeability and absorption. For a diverse set of 17 compounds, van de Waterbeemd and Camenisch showed that the log of the apparent permeability ($\log P_{app}$) measured from Caco-2 monolayers correlates well with the molecular weight and the PSA obtained for a single conformation [36]. By considering multiple conformations for the β-adrenoreceptor antagonists series, a sigmoidal relationship between the dynamic PSA and the fractional absorption (%HIA) was derived [37]. Similarly, the Boltzmann-averaged dynamic solvent-accessible surface area obtained from molecular dynamics simulations was related to the intestinal absorption [22].

Artursson and coworkers [38] extended the PSA model [17] by including the dynamic non-polar surface area (NPSA) component and showed a good correlation with the intestinal permeability of 19 oligopeptides. However, the PSA, NPSA and H-bond atom counts were not determined to be the critical elements responsible for the observed cellular permeability of the 21 peptide and peptidomimetic compounds considered by a different group [39]. This suggests that QSPR models based on logP, PSA or NPSA, while helpful for deriving meaningful correlations within a narrow structural class, may not extend universally. Including other physicochemical variables in the QSPR model would remove the over-dependence of P_{app} on PSA. Alternatively, descriptors replacing PSA can be mapped to HIA so that the intestinal absorption of both polar and apolar molecules can be modeled simultaneously. QSPR efforts along these directions reveal similar or better performances compared to models obtained using PSA as the response variable. In fact, the statistical results obtained by employing multivariate partial least square fitting (PLS) protocols with several molecular descriptors excluding PSA [40] and also by using molecular hash keys [41] was comparable to that derived using PSA [37] for the same set of compounds (Table 1). Very recently, a quantitative model for predicting the %HIA was proposed by combining a genetic algorithm and a neural network scoring function [42]. The significance of the model can be appreciated from the small (9.4%) root-mean-square error (rmse) obtained for a training set of 67 compounds. However, the major drawback rests in the compound selection since a single model was used to describe both passive and carrier-mediated absorption mechanisms. In addition, with 52/67 compounds exhibiting >75% HIA (strongly absorbed), the training set data primarily contains compounds with high absorption and therefore may result in a biased model.

One of the primary goals of this study is to develop QSPR models for estimating both $\log P_{app}$ and %HIA following the procedure outlined in Fig. 1. As a first step, experimental P_{app} measurements and %HIA data reported in literature were compiled. Subsequently, actively transported compounds were identified and separated from the dataset. The remaining molecules were then divided into training and prediction sets. The training set was composed such that the experimental value was distributed uniformly with respect to the measured P_{app} and %HIA. Using these evenly distributed datasets, QSPR models for computing $\log P_{app}$ and % HIA are derived independently. By deriving both simple and

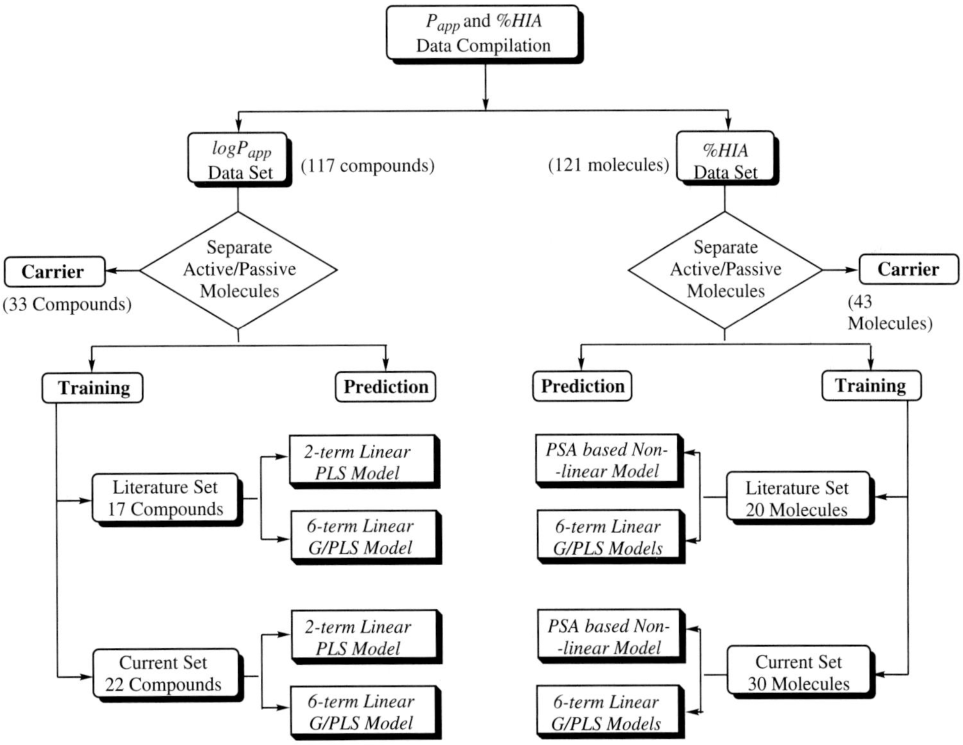

Fig. 1 Illustration of data compilation. The datasets were collected from the literature and compounds were divided into passively and actively transported sets. Further separations into training and prediction sets are described in the text. Regression analysis was only performed on passively absorbed compounds

complex QSPR equations, the influence of certain independent variables was assessed. The models were then extended to predict the $\log P_{app}$ and %HIA for an external validation set of passively absorbed compounds and compared to reported QSPR models [32, 36, 37, 40, 42]. Deriving similar mathematical relations for $\log P_{app}$ and %HIA and comparing the results obtained with the present training set data demonstrate the limitations with the oft-used training set compounds in the literature. As a next step, the $\log P_{app}$ and %HIA are predicted for molecules absorbed through active transport processes to see if these external influences are manifested in the molecular descriptors identified. This will also help determine if simple relationships exist between the active and passive absorption mechanisms. The quantitative $\log P_{app}$ and %HIA results are then extended to categorize the degree of intestinal permeability and absorption and compared to literature predictions [31, 32].

Dataset

Experimental P_{app} values determined using Caco-2 cell lines were pooled from several literature sources resulting in 117 compounds [16–23]. The data revealed that the P_{app} measurements reported by different laboratories for the same compounds varied significantly in magnitude. For example, the experimental $\log P_{app}$ for alprenolol ranges from −3.9 to −5.8, while that of atenolol varies between −5.7 and −7.5. For compounds where inter-laboratory experimental determinations were available, the values most consistent among the reported ones were used (see supporting information). The drawback of models derived based on the Caco-2 P_{app} values determined from a single laboratory is the potential bias in the results that limits extension to other compound sets reported by different groups. Alternatively, by using experimental P_{app} values that are consistent across different groups, realistic estimates of the experimental uncertainty in the various measurements can be ascertained for a larger set of compounds. This also provides a confidence limit for the experimental values reported for compounds from that group only. By following this procedure, some reported P_{app} values [43] were observed to deviate considerably from other literature reports for the same compounds and so were not included while generating the final training set. However, some of these compounds are used in the second $\log P_{app}$ prediction set but are not discussed in the text. The chemical structures used in this work are available from the authors upon request.

We believe that it is important to use compounds that permeate through passive absorption process in the dataset, since actively transported molecules are influenced by external environments that may not be modeled accurately. Prior to generating the final dataset, molecules reported to be substrates for various transport mechanisms or carrier-mediated processes (P-gp, peptide, nucleoside, etc.) were separated from those that are likely to undergo passive intestinal permeation. This division of compounds (Fig. 1) results in a dataset of only 59 compounds that most likely permeate through diffusion-controlled processes. A similar protocol was employed to gather the %HIA data from literature [19–21, 37, 42–45]. Although most of the fractional intestinal absorption data had been compiled [42], additional %HIA values reported in literature [43, 44] were also included, resulting in 121 compounds of which 76 molecules are probably passively absorbed. The training and prediction sets were chosen so that the $\log P_{app}$ and %HIA values span the whole range. Also, in the case of %HIA, compounds with values of 100% or 0% were used only in the prediction set.

Materials and methods

All the compounds were energy minimized using the universal force field [46] implemented in the Cerius^2 molecular modeling package (Accelrys, San Diego, CA) and imported into a Cerius^2 study table. Structural, constitutional, topological, and other calculated descriptors available through the Cerius^2 QSAR module were then included. In addition, the computed PSA [47] and the cube root of the gravitational index (GRAVIND) [42] were added. Since the passive transport of the compound across the intestinal epithelium is largely diffusion-controlled, and since the diffusion coefficient is inversely proportional to the square (SQINMW) or the cube root (CBINMW) of the molecular weight [48], descriptors representing such features were also included. By inspecting the independent variables in the study table, descriptors that did not have sufficient non-zero values or adequate variation were identified and removed.

A structurally heterogeneous $\log P_{app}$ dataset containing 22 compounds was used as the training set for predicting the apparent permeability of the remaining 19 molecules. Similarly, the training set for %HIA consisted of 30 compounds, leading to an external prediction set consisting of the remaining 46 molecules. The training set for modeling %HIA was selected such that molecules with 0 or 100 %HIA were not included. We believe that these limiting values do not correspond to the actual measurements, but reflect on the compound having crossed a threshold value. Therefore, using these compounds in the training set will affect the overall performance of the derived QSPR model. Hence, molecules with 0% and 100% HIA values are used in the prediction set and not in our training set. To have a direct comparison with reported QSPR models, additional equations were derived for the 17 and 20 compound training sets used in the literature for predicting $\log P_{app}$ and %HIA, respectively (Fig. 1) and the results are provided in the supporting information accompanying the paper.

Several studies in the literature have pointed out that P_{app} and %HIA are not likely to be linearly related to the computed descriptors [24–29, 36]. Therefore, the experimental P_{app} and the reported %HIA were transformed to logarithmic units. Since the %HIA has a closed scale with limiting values, fitting the data using linear approximations may lead to a statistically incorrect model. As in previous studies [40], a logit transformation was performed on the

Table 2 Experimental and computed $\log P_{app}$ for training set, prediction set, and compounds permeating through carrier-mediated processes[a]

Molecule	Expt	Equation (11)	Equation (3)
Training Set			
H216	−6.983	−5.929	−6.973
doxorubicin	−6.796	−7.182	−6.772
dnalaprilate	−6.208	−5.507	−5.661
benzylpenicillin	−5.708	−5.264	−5.398
practolol	−5.613	−4.811	−6.047
H95	−5.426	−4.807	−5.287
hydrocortisone	−4.914	−5.309	−4.595
imipramine	−4.851	−3.978	−4.601
desipramine	−4.666	−4.096	−4.627
naloxone	−4.550	−4.973	−4.885
clonidine	−4.521	−4.345	−4.094
pindolol	−4.460	−4.619	−4.752
propranolol	−4.378	−4.337	−4.010
betaxolol	−4.314	−4.627	−4.546
tenidap	−4.291	−5.073	−4.584
ibuprufen	−4.280	−4.153	−3.969
oxprenolol	−4.184	−4.512	−4.240
alprenolol	−4.125	−4.353	−4.234
caffeine	−4.074	−4.375	−4.035
betaxolol ester	−4.015	−4.842	−4.270
propranolol ester	−3.983	−4.596	−4.452
naproxen	−3.678	−4.329	−3.985
Prediction Set			
G6264	−6.854	−6.194	−7.530
G6700	−6.854	−6.114	−6.678
G6249	−6.745	−6.058	−6.681
G6191	−6.456	−6.297	−7.557
G6262	−6.367	−6.092	−6.848
G6703	−6.244	−6.228	−6.805
sulpiride	−6.160	−5.527	−5.356
atenolol	−5.936	−5.092	−6.494
acyclovir	−5.699	−5.486	−5.353
sumatriptan	−5.523	−4.933	−6.193
G6203	−5.509	−6.186	−6.072
H244	−5.220	−4.881	−5.470
ziprasidone	−4.910	−5.077	−2.741
tiacrilast	−4.900	−4.754	−3.245
mibefradil	−4.870	−5.382	−4.316
fleroxacin	−4.810	−5.033	−3.928
nitrendipine	−4.770	−5.502	−5.443
guanoxan	−4.710	−4.900	−5.597
chloramphenicol	−4.686	−5.559	−4.831
felodipine	−4.644	−4.914	−4.940
fluconazole	−4.526	−5.137	−3.518
trovaflaxicin	−4.520	−5.646	−4.134
warfarin	−4.417	−4.694	−3.980
timolol	−4.354	−5.068	−4.370
theophylline	−4.350	−4.603	−4.181
antipyrine	−4.310	−3.928	−3.745
diltiazem	−4.310	−4.945	−4.625
testosterone	−4.286	−4.382	−3.400

Table 2 (continued)

Molecule	Expt	Equation (11)	Equation (3)
corticosterone	−4.263	−5.065	−4.273
lidocaine	−4.210	−4.184	−5.512
diazepam	−4.149	−4.310	−3.954
guanabenz	−6.000	−4.879	−3.908
coumarin	−4.110	−3.803	−1.792
timolol ester	−4.103	−5.331	−3.712
metoprolol	−4.036	−4.491	−4.978
oxprenolol ester	−4.012	−4.756	−4.369
alprenolol ester	−3.967	−4.610	−4.671
2nd Prediction Set			
olsalazine	−6.959	−5.859	−3.668
terbutaline	−6.420	−4.728	−4.394
epinephrine	−6.020	−4.659	−4.176
acetaminophen	−6.000	−4.203	−4.896
acrivastine	−7.721	−4.696	−4.390
bupropion	−5.824	−4.149	−4.572
cefuroxime	−8.420	−6.647	−5.477
chlorothiazide	−8.495	−5.650	−4.691
fluparoxan	−5.699	−4.132	−3.476
hydrochlorothiazide	−8.036	−5.705	−4.238
ketoprofen	−6.032	−4.515	−4.242
labetalol	−6.119	−5.421	−6.599
lamotrigine	−5.959	−5.233	−3.778
nadolol	−8.409	−5.109	−5.309
netivudine	−8.168	−5.558	−6.106
penicillin V	−8.770	−5.436	−5.290
phenytoin	−5.796	−4.680	−4.889
progesterone	−6.009	−4.398	−4.099
propylthiouracil	−6.018	−4.190	−4.608
sotalol	−7.377	−4.999	−4.850
trimethoprim	−6.061	−5.415	−5.601
Active Processes			
furosemide	−8.854	−5.688	−5.124
lisinopril	−8.658	−6.159	−6.689
loracarbef	−8.620	−5.650	−6.579
cefatrizine	−8.119	−7.055	−6.552
sulfasalazine	−6.886	−6.204	−3.679
methylprednisolone	−6.602	−5.375	−5.340
zidovudine	−6.553	−5.727	−4.000
amoxicillin	−6.481	−6.029	−6.380
L-leucine	−6.301	−4.320	−4.785
mannitol	−6.187	−5.253	−5.998
sucrose	−6.149	−6.412	−7.691
L-glutamine	−6.071	−5.065	−6.409
L-dopa	−6.000	−5.143	−5.474
azithromycin	−5.983	−7.163	−7.033
ondansetron	−5.959	−4.410	−4.817
D-mannitol	−5.932	−5.200	−5.173
methotrexate	−5.921	−7.552	−7.272
acetylsalicylate	−5.620	−4.300	−1.296
cephalexin	−5.570	−5.631	−6.384
cimetidine	−5.514	−5.284	−5.524
erythromycin	−5.428	−7.357	−6.789
taurocholate	−5.396	−6.500	−4.240

Table 2 (continued)

Molecule	Expt	Equation (11)	Equation (3)
acebutolol	−5.351	−5.206	−6.635
L-phenylalanine	−5.161	−4.478	−4.822
salicylic acid	−4.924	−4.228	−3.390
quinidine	−4.690	−4.678	−3.669
dexamethasone	−4.631	−5.423	−5.100
D-glucose	−4.602	−5.094	−3.323
verapamil	−4.580	−5.053	−4.566
acetylsalicylic acid	−4.513	−4.381	−3.678
taurocholic acid	−4.462	−6.487	−6.922
prazosin	−4.361	−5.536	−5.204
valproic acid	−4.319	−3.989	−4.595
glycine	−4.097	−4.236	−4.099
gabapentin	−4.046	−4.468	−5.231

[a]underlined values in the prediction set represent compounds that exhibited more than 1 log unit rms error and so were not included in the calculations of the prediction subset in Table 1

dependent variable (%HIA) using Eq. (1). The computed logitHIA was transformed back to %HIA using Eq. (2), so that direct comparisons with experimental values could be achieved.

$$\log\mathrm{it(HIA)} = \ln\left[(\%\mathrm{HIA} + 10)/(110 - \%\mathrm{HIA})\right] \quad (1)$$

$$\%\mathrm{HIA} = \left[(110 * \exp(\mathrm{logitHIA}) - 10)/(1 + \exp(\log\mathrm{itHIA}))\right] \quad (2)$$

Using Cerius^2, the training-set molecules were initially subjected to a Genetic/Partial Least Squares (G/PLS) statistical fitting procedure [49, 50] to fit the dependent function (logP_{app} or logitHIA) using 15 variables and up to four PLS components. In this way, a plausible subset of descriptors that best describe logP_{app} and logitHIA were obtained from the pool of independent variables. Subsequent G/PLS runs on the same training-set data used only the descriptor subset and seven terms in the regression equation. The initial random population of 100 equations was evolved for 20,000 generations, resulting in a final set of 100 model equations that characterizes the observed logP_{app} and logitHIA through the square of the correlation coefficient (r^2) fitness function. All the variables involved in the G/PLS regression were scaled and a maximum of three principal components were allowed. To determine the role of certain variables like PSA, Jurs-terms [51], electrotopological indices [52], and AlogP [53], four different models were obtained for logitHIA by including or removing some descriptors from the initial descriptor subset used in the G/PLS run. In addition, QSPR models with varying complexity and for the standard training set used in the literature were obtained for direct comparison with reported models (Table 1).

The internal cross-validated correlation coefficient, r^2(cv) (using the leave-one-out method) was used to ascertain the statistical quality of the derived functions. The accuracy of the estimations was evaluated by comparing the results obtained using the best model equation against the experimental logP_{app} and logitHIA values for each of the prediction sets. The quantitative logP_{app} and %HIA results were also used to categorize the molecules based on the extent of intestinal permeability and absorption using the following threshold values. Thus, compounds with computed logP_{app}<−6 and<−5 are classified as *poorly* and *moderately* permeable, respectively. All the remaining molecules are considered to possess *good* intestinal permeabilities. Similarly, a %HIA of <30 suggests poor intestinal absorption, while molecules exhibiting HIA >70% possess good intestinal absorption. Compounds with medium HIA range between 30 and 70%.

Results

Table 1 summarizes the performances of the various QSPR models derived using the experimental logP_{app} and %HIA for the passively absorbed compounds listed in Tables 2 and 3. The computed logP_{app} and %HIA are also given for molecules absorbed through various transport processes but were not used as part of the training set.

Statistical results for logP_{app}

The G/PLS analysis on the 22 training-set molecules (Table 2) resulted in Eq. (3) with an r^2 of 0.89 for logP_{app} and an r^2(cv) of 0.54. The 0.3 log unit root-mean-square error (rmse) for the training set is within experimental error limits.

$$\begin{aligned}\log\mathrm{P_{app}} = &- 4.628 + 0.399 * \mathit{Hbond\ acceptor} + 0.450 \\ &* \mathit{AlogP} - 0.329 * \mathit{Hbond\ donor} + 0.698 \\ &* \mathit{Jurs\_RPCS} - 0.166 * \mathit{Kappa\_1} \\ &+ 0.00161 * \mathit{Jurs-WNSA\_1}\end{aligned} \quad (3)$$

The computed r^2(pr) and rmse obtained for the 37 prediction-set compounds suggest a satisfactory performance of the model when applied to an external dataset, but with a significant rms error. However, closer examination reveals that the predicted logP_{app} is greater than one log unit when compared with the experimental values for six compounds identified in Table 2. Since the measured apparent permeabilities are available from one literature source only (see supporting information), it is difficult to ascertain if these compounds possess large uncertainty in the observed logP_{app} values or should be treated as the prediction-set outliers. Assuming the latter, significant improvement in the predicted r^2 and rmse is achieved for the remaining 31 molecules in the prediction subset

Table 3 Experimental and computed %HIA values for the training set, prediction set, and compounds absorbed through carrier-mediated mechanisms

Molecule	Expt	Equation (4)	Equation (5)	Equation (6)	Equation (7)	Equation (12)
Training Set						
cromolyn	0.5	−1.7	7.7	−1.2	3.4	23.2
olsalazine	2.3	0.1	8.4	6.7	19.4	27.5
cefuroxime	5.0	4.4	1.9	9.4	0.9	14.0
enalaprilat	10.0	18.1	12.5	12.6	50.3	58.1
benzyl penicillin	30.0	58.0	32.6	67.1	58.4	67.7
norfloxacin	35.0	33.3	22.5	67.7	78.6	77.0
Phenoxymethylpenecillinic acid	45.0	56.0	43.6	61.7	44.7	60.5
ziprasidone	60.0	85.1	64.7	69.0	88.1	85.7
hydorchlorothiazide	67.0	66.4	87.4	74.8	70.1	36.4
lamotrigine	70.0	74.6	79.8	61.4	67.2	56.4
guanabenz	75.0	67.5	82.9	73.0	82.8	71.7
propylthiouracil	75.0	86.7	80.5	81.0	76.8	89.3
sorivudine	82.0	80.3	66.3	47.2	57.4	46.6
bupropion	87.0	92.3	81.1	86.4	93.3	93.9
acrivastine	88.0	80.0	83.5	95.5	89.8	89.7
betaxolol	90.0	88.1	93.3	88.5	94.0	89.3
oxprenolol	90.0	93.8	96.9	86.9	88.1	87.9
pindolol	90.0	92.2	85.6	92.7	92.3	83.5
scopolamine	90.0	95.7	90.5	100.4	82.5	85.7
timolol	90.0	91.2	87.4	88.6	64.7	72.6
naloxone	91.0	89.1	83.0	83.8	87.5	80.4
progesterone	91.0	91.6	96.1	84.1	87.9	93.2
clonidine	95.0	92.7	91.2	87.7	95.6	89.8
imipramine	95.0	89.7	88.6	98.2	103.1	96.9
metoprolol	95.0	93.9	99.6	101.3	88.1	88.2
sotalol	95.0	89.2	85.8	90.2	85.5	73.3
trimethoprim	97.0	87.9	101.5	81.6	61.9	52.4
warfarin	98.0	80.9	98.7	91.2	87.8	87.2
diltiazem	99.0	94.4	86.5	79.2	71.8	88.1
nordiazepam	99.0	98.6	93.7	92.8	92.2	89.7
Prediction Set						
ceftriaxone	1.0	−4.2	−6.6	−7.6	−8.9	3.0
ganciclovir	3.8	62.9	62.0	45.0	27.4	25.3
doxorubicin	5.0	29.3	39.8	−4.0	5.6	8.3
chlorothiazide	13.0	69.6	83.5	76.8	58.2	39.2
netivudine	28.0	62.2	49.0	57.9	46.5	43.0
acyclovir	30.0	56.0	66.9	66.8	31.1	38.6
nadolol	34.5	95.4	87.8	84.6	86.3	71.6
sulpiride	36.0	64.2	75.4	74.2	59.8	55.0
atenolol	50.0	81.9	89.2	89.0	75.9	67.9
metolazone	64.0	75.3	77.4	60.5	84.6	64.6
ciprofloxacin	69.0	30.4	21.2	73.2	81.4	76.5
terbutaline	73.0	83.6	73.6	75.2	86.0	77.5
sumatriptan	75.0	93.4	83.6	97.9	92.9	82.5
acetaminophen	80.0	56.0	84.7	100.5	57.5	87.7
bromazepam	84.0	101.9	95.0	96.4	82.6	84.6
trovafloxacin	88.0	23.7	1.1	18.9	73.0	60.6
chloramphenicol	90.0	63.3	71.3	39.6	36.3	50.5
ephedrine	90.0	75.0	86.3	103.0	79.9	92.1
phenytoin	90.0	60.6	92.6	94.1	87.1	82.8
propranolol	90.0	93.7	85.3	91.4	97.5	90.8
tenidap	90.0	62.5	53.7	75.8	94.8	75.2

Table 3 (continued)

Molecule	Expt	Equation (4)	Equation (5)	Equation (6)	Equation (7)	Equation (12)
hydrocortisone	91.0	88.6	86.1	48.3	71.6	68.7
terazosin	91.0	89.4	95.9	77.8	38.3	58.2
alprenolol	93.0	89.4	89.9	96.4	92.8	90.8
fluconazole	95.0	95.3	67.7	84.3	43.1	67.3
labetalol	95.0	64.6	63.6	56.0	92.2	63.8
oxazepam	97.0	87.6	86.4	81.7	87.5	81.7
theophylline	98.0	46.6	40.1	82.6	35.2	77.7
prednisolone	98.8	87.5	75.6	55.6	72.4	68.7
naproxen	99.0	66.1	88.5	95.4	85.3	90.1
tiacrilast	99.0	59.1	49.7	78.4	63.9	79.3
antipyrine	100.0	99.6	100.2	102.4	84.1	94.5
bumetanide	100.0	28.6	79.1	47.5	63.3	45.4
caffeine	100.0	64.8	42.2	87.9	48.5	86.6
corticosterone	100.0	90.8	95.3	74.6	73.9	78.7
coumarin	100.0	49.5	87.3	101.2	67.0	94.6
desipramine	100.0	84.0	96.0	104.9	102.6	95.1
diazepam	100.0	102.4	96.4	96.7	93.3	93.2
felodipine	100.0	104.8	104.1	79.8	80.0	85.9
fluparoxan	100.0	87.4	81.1	89.7	89.8	91.9
fluvastatin	100.0	74.1	64.4	42.0	81.7	74.2
ibuprofen	100.0	45.8	75.0	91.7	76.5	92.0
ketoprofen	100.0	52.2	75.7	90.0	78.8	87.6
lormetazepam	100.0	94.2	85.2	75.2	88.8	87.4
practolol	100.0	92.8	92.6	90.0	84.1	79.9
testosterone	100.0	90.6	100.6	99.5	94.1	92.1
Active Processess						
gentamicin	0.0	102.5	105.7	107.8	83.7	10.2
raffinose	0.3	99.5	79.8	−8.2	−1.2	1.1
lactulose	0.6	80.1	54.0	2.4	16.6	7.3
sulfasalazine	13.0	18.5	31.3	17.4	12.5	25.0
mannitol	15.0	83.0	79.1	55.8	45.3	46.4
foscarnet	17.0	−9.2	54.5	−8.8	−4.1	68.4
lisinopril	25.0	7.5	21.1	−2.6	31.6	31.6
pravastatin	34.0	73.4	79.6	10.4	30.0	44.3
erythromycin	35.0	56.3	104.0	3.6	−9.1	13.0
azithromycin	35.0	82.8	105.9	4.1	−8.6	18.1
defuroxime axetil	36.0	2.0	12.8	8.3	−6.9	12.7
sucrose	42.0	98.4	71.2	69.2	34.2	9.7
etoposide	50.0	92.7	108.9	21.7	−2.3	23.6
ranitidine	50.0	89.9	70.4	76.2	55.7	70.8
gabapentin	50.0	21.9	34.2	81.0	55.0	81.3
tranexamic acid	55.0	20.2	27.1	75.0	46.5	81.0
L-glutamine	60.0	6.6	25.7	51.5	14.2	49.0
taurocholic acid	60.0	50.7	51.7	38.4	25.0	29.2
furosemide	61.0	44.9	56.5	58.4	70.1	44.7
cimetidine	85.0	82.4	67.4	66.2	34.2	53.8
captopril	67.0	57.1	42.2	76.8	75.5	86.0
cefatrizine	76.0	35.9	12.6	6.0	2.6	7.3
quinidine	80.0	95.0	89.6	99.7	89.2	88.4
methylprednisolone	82.0	88.4	79.0	29.6	65.5	66.4
acebutolol	89.5	86.9	94.3	77.5	64.1	70.0
amoxicillin	93.5	37.4	14.2	22.4	45.5	30.4
verapamil	95.0	92.4	104.8	85.0	46.8	86.9
cephalexin	98.0	65.6	30.6	54.9	66.3	49.6

Table 3 (continued)

Molecule	Expt	Equation (4)	Equation (5)	Equation (6)	Equation (7)	Equation (12)
acetylsalicylic acid	100.0	24.1	63.0	87.5	48.0	85.4
D-glucose	100.0	79.6	52.2	52.1	49.0	54.1
D-Phe-L-Pro	100.0	48.7	42.5	80.8	71.7	71.9
dexamethasone	100.0	85.6	70.7	55.7	72.2	68.6
glycine	100.0	−7.1	2.4	65.5	−6.9	80.3
L-dopa	100.0	29.8	37.1	42.8	54.9	53.0
L-leucine	100.0	13.6	37.9	67.4	25.4	82.7
L-phenylalanine	100.0	30.3	52.9	88.2	58.6	80.9
loracarbef	100.0	47.4	16.0	43.5	60.8	48.4
methotrexate	100.0	−0.6	6.7	−6.9	−6.7	2.7
ondansetron	100.0	95.5	81.9	98.9	90.5	92.0
prazosin	100.0	89.9	85.1	78.2	48.5	60.7
salicylic acid	100.0	14.6	62.4	92.0	43.4	86.3
valproic acid	100.0	13.0	57.0	73.6	32.1	91.8
zidovudine	100.0	38.6	65.6	75.0	−4.5	22.0

[a]The underlined %HIA values correspond to compounds removed in the statistical analysis of the prediction subset (Table 1)

(Table 1). Clearly, there are many other interpretations available for the outliers in this case, including structural diversity outside the training set or alternate absorption mechanisms in these compounds.

Statistical results for logitHIA

The statistical quality of logitHIA (Eqs. 4, 5, 6, 7) and the quantitative estimations of %HIA for the 30 training-set compounds (Table 3) show that good QSPR models are achieved for %HIA. Among the four models, the best results are obtained through Eq. (4) with an r^2(tr) of 0.907 while the logitHIA estimations obtained using Eq. (7) account for only 67% of the variance in the training set (Table 1). This is also reflected in the larger rmse obtained using Eq. (7) compared to the 9% error in HIA associated with Eq. (4). The subtle changes in the r^2(tr) obtained through Eqs. (4, 5, 6 and 7) are attributed to the effect of certain independent variables like PSA, Jurs term, etc. in modeling the intestinal absorption.

$$\begin{aligned}\text{logitHIA} = {} & 10.342 + 0.783 * S\_dssC - 104.493 \\ & * SQINMW - 0.0262 * S\_dO \\ & - 0.022 * PSA - 0.164 * S\_ssCH2 \\ & - 0.366 * A\log P\end{aligned} \tag{4}$$

$$\begin{aligned}\log \text{itHIA} = {} & 3.849 + 10.991 * Jurs\_RNCG \\ & + 0.686 * S\_dssC + 0.114 * S\_ssO \\ & + 0.883 * JX - 95.220 * SQINMW \\ & - 0.232 * Hbond\ acceptor\end{aligned} \tag{5}$$

$$\begin{aligned}\log \text{itHIA} = {} & 11.324 - 0.015 * PSA - 55.141 * CBINMW \\ & + 1.032 * Jurs\_FNSA\_2 + 0.0234 \\ & * Jurs\_WNSA\_3 + 11.329 * Jurs\_RNCG \\ & - 0.171 * A\log P\end{aligned} \tag{6}$$

$$\begin{aligned}\log \text{itHIA} = {} & 8.029 - 0.00379 * Area + 0.735 \\ & * GRAVIND - 0.0577 * PSA \\ & - 61.408 * CBINMW + 0.724 \\ & * Hbond\ donor - 0.0142 * MW\end{aligned} \tag{7}$$

The computed r^2(pr) and rmse associated with logitHIA for the 46 prediction-set compounds suggest a weaker performance of the above models for quantitative logitHIA predictions. In striking contrast to the training set results, the predictions obtained through Eqs. (6) and (7) are the best among the four G/PLS models derived here for estimating %HIA. The poor performance of Eqs. (4) and (5) is attributed to the use of certain electrotopological indices in these models that are not represented in some of the validation-set compounds. For example, the absence of methylene carbons (>CH_2) in ketoprofen, ibuprofen, and coumarin is partially responsible for the >40% rmse observed for these molecules by using Eq. (4) (Table 3). Similarly, the lack of ether-like oxygen (–O–) in ticrilast, caffeine, and ciprofloxacin contributes to the smaller r^2 observed by using Eq. (5), while compounds lacking a carbonyl carbon (nadolol, chlorothiazide, etc.) are predicted poorly by Eqs. (4) and (5). When molecules predicted with large rmse (>40%) are excluded from the validation set, the correlation coefficient for the validation subset is improved significantly along with a substantial decrease in the rmse (~10%). Considering the large deviations in the reported inter-laboratory %HIA values for certain compounds (see supporting information), the predictions are still quantitative and within acceptable experimental error limits.

Discussion

The G/PLS results presented above reveal that reliable models have been obtained for estimating $\log P_{app}$ and %HIA across a heterogeneous dataset (Tables 2 and 3). However, the choice of the training-set data and the complexity of the regression equations (Eqs. 3, 4, 5, 6, 7) obtained in this study do not allow a direct comparison of the present results with reported QSPR models. Consequently, additional models were derived for the standard 17 and 20 training-set compounds used in the literature for estimating $\log P_{app}$ and %HIA and compared with the existing models (Table 1) derived using 22 and 30 compound sets, respectively (see supporting information).

Comparison with reported $\log P_{app}$ models

Palm et al. reported one of the earliest $\log P_{app}$ models by correlating the dynamic PSA for six β-adrenoreceptor-blocking agents [17, 23] and subsequently extended to model nine homologues [23] with considerable success. Our initial attempts to obtain a linear relationship between the computed PSA and $\log P_{app}$ for the 22 training-set compounds yielded an r^2 of 0.583, considerably lower than the results obtained with Eq. (3) (Table 1). This reveals the limitation on the use of a single variable for quantitative $\log P_{app}$ predictions across a diverse dataset. However, a statistically significant correlation coefficient (0.833) was obtained for the literature training set of 17 compounds when $\log P_{app}$ was modeled using molecular weight and PSA (Eq. 8) [32]. This equation also agrees with the general notion that intestinal permeability is enhanced by increasing the molecular weight and by decreasing PSA. Since Eq. (8) was not validated through predictions for an external dataset, the generality of the model could not be assessed. Therefore, we refit a model using these terms to verify its extension for all the molecules considered in this study (Table 2). The comparable statistical quality (Table 1) and the coefficient on the variables obtained through Eq. (9) for the same 17 training-set compounds reflect the similarity of our model with Eq. (8). The marginal difference in the coefficients in Eqs. (8) and (9) is attributed to the use of slightly different PSA values and also to the use of consensus $\log P_{app}$ values (see supporting information), rather than the experimental values reported in the previous work [9].

$$\log P_{app} = 0.008 * MW - 0.043 * PSA - 5.165 (n = 17) \tag{8}$$

$$\log P_{app} = 0.00443 * MW - 0.0288 * PSA - 4.459 (n = 17) \tag{9}$$

$$\log P_{app} = -3.238 - 0.695 * CHI\_3\_C + 0.00129 * Jurs\_PPSA\_2 - 0.309 * Kappa\_3 - 0.00730 * Jurs\_TPSA - 0.149 * Hbond\ donor + 0.0655 * Rotlbonds \tag{10}$$

$$\log P_{app} = -0.00291 * MW - 0.0150 * PSA - 3.052 (n = 22) \tag{11}$$

The simplicity and the easily computable variables suggest Eqs. (8) or (9) to be versatile in modeling the apparent intestinal permeability of virtual compound-libraries. Applying Eq. (9) to a validation set of 50 molecules resulted in an inferior performance with an r^2(pr) of 0.455 and an rmse of 0.73 log units. To determine if the predictions are affected by the choice of descriptors, additional regressions were performed and compared with a reported PLS model [54] (r^2=0.91 versus $\log P_{app}$ and using nine variables). The G/PLS model represented by Eq. (10) for the same 17 compounds demonstrates a much better performance with fewer variables (Table 1). In contrast, the correlation coefficient obtained for the 50 validation-set molecules clearly invalidates the extension of Eq. (10) for $\log P_{app}$ predictions on unknowns.

In comparing previous QSPR models of $\log P_{app}$ and Eq. (10), it is clear that neither the choice of descriptors nor the number of terms in the QSPR equation is a major contributor for the weaker correlations of the models. One of the key reasons for the failure of Eqs. (9) and (10) derived using the standard 17 compounds in the literature arises due to the lack of structural diversity in the training set. Molecules like alprenolol, atenolol, practolol, and metaprolol fall within a homologous series, as do the steroids corticosterone, hydrocortisone, and testosterone. In addition, recent pharmacokinetic studies show some of the literature training-set compounds, like dexamethasone and sulfasalazine, to be transported through efflux mechanisms. [43] Since our attempt was to derive a QSPR model for passively permeable compounds, a training set that spans more of the diversity space was essential. The 22 training-set compounds (Table 2) used for deriving the $\log P_{app}$ model in the present study overcome some of the contaminations in the dataset and represent diverse compounds that appear to permeate predominantly through passive processes.

Equation (11) derived from 22 training-set compounds yielded weaker correlations with an increased rmse compared to equations obtained with the 17-compound training set (Table 1). However, the contrasting interpretations of the effect of molecular weight (compare the coefficients in Eqs. (9) and (11) suggest the sensitivity of the independent variables to the number and choice of the compounds in the dataset. On the other hand, the comparable statistical quality of Eq. (3) with the reported PLS model [54] (Table 1) clearly demonstrates that the lack of predictive ability of Eq. (11) is not due to the different selection of the training set molecules but caused by the incomplete representation of the descriptors. This incomplete representation is further substantiated by the fact that models of HIV-protease-inhibitor uptake by Caco-2 monolayers required multivariate regressions with proper choice of independent variables [55].

Comparison with reported %HIA models

Similar to the $\log P_{app}$ models, QSPR efforts in modeling %HIA involved a standard dataset of 20 molecules in the literature (see supporting information). Using only PSA as the independent variable, a simple non-linear model was obtained with an r^2 of 0.90 [23, 29]. Similarly, a linear model was derived for logitHIA by using multivariate statistics [40]. In spite of a good correlation [r^2(tr)=0.916] for the 20 training-set compounds, neither of these models has been used for quantitative %HIA prediction on an external validation set. Our attempts to model the %HIA using the Boltzmann sigmoidal curve function (Eq. 12) and PSA for the same 20 compounds yielded a correlation coefficient of 0.941, in close agreement to reported QSPR models (Table 1).

$$\%\mathrm{HIA} = 100/[1 + \exp((\mathrm{PSA}_{50} - \mathrm{PSA})/\mathrm{slope}) \text{ where } \mathrm{PSA}_{50} \text{ is the PSA value at which HIA} = 50\% \quad (12)$$

However, the use of both actively transported and passively diffused compounds in the literature %HIA and logitHIA models can affect the predictive ability. Although stringent conditions were used in obtaining the 20 training-set compounds in the literature [23], pharmacokinetic studies reveal foscarnet and sulfsalazine to be absorbed through various transport mechanisms [7, 8, 43]. Consequently, Eq. (12) was fitted against the experimental %HIA values for the 30 training-set compounds used in this study (Table 3) and compared with the G/PLS results obtained through Eqs. (4, 5, 6 and 7). In spite of explaining only 55% of the variance in the training-set data, this single descriptor non-linear model outperforms all the G/PLS models in its predictive ability. The reversal in the r^2 and rmse trends between the training and prediction sets (Table 1) for the %HIA models also demonstrates that the G/PLS procedure perhaps overfits the 30 training-set compounds considerably and more so when electrotopological descriptors are used (compare Eqs. 4 and 5 with 6 and 7). Given the similar r^2(ss) and rmse obtained through Eqs. (6) and (7) with that of Eq. (12), it is tempting to use the PSA model for the sake of simplicity. However, this non-linear model breaks down when apolar molecules are considered and while predicting the %HIA of a congeneric series where hydrophobic substitutions ($-CH_3$, $-C_2H_5$, $-$Ph, etc.) are made. In both these cases, no change in the computed %HIA will be necessarily observed, although the intestinal absorption process could be affected significantly. Again, by virtue of the sigmoidal relationship of PSA with %HIA, the PSA_{50} and the slope values in Eq. (12) are highly sensitive to the number of compounds used in the training set (Table 1).

Interpretation of the physicochemical descriptors

Several studies have demonstrated that the intestinal absorption and permeability are governed by a number of factors including the lipophilicity, molecular size and shape, and hydrogen-bonding capabilities [27]. The G/PLS models derived here (Eqs. 3, 4, 5, 6, 7 and 10) utilize all these variables in the regression equations and illustrate that random descriptors are not included in explaining the dependent property. For example, three out of the five variables considered by Lipinski [31] as features consistent with drug-like compounds for classifying the intestinal absorption process are manifested through Equation (3). The diversity in the physicochemical variables describing

Table 4 Total and percent correct classification (%, in parenthesis) of HIA based on Eqs. (4, 5, 6, 7, and 12) for the training set, prediction set, and the molecules considered to be absorbed through carrier-mediated processes

Dataset	N[a]	Eq. (3)	Eq. (11)	N[a]	Eq. (4)	Eq. (5)	Eq. (6)	Eq. (7)	Eq. (12)
Training	22	20 (90.9)	16 (72.7)	30	28 (93.3)	27 (90.0)	27 (90.0)	22 (73.3)	24 (80.0)
Prediction	19	17 (89.5)	10 (52.6)	46	26 (56.5)	29 (63.0)	32 (69.6)	31 (67.4)	34 (73.9)
Transport	34	15 (44.1)	15 (44.1)	43	15 (34.9)	13 (30.2)	21 (48.8)	14 (32.6)	19 (44.2)

[a]Number of molecules in the dataset

Table 5 Total and percent correct consensus prediction and success rate (%, in parenthesis) based on Eqs. (4, 5, 6, 7 and 12) for the training set, prediction set, and the molecules considered to be absorbed through other mechanisms (Transport Set)

Confidence level	N[a]	100%	80%	60%	40%	20%	0%
Equation's (4, 5, 6, and 7, 12)							
Training Set	30	21(70.0)	2(6.7)	2(6.7)	4(13.3)	1(3.3)	0(0.0)
Prediction Set	46	18(39.1)	5(10.9)	8(17.4)	6(13.0)	6(13.0)	3(6.6)
Transport Set	43	2(4.6)	5(11.6)	8(18.6)	10(23.3)	8(18.6)	10(23.3)

[a]Number of molecules in the dataset

the logP_{app} also shows that the H-bonding and the Jurs' terms in Eq. (3) capture the effects of PSA in Eq. (11). Eq. (3) further reveals that H-bond acceptor atoms and lipophilic substitutions improve the intestinal permeability while H-bond donors decrease permeability.

The multiple QSPR models for logitHIA substitute the Jurs descriptors (compare Eq. 4 with Eq. 5) and the substitution of electrotopological indices through components of Jurs-terms, gravitational index, and area (compare Eqs. 4 and 5 with Eqs. 6 and 7). The use of molecular weight in all the G/PLS equations agrees with the general notion that molecular weight and diffusion are interrelated.

Qualitative predictions of intestinal permeability and absorption

Although quantitative estimates of logP_{app} and %HIA are effective for comparisons across a homologous series, such accuracies are not essential for screening large compound libraries. Furthermore, given the considerable uncertainty in the experimental %HIA measurements and the sizeable errors in the computed values, it would be beneficial if the QSPR models reported in this study reproduce qualitative features correctly. The coarse filters (poor, medium, good) used to define the extent of intestinal permeability (see Materials and methods) demonstrate that reasonable hit rates are observed when the logP_{app} values obtained from multivariate analysis are used (Table 4; Eq. 3). The results also show that the 2-term logP_{app} model is approximately 20% less accurate in qualitatively classifying the apparent permeability of compounds in the training and the prediction sets. In spite of the relatively poor r^2(pr) observed for the four %HIA models (Eqs. 4, 5, 6 and 7), the consistent classification of the degree of intestinal absorption adds confidence in extending these equations for qualitative %HIA predictions of molecules that have not been considered in the prediction set. Although the classifications obtained through the non-linear %HIA model (Eq. 12) perform better for the prediction-set compounds, Eqs. (6) and (7) perform well when outliers are removed. Also, Eq. (12) may not perform well for a homologous series where the analogues differ by their hydrophobic substitutions ($-CH_3$, $-C_2H_5$, $-CH(CH_3)_2$, etc.) while Eqs. (6) and (7) should predict trends in these series more consistently. The <50% hit rates obtained in correctly classifying compounds that are carrier-mediated (Table 4) suggests that Eqs. (6), (7) and (12) discriminate molecules transported through different transport mechanisms from those that are passively diffused. In addition, the QSPR models derived here indirectly suggest that the intestinal permeability and absorption of actively transported compounds are modeled reasonably, except for the inability in accounting for the factors contributed by the solute environment and the kinetics of the mediators.

Since the five QSPR models derived for predicting the %HIA perform only moderately well for the prediction-set data, a consensus approach was designed to minimize the number of false positives in the classification scheme. Alternatively, comparing the results obtained using all five %HIA models provides an indirect mechanism of assessing the reliability of computed predictions. The results in Table 5 show that 21 of the 32 compounds in the training-set compounds are classified correctly by all the models, resulting in a 100% confidence in the computed values. The predictions are 80% accurate if four out of the five models classify a molecule similarly. Alternatively, borderline cases are identified when three out of the five models predict similarly. These results show an improved performance when the results of all the five models are used collectively (Table 5) over the %HIA models considered individually (Table 4) for qualitative classification.

Conclusions

To derive the logP_{app} and logitHIA models, linear relationships were assumed to exist with the descriptors investigated. However, a non-linear behavior is illustrative when %HIA is modeled using PSA. In comparing the QSPR models derived for logP_{app} and %HIA, we believe that the multivariate-statistics approach performs the best for estimating the P_{app} while a single variable like PSA describes the human intestinal absorption process satisfactorily. The logP_{app} models derived in the present study also demonstrate that the intestinal permeability is governed by several parameters. The varied dataset used in the reported QSPR models restricts a direct comparison on the performance of the present models to that proposed in literature for modeling logP_{app} and HIA. However, QSPR models derived using the standard training set used in the literature resulted in weaker correlations when extended to an external prediction set. This is attributed to the lack of structural diversity and the inclusion of actively transported compounds in the literature training set and to the use of non-passively absorbed molecules. In contrast, the use of training-set data containing more passively absorbed compounds resulted in models with good predictive ability. In the absence of a quantitative %HIA comparison for the models derived here with reported QSPR results, qualitative features on the extent of intestinal absorption can be used as an index to assess the performance.

We believe that Eqs. (4, 5, 6, 7) describe %HIA and Eqs. (9, 10 and 11) describe $\log P_{app}$ well. The complexity of both experiments implies that careful use of any models for these properties is necessary. We have attempted to eliminate actively transported compounds from the data. Further work on computational models will be required as additional measurements are reported. We do find that we can obtain predictive models without large numbers of variables and that these models are predictive outside the training set.

The fact that coefficients on common variables in the two equation sets are not always consistent is problematic. P_{app} is often used as a surrogate test for HIA, yet some of our equations imply the physical processes involved in the two experiments are not the same. Because the correlations and validations are better for P_{app} and that it is a simpler experiment, we believe that the P_{app} models are more likely to extend to new compounds.

References

1. Smith DA, van de Waterbeemd H (1999) Curr Opin Chem Biol 3:373–378
2. Clark DE, Pickett SD (2000) Drug Discovery Today 5:49–58
3. Fecik RA, Frank KE, Gentry EJ, Menon SR, Mitscher LA, Telikepalli H (1998) Med Res Rev 18:149–185
4. Tarbit MH, Berman J (1998) Curr Opin Chem Biol 2:411–416
5. Navia MA, Chaturvedi PR (1996) Drug Discovery Today 1:179–189
6. Chan OH, Stewart BH (1996) Drug Discovery Today 1:461–473
7. Tsuji A, Tamai I (1996) Pharm Res 13:963–977
8. Lin JH, Lu AYH (1997) Pharmacol Rev 49:403–449
9. Stewart BH, Chan OH, Lu RH, Reyner EL, Schmid HL, Hamilton HW, Steinbaugh BA, Taylor MD (1995) Pharm Res 12:693–699
10. Stewart BH, Chan OH, Jezyk N, Fleisher D (1997) Advanced Drug Delivery Reviews 23:27–45
11. Barthe L, Woodley J, Houin G (1999) Fundam Clin Pharm 13:154–168
12. Pinto M, Robine-Leon S, Appay M-D, Kedinger M, Triadou N, Dussaulx E, LaCroix B, Simon-Assmann P, Haffen K, Fogh J, Zweibaum A (1983) Biol Cell 47:323–330
13. Hidalgo IJ, Raub TJ, Borchardt RT (1989) Gastroenterology 96:736–749
14. Wilson G, Hassan IF, Dix CJ, Williamson I, Shah R, Mackay M (1990) J Controlled Release 11:25–40
15. Delie F, Rubas WA (1997) Crit Rev Ther Drug Carrier Syst 14:221–286
16. Artursson P, Karlsson J (1991) Biochem Biophy Res Commun 175:880–885
17. Palm K, Luthman K, Ungell A-L, Strandlund G, Artursson P (1996) J Pharm Sci 85:32–39
18. Rubas W, Cromwell MEM (1997) Advanced Drug Delivery Rev 23:157–162
19. Yee S (1997) Pharm Res 14:763–766
20. Lennernäs H (1998) J Pharm Sci 87:403–410
21. Grès M-C, Julian B, Bourrié M, Meunier V, Roques C, Berger M, Boulenc X, Berger Y, Fabre G (1998) Pharm Res 15:726–733
22. Krarup LH, Christensen IT, Hovgaard L, Frokjaer S (1998) Pharm Res 15:972–978
23. Palm K, Luthman K, Ungell A-L, Strandlund G, Beigi F, Lundahl P, Artursson P (1998) J Med Chem 41:5382–5392
24. Martin YC (1981) J Med Chem 24:229–237
25. Nook T, Doelker E, Buri P (1988) Int J Pharmaceut 43:119–129
26. Merino V, Freixas J, Val Bermejo MD, Garrigues TM, Moreno J, Plá-Delfina JM (1995) J Pharm Sci 84:777–782
27. Camenisch G, Folkers G, van de Waterbeemd H (1996) Pharmaceutica Acta Helvetiae 71:309–327
28. Testa B, Carrupt P-A, Gaillard P, Billois F, Weber P (1996) Pharm Res 11:335–343
29. Camenisch G, Folkers G, van de Waterbeemd H (1998) Eur J Pharm Sci 6:325–333
30. Camenisch G, Alsenz J, van de Waterbeemd H, Folkers G (1998) Europ J Pharm Sci 6:313–319
31. Lipinski CA, Christopher AL (1997) Advanced Drug Delivery Rev 23:3–25
32. Clark DE (1999) J Pharmaceutical Sci 88:807–814
33. Sugawara M, Takekuma Y, Yamada H, Kobayashi M, Iseki K, Miyazaki K (1998) J Pharmaceutical Sci 87:960–966
34. Winiwarter S, Bonham NM, Ax F, Hallberg A, Lennernäs H, Karlén A (1998) J Med Chem 41:4939–4949
35. Bermejo M, Merino V, Garrigues TM, Plá-Delfina JM, Mulet A, Vizet P, Trouiller G, Mercier C (1999) J Pharm Sci 88:398–405
36. van de Waterbeemd H, Camenisch G (1996) Quant Struct Act Relat 15:480–490
37. Palm K, Stenberg P, Luthman K, Artursson P (1997) Pharm Res 14:568–571
38. Stenberg P, Luthman K, Artursson P (1999) Pharm Res 16:205–212
39. Goodwin JT, Mao B, Vidmar TJ, Conradi RA, Burton PS (1999) J Peptide Res 53:355–369
40. Norinder U, Österberg T, Artursson P (1999) European J Pharm Sci 8:49–56
41. Ghuloum AM, Sage CR, Jain AN (1999) J Med Chem 42:1739–1748
42. Wessel MD, Jurs PC, Tolan JW, Muskal SM (1998) J Chem Inf Comp Sci 38:726–735
43. Irvine JD, Takahashi L, Lockhart K, Cheong J, Tolan JW, Selick HE, Grove JR (1999) J Pharm Sci 88:28–33
44. Kansy M, Senner F, Gubernator K (1998) J Med Chem 41:1007–1010
45. Pade V, Stavchansky S (1998) J Pharmaceutical Sci 87:1604–1607
46. Rappe AK, Casewit CJ, Colwell KS, Goddard WA, Skiff WM (1992) J Am Chem Soc 114:10024–10035
47. The PSA arising from the oxygen, nitrogen, and halogen atoms, and the –OH, –NH2 groups was calculated using the solvent accessible surface area in Cerius2 with the van der Waals radii reported in reference 37.
48. Herman RA, Veng-Pedersen P (1994) J Pharm Sci 83:423–428
49. Rogers D, Hopfinger AJ (1994) J Chem Inf Comp Sci 34: 854–866
50. Genetic partial least squares in QSAR: Rogers D (1996) In: Devillers J (ed) Genetic algorithms in molecular modeling. Academic, London
51. Stanton DT, Jurs PC (1990) Anal Chem 62:2323–2329
52. Hall LH, Kier LB, Brown BB (1995) J Chem Inf Comp Sci 35:1074–1080
53. Viswanadham VN, Ghose AK, Revankar GR, Robins RK (1989) J Chem Inf Comput Sci 29:163–172
54. Norinder U, Österberg T, Artursson P (1997) Pharm Res 14:1786–1791
55. Stewart BH, Chung FY, Tait B, Blankley CJ, Chan OH, John C (1998) Pharm Res 15:1401–1406

Scheme 1 Umpolung catalysts for conversions of aldehydes (Ef = H) and acylsilanes (Ef = SiR_3') to d[1]-acyl anion equivalents, which couple with aldehydes or Michael-systems to acyloin- or Stetter-adducts, respectively. Pyruvate decarboxylation (Ef = CO_2 and elimination) is in vivo initiated by thiamin (vitamin B_1)

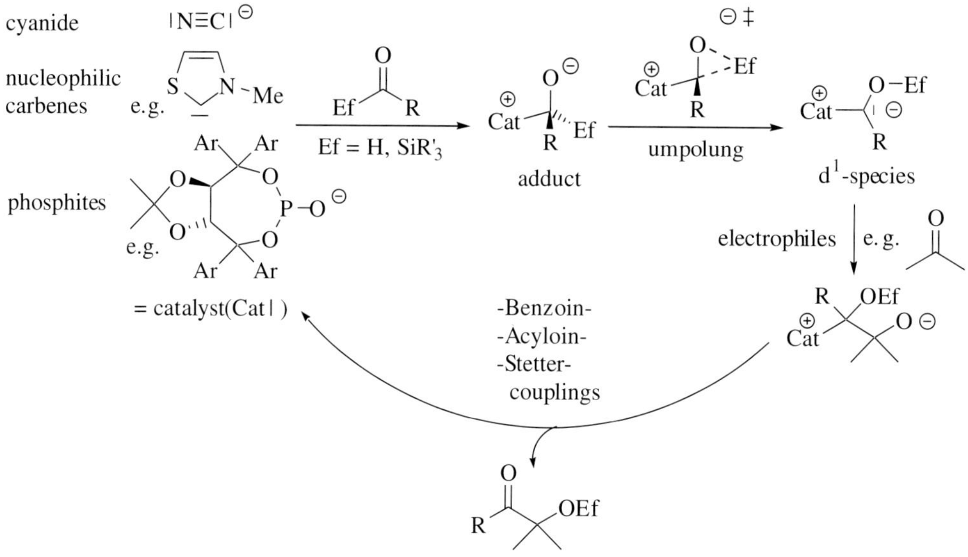

Scheme 2 Relative electronic energies for the cyanide-induced umpolung of formaldehyde (THF solvent, CPCM-B3LYP/6-31G*) referring to isolated reactants (cf. Table 1) and the electronic activation energy E_a

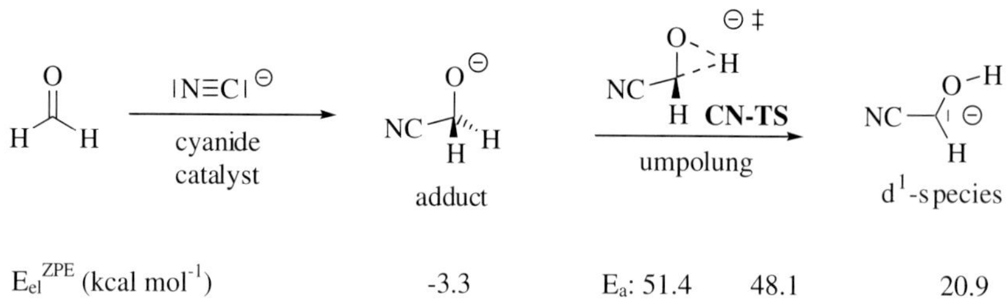

solvent environments (here THF) are more suitable than gas-phase computations. Additional diffuse functions provide only small corrections (Tables 1, 2). Relative to isolated reactants, a stable adduct forms from cyanide with formaldehyde (−3.3 kcal mol^{-1}, Scheme 2), but formation of this adduct is endergonic if entropy is considered (+5.9 kcal mol^{-1}, Table 1). The adduct transforms via 1,2-H-migration in transition structure CN-TS (Scheme 2, Fig. 1) to the unstable (20.9 kcal mol^{-1}) carbanionic d[1]-species (Scheme 2, Table 1).

Substituted (R) aldehydes form less stable adducts with cyanide than formaldehyde (Scheme 3, Tables 3, 4). Because of these thermodynamically instable adducts with R ≠ H, even lower activation barriers for the umpolung step relative to formaldehyde (E_a = 51.4 kcal mol^{-1}) are apparent for benzaldehyde

Table 1 Relative energies (kcal mol^{-1}) for the cyanide-induced umpolung of formaldehyde (Scheme 2)

	Adduct formation	TS	d[1]-species formation
E_{el}^{ZPE}	−3.3, −4.3,[a] −9.3[b]	48.1, 50.4,[a] 39.3[b]	20.9, 19.5,[a] 11.8[b]
ΔH	−4.6, −5.7,[a] −10.5[b]	46.8, 49.1,[a] 38.0[b]	19.4, 18.5,[a] 10.7[b]
ΔG	5.9, 4.9,[a] −0.2[b]	57.3, 59.6,[a] 48.5[b]	30.2, 28.3,[a] 20.8[b]

CPCM optimization in THF, B3LYP/6-31G*, and ZPE (scaled by 0.9806) corrected electronic, enthalpic, and free energies (298 K, 1 bar), relative to isolated adducts (cf. Table 2)
[a]CPCM-THF B3LYP/6-31+G*
[b]Gas-phase B3LYP/6-31+G*

and acrolein (E_a = 46.0 and 45.3 kcal mol^{-1}, Scheme 3, Table 3). The carbanionic d[1]-species are also less destabilized for phenyl (only 12.1 kcal mol^{-1}) and especially vinyl substituents (only 9.8 kcal mol^{-1}) relative to formaldehyde (20.9 kcal mol^{-1}, Scheme 3) (Glorius et al. recently reported successful umpolung reactions with acrolein derivatives using a nucleophilic carbene catalyst: [42, 43]). The methyl substitution in acetaldehyde increases the kinetic barrier (E_a = 52.3 kcal mol^{-1}) and the thermodynamic destabilization of the d[1]-species (25.9 kcal mol^{-1}, Scheme 3, Table 3).

A comparison among the three types of catalysts, i.e., cyanide, *N*-methylthiazol−2-ylidene, as an exemplary nucleophilic carbene, and a glycol-based phosphite (phosphonate), emphasizes the extraordinary quality of carbenes for the umpolung of aldehydes (Scheme 4, Tables 5, 6) (Glorius et al. recently reported successful umpolung reactions with acrolein derivatives using a nucleophilic carbene catalyst: [42, 43]). The thiazole-2-ylidene catalyst exhibits by far the strongest tendency for adduct formation with formaldehyde (−12.2 kcal mol^{-1}, Scheme 4). Despite the high stability of this carbene adduct, the activation barrier (E_a = 41.4 kcal mol^{-1}) is much lower than for cyanide (51.4 kcal mol^{-1}) or for the phosphite catalyst (52.4 kcal mol^{-1}, Scheme 4). Furthermore, the crucial d[1]-species is excellently stabilized by the carbene catalyst forming a neutral, electron-rich alkene (−11.8 kcal mol^{-1}). Cyanide is slightly superior to the phosphite with respect to thermodynamic

Table 2 Total energies (a.u.) for the cyanide-induced umpolung of formaldehyde (Scheme 2)

	$O=CH_2$	CN^-	Adduct	TS	d^1-species
E_{el}^{ZPE}	−114.48361	−92.93493	−207.42374	−207.34183	−207.38523
H	−114.47979	−92.93163	−207.41868	−207.33677	−207.38051
G	−114.50528	−92.95398	−207.44991	−207.36792	−207.41107
E_{el}^{ZPEa}	−114.49402	−92.95992	−207.46082	−207.37360	−207.42293
H^a	−114.49020	−92.95661	−207.45582	−207.36849	−207.41730
G^a	−114.51570	−92.97896	−207.48692	−207.39963	−207.44964
E_{el}^{ZPEb}	−114.48259	−92.86068	−207.35803	−207.28068	−207.32449
H^b	−114.47878	−92.85738	−207.35285	−207.27561	−207.31908
G^b	−114.50426	−92.87974	−207.38429	−207.30677	−207.35078

CPCM optimization in THF, B3LYP/6-31G*, and ZPE (scaled by 0.9806) corrected electronic, enthalpic, and free energies (298 K, 1 bar)
[a]CPCM-THF B3LYP/6-31 + G*
[b]Gas-phase B3LYP/6-31 + G*

and kinetic criteria of the umpolung sequence (Scheme 4).

The similarity between cyanide- and phosphite-catalysts is not only apparent from similar activation and reaction energies but also from similar geometries of the transition structures for the 1,2-H-migrations. Regarding the transfer of the proton from carbon to oxygen, the carbene transition structure SN-TS is early on the reaction coordinate with short C–H (1.205 Å) and long O–H distances (1.296 Å, Fig. 2). In contrast, CN-TS (Fig. 1) and PO-TS (Fig. 3) represent late transition structures with longer C–H (1.222, 1.219 Å) and shorter O–H distances (both 1.227 Å). Despite the early nature of SN-TS, its crucial carbon atom is much more planar coordinated by the catalyst, the hydrogen (fixed), and the oxygen atoms (angle sum: 353.8°) than in CN-TS (angle sum: 345.2°) or PO-TS (angle sum: 344.3°). This points to very strong conjugation and double bond formation in SN-TS and explains both the low activation barrier and the high stability of the d^1-species.

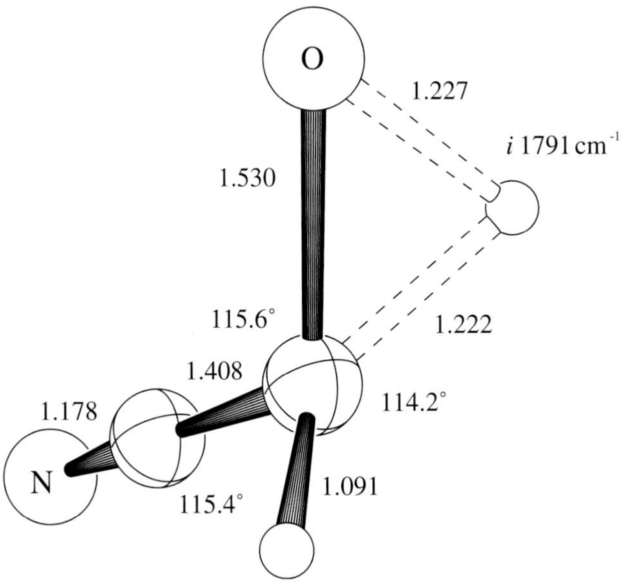

Fig. 1 CN-TS, transition structure of cyanide-induced umpolung of formaldehyde (CPCM-THF-B3LYP/6-31G*, bond distances in Angstrom, angles at carbon refer to fixed substituents)

Table 3 Relative energies (kcal mol^{-1}) for the cyanide-induced umpolung of different aldehydes (Scheme 3)

R =	Energy (kcal mol^{-1})	Adduct formation	TS	d^1-species formation
H	E_{el}^{ZPE}	−3.3	48.1	20.9
	ΔH	−4.6	46.8	19.4
	ΔG	5.9	57.3	30.2
Methyl	E_{el}^{ZPE}	3.3	55.6	25.9
	ΔH	2.2	54.6	25.1
	ΔG	13.2	65.4	34.8
Phenyl	E_{el}^{ZPE}	5.2	51.2	12.1
	ΔH	4.4	50.4	11.6
	ΔG	15.0	61.2	21.9
Vinyl	E_{el}^{ZPE}	4.9	50.2	9.8
	ΔH	4.0	49.2	9.2
	ΔG	14.7	60.1	19.3

CPCM optimization in THF, B3LYP/6-31G*, and ZPE (scaled by 0.9806) corrected electronic, enthalpic, and free energies (298 K, 1 bar), relative to isolated adducts (cf. Table 4)

Scheme 3 Relative electronic energies (E_{el}^{ZPE}, kcal mol^{-1}) for the cyanide-induced umpolung of different aldehydes (THF solvent, CPCM-B3LYP/6-31G*) referring to isolated reactants (cf. Table 3) and the electronic activation energies E_a

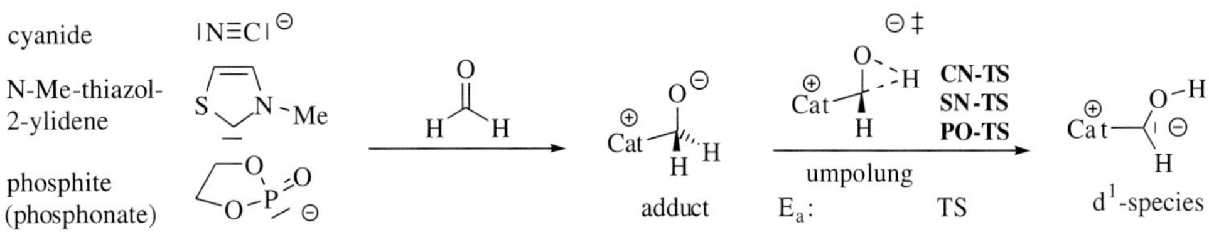

		adduct	E_a	TS	d^1-species
Cat =	Cyanide	-3.3	51.4	48.1	20.9
	Carbene	-12.2	41.4	29.2	-11.8
	Phosphite	-0.2	52.4	52.2	21.1

Table 4 Total energies (a.u.) for the cyanide-induced umpolung of different aldehydes (Scheme 3)

	Formaldehyde	Adduct	TS	d[1]-species
R = H				
E_{el}^{ZPE}	−114.48361	−207.42374	−207.34183	−207.38523
H	−114.47979	−207.41868	−207.33677	−207.38051
G	−114.50528	−207.44991	−207.36792	−207.41107
R = methyl	Acetaldehyde			
E_{el}^{ZPE}	−153.78571	−246.71535	−246.63201	−246.67937
H	−153.7809	−246.70902	−246.62557	−246.67258
G	−153.81059	−246.74357	−246.66039	−246.70911
R = phenyl	Benzaldehyde			
E_{el}^{ZPE}	−345.47650	−438.40311	−438.32983	−438.39216
H	−345.46918	−438.39372	−438.32053	−438.38239
G	−345.50702	−438.43713	−438.36349	−438.42614
R = vinyl	Acrolein			
E_{el}^{ZPE}	−191.86247	−284.78956	−284.71741	−284.78174
H	−191.85714	−284.78243	−284.71036	−284.77406
G	−191.88875	−284.81938	−284.74698	−284.81202

CPCM optimization in THF, B3LYP/6-31G*, and ZPE (scaled by 0.9806) corrected electronic, enthalpic, and free energies (298 K, 1 bar). Cyanide E_{el}^{ZPE}: −92.93493 a.u., enthalpy: −92.93163 a.u., free energy: −92.95398 a.u. (cf. Table 1)

Table 5 Relative energies (kcal mol^{-1}) for umpolung transformations of formaldehyde with different catalysts (Scheme 4)

Catalyst =		Adduct formation	TS (activation energy)	d[1]-species formation
Cyanide	E_{el}^{ZPE}	−3.3	48.1	20.9
ΔH	−4.6	46.8	19.4	
ΔG	5.9	57.3	30.2	
Carbene	E_{el}^{ZPE}	−12.2	29.2	−11.8
ΔH	−13.1	28.3	−12.1	
ΔG	−0.8	40.6	−1.2	
Phosphite	E_{el}^{ZPE}	−0.2	52.2	21.1
ΔH	−1.1	51.3	20.4	
ΔG	11.1	63.7	32.5	

CPCM optimization in THF, B3LYP/6-31G*, and ZPE (scaled by 0.9806) corrected electronic, enthalpic, and free energies (298 K, 1 bar), relative to isolated adducts (cf. Table 6)

Table 6 Total energies (a.u.) for umpolung transformations of formaldehyde with different catalysts (Scheme 4)

Catalyst =	Cyanide	Adduct	TS	d[1]-species
E_{el}^{ZPE}	−92.93493	−207.42374	−207.34183	−207.38523
H	−92.93163	−207.41868	−207.33677	−207.38051
G	−92.95398	−207.44991	−207.36792	−207.41107
Catalyst =	Carbene			
E_{el}^{ZPE}	−608.24051	−722.74359	−722.67765	−722.74285
H	−608.23395	−722.73466	−722.66858	−722.73296
G	−608.26981	−722.77631	−722.71034	−722.77699
Catalyst =	Phosphite			
E_{el}^{ZPE}	−645.74505	−760.22901	−760.1455	−760.19507
H	−645.73836	−760.21986	−760.13644	−760.18565
G	−645.77471	−760.26225	−760.17854	−760.22812

CPCM optimization in THF, B3LYP/6-31G*, and ZPE (scaled by 0.9806) corrected electronic, enthalpic, and free energies (298 K, 1 bar). Formaldehyde E_{el}^{ZPE}: −114.48361 a.u., enthalpy: −114.47979 a.u., free energy: −114.50528 a.u. (cf. Table 1)

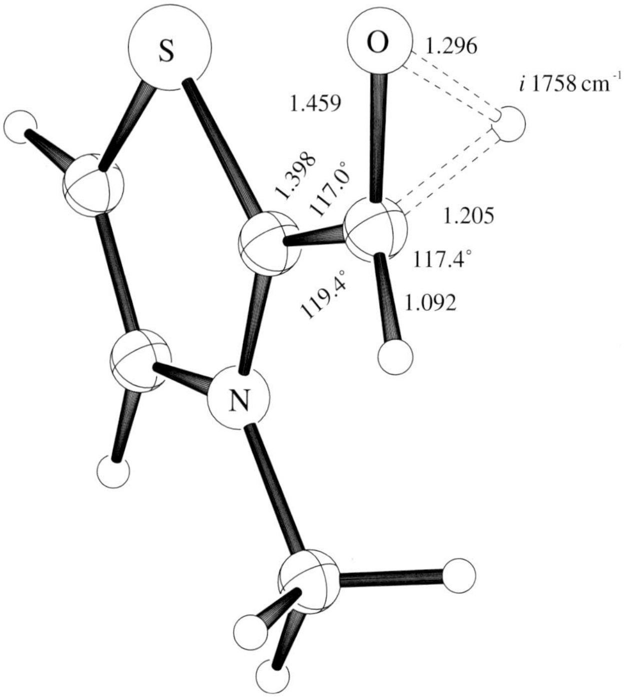

Fig. 2 SN-TS, transition structure of carbene-induced umpolung of formaldehyde (CPCM-THF-B3LYP/6-31G*, bond distances in Angstrom)

Conclusions

Aldehyde substrates with π-conjugation, i.e., phenyl or vinyl substituents, form the crucial carbanionic d[1]-intermediate more easily than formaldehyde or aldehydes with alkyl substituents. Phosphites, the new class of umpolung catalysts, resemble more cyanide than nucleophilic carbenes in umpolung reactions due to

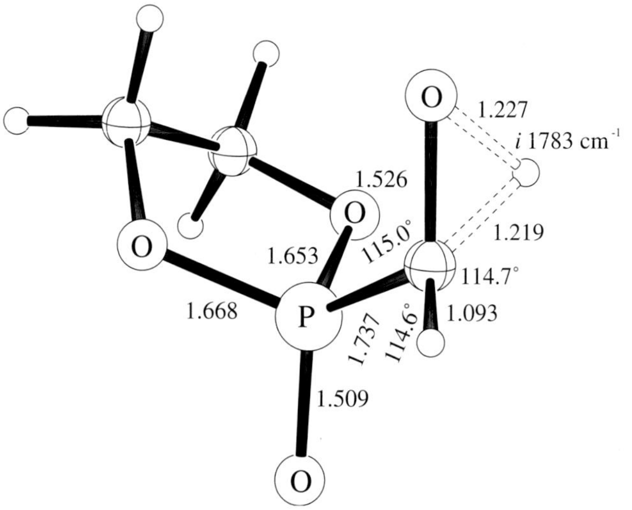

Fig. 3 PO-TS, transition structure of phosphite-induced umpolung of formaldehyde (CPCM-THF-B3LYP/6-31G*, bond distances in Angstrom, angles at carbon refer to fixed substituents). A *syn* OPCO conformer is less stable

kinetic, thermodynamic, and geometric assessments. *N*-methylthiazol-2-ylidene, was found to be by far the strongest umpolung catalyst, both with regard to nucleophilic attack on the aldehyde as well as kinetic and thermodynamic stabilization of 1,2-H-migration and of the d^1-intermediate. Further studies with focus on the role of counterions and other electrofuges (e.g., SiR_3 in Brook rearrangements) will provide an even better understanding of umpolung catalysis.

Acknowledgments We are grateful to the Fonds der Chemischen Industrie for financial support as well as for a Dozenten-Stipendium to BG. We especially thank the Deutsche Forschungsgemeinschaft (DFG) for funding (GO-930/9-1, GO-930/7-1, and GO-930/5-1) as well as the Bayer AG, the BASF AG, the Wacker AG, the Degussa AG, the Raschig GmbH, the Symrise GmbH, the Solvay GmbH, and the OMG AG for generous support.

References

1. Wöhler F, Liebig J (1832) Ann Pharm 3:249–287
2. Enders D, Balensiefer T (2004) Acc Chem Res 37:534–541
3. Kluger R (1997) Pure Appl Chem 69:1957–1967
4. Lapworth A (1903) J Chem Soc 83:995–1005
5. Seebach D (1979) Angew Chem 91:259–278
6. Seebach D (1979) Angew Chem Int Ed 18:239–258
7. Hase TA (1987) Umpoled synthons. Wiley, New York
8. Dugas H (1989) Bioorganic chemistry. Springer, Berlin Heidelberg New York, p 563
9. Breslow R (1958) J Am Chem Soc 80:3719–3726
10. Pohl M, Lingen B, Müller M (2002) Chem Eur J 8:5288–5295
11. Enders D, Kallfass U (2002) Angew Chem 114:1822–1824
12. Enders D, Kallfass U (2002) Angew Chem Int Ed Engl 41:1743–1745
13. Doudding T, Houk KN (2004) Proc Natl Acad Sci USA 5770–5775
14. Enders D, Breuer K, Kallfass U, Balensiefer T (2003) Synthesis 1292–1295
15. Kerr MS, Read J, de Alaniz R, Rovis T (2002) J Am Chem Soc 124:10298–10299
16. Enders D, Breuer K (1999) Addition of acyl carbanion equivalents to carbonyl groups and enones. In: Jacobsen EN, Pfaltz A, Yamamoto H (eds) Comprehensive asymmetric catalysis, chapter 30. Springer, Berlin Heidelberg New York
17. Stetter H, Kuhlmann H (1991) Org React 40:407–496
18. Stetter H (1976) Angew Chem 88:695–704
19. Stetter H (1976) Angew Chem Int Ed 15:639–648
20. Stetter H (1973) Angew Chem 85:89
21. Linghu X, Potnick JR, Johnson JS (2004) J Am Chem Soc 126:3070–3071
22. Takeda K, Tanaka T (1999) Synlett 705–708
23. Koenigkramer RE, Zimmer H (1980) Tetrahedron Lett 21:1017–1020
24. Linghu X, Bausch CC, Johnson JS (2005) J Am Chem Soc 127:1833–1840
25. Schwartz AW, de Graaf RM (1993) J Mol Evol 36:101–106
26. Saimoto H, Sugiyama N, Sashiwa H, Shigemasa Y (1993) Chem Express 8:737–740
27. Huskey WP, Epstein IR (1989) J Am Chem Soc 111:3157–3163
28. Shigemasa Y, Matsuba A, Ueda N, Nakashima R, Harada K, Takeda N, Suzuki M, Saito S (1984) Carbohydr Res 134:C4–C6
29. Harsch G, Bauer H, Voelter W (1984) Liebigs Ann Chem 4:623–635
30. Harsch G, Harsch M, Voelter W (1983) Z Naturforsch B 38B:1257–1268
31. Harsch G, Harsch M, Bauer H, Voelter W (1983) Z Naturforsch B 38B:1269–1280
32. Frisch MJ, Trucks GW, Schlegel HB, Scuseria GE, Robb MA, Cheeseman JR, Montgomery JA Jr, Vreven T, Kudin KN, Burant JC, Millam JM, Iyengar SS, Tomasi J, Barone V, Mennucci B, Cossi M, Scalmani G, Rega N, Petersson GA, Nakatsuji H, Hada M, Ehara M, Toyota K, Fukuda R, Hasegawa J, Ishida M, Nakajima T, Honda Y, Kitao O, Nakai H, Klene M, Li X, Knox JE, Hratchian HP, Cross JB, Adamo C, Jaramillo J, Gomperts R, Stratmann RE, Yazyev O, Austin AJ, Cammi R, Pomelli C, Ochterski JW, Ayala PY, Morokuma K, Voth GA, Salvador P, Dannenberg JJ, Zakrzewski VG, Dapprich S, Daniels AD, Strain MC, Farkas O, Malick DK, Rabuck AD, Raghavachari K, Foresman JB, Ortiz JV, Cui Q, Baboul AG, Clifford S, Cioslowski J, Stefanov BB, Liu G, Liashenko A, Piskorz P, Komaromi I, Martin RL, Fox DJ, Keith T, Al-Laham MA, Peng CY, Nanayakkara A, Challacombe M, Gill PMW, Johnson B, Chen W, Wong MW, Gonzalez C, Pople JA (2004) Gaussian 03, Revision C.02. Gaussian Inc., Wallingford
33. Ditchfield R, Hehre WJ, Pople JA (1971) J Chem Phys 54:724–728
34. Rassolov VA, Ratner MA, Pople JA, Redfern PC, Curtiss LA (2001) J Comput Chem 22:976–984
35. Becke AD (1993) J Chem Phys 98:5648–5652
36. Stephens PJ, Devlin FJ, Chabalowski CF, Frisch MJ (1994) J Phys Chem 98:11623–11627
37. Lee C, Yang W, Parr RG (1988) Phys Rev B 37:785–789
38. Miehlich B, Savin A, Stoll H, Preuss H (1989) Chem Phys Lett 157:200–206
39. Scott AP, Radom L (1996) J Phys Chem 100:16502–16513
40. Barone V, Cossi M (1998) J Phys Chem A 102:1995–2001
41. Cossi M, Rega N, Scalmani G, Barone V (2003) J Comput Chem 6:669–681
42. Burstein C, Glorius F (2004) Angew Chem 116:6331–6334
43. Burstein C, Glorius F (2004) Angew Chem Int Ed 43:6205–6208

J Mol Model (2006) 12: 597–609
DOI 10.1007/s00894-005-0037-3

ORIGINAL PAPER

Farooq A. Kiani · Matthias Hofmann

Periodic trends and easy estimation of relative stabilities in 11-vertex *nido-p*-block-heteroboranes and -borates

Received: 22 April 2005 / Accepted: 26 July 2005 / Published online: 29 October 2005

Abstract Density functional theory computations were carried out for 11-vertex *nido-p*-block-hetero(carba)boranes and -borates containing silicon, germanium, tin, arsenic, antimony, sulfur, selenium and tellurium heteroatoms. A set of quantitative values called "estimated energy penalties" was derived by comparing the energies of two reference structures that differ with respect to one structural feature only. These energy penalties behave additively, i.e., they allow us to reproduce the DFT-computed relative stabilities of 11-vertex *nido*-heteroboranes in general with good accuracy and to predict the thermodynamic stabilities of unknown structures easily. Energy penalties for neighboring heteroatoms (HetHet and HetHet′) decrease down the group and increase along the period (indirectly proportional to covalent radii). Energy penalties for a five- rather than four-coordinate heteroatom, [$Het_{5k}(1)$ and $Het_{5k}(2)$], generally, increase down group 14 but decrease down group 16, while there are mixed trends for group 15 heteroatoms. The sum of HetHet′ energy penalties results in different but easily predictable open-face heteroatom positions in the thermodynamically most stable mixed heterocarbaboranes and -borates with more than two heteroatoms.

Keywords Structural features · Structural increments · Energy penalties · Density functional theory · 11-Vertex *nido*-heteroboranes and heteroborates · 11-Vertex *nido*-heterocarbaboranes and heterocarbaborates

Electronic Supplementary Material Supplementary material is available for this article at http://dx.doi.org/10.1007/10.1007/s00894-005-0037-3 and is accessible for authorized users.

Dedicated to Professor Dr. Paul von Ragué Schleyer on the occasion of his 75th birthday

F. A. Kiani · M. Hofmann (✉)
Anorganisch-Chemisches Institut,
Ruprecht-Karls-Universität Heidelberg,
Im Neuenheimer Feld 270, 69120 Heidelberg, Germany
E-mail: matthias.hofmann@aci.uni-heidelberg.de

Introduction

The 11-vertex *nido*-cluster represents the most diverse family of heteroboranes and -borates. Many reactions are known [1–3] to incorporate a hetero-fragment into a smaller *nido*- or *arachno*-cluster, leading to 11-vertex *nido*-heteroboranes. Removal of one vertex from a 12-vertex *closo*-heteroborane cluster also leads to 11-vertex *nido*-heteroboranes and -borates [2, 4, 5]. Experimentally known 11-vertex *nido*-heteroborane and -borate clusters include: group 14 heteroatoms, i.e., carbon [6–9], silicon [10–13], germanium [14–17] and tin [6–8, 18–21]; group 15 heteroatoms, i.e., nitrogen, phosphorus [1, 2], arsenic [5, 22–32] and antimony [33]; group 16 heteroatoms, i.e., sulfur [34], selenium [35–41] and tellurium [35–37, 42, 53]. Williams' qualitative rules predict isomers with low-coordinate heteroatoms and separated heteroatoms to be preferred [6, 54, 55]. While these rules suffice to select the most stable *closo*-heteroboranes, the presence of additional endo-hydrogen atoms, the large number of isomers and possibly irresolvable conflicts ask for more sophisticated rules to predict the most favorable isomer in the case of *nido*-clusters.

A set of quantitative rules was presented that reproduced the stability order of 6-vertex *nido*-carboranes on the basis of 15 structural increments [56]. Disfavoring structural features, e.g., neighboring carbon atoms, were identified and the so-called energy penalties were derived by a statistical fitting procedure. Applying these energy penalties additively, the stability order of isomeric 6-vertex *nido*-(carba)boranes and -borates can easily be derived by a paper-and-pencil approach. With only nine such fitted quantitative rules, the relative stability order of numerous 11-vertex *nido*-(carba)boranes and -borates [57] was reproduced successfully. The approach was applied to the 10-vertex *nido*-(carba)boranes and -borates [58], and to the 11-vertex *nido*-mixed hetero(carba)boranes and -borates [59] with H–C, P, H–P, N and H–N heteromoieties. Our work [56–59] quantified Williams' rules [6, 54, 55] by corresponding energy penalties for each heteroatom and introduced some more rules due to open-face hydrogen characteristics of the *nido*-cluster. These quantitative rules allow us not only to predict the thermodynamically most stable isomer but also to estimate a stability order of various isomers easily [56–59]. Furthermore, these energy penalties successfully elaborate which two heteroatoms are more favorable choices for adjacent positions in the thermodynamically most stable mixed *nido*-heteroboranes. For example, quantitative rules indicate 7,8,10- rather than 7,8,9-, 7,9,10- and 7,9,8-positions for the heteroatoms in *nido*-$[P_2CB_8H_9]^-$ to be thermodynamically most stable [59].

In our previous work [56–59], energy penalties (E_{inc}) were determined by statistical fitting to a large number of structures. This procedure gives accurate values but requires extensive computations. Estimated energy penalties, (E_{inc}'), which are the energy difference of two suitable reference structures differing with respect to one structural feature only, are usually very close to the energy penalties arising from statistical fitting to a large number of isomers [59]. This is to be expected when structural features behave additively. For instance, the estimated energy penalty for adjacent carbon atoms, i.e., the energy difference of 7,8-$C_2B_9H_{11}^{2-}$ and 7,9-$C_2B_9H_{11}^{2-}$ is 16.3 kcal mol^{-1}, very close to the statistically fitted value (16.0 kcal mol^{-1}) derived from 20 carboranes [57–59]. Here, we present the relative stability order ($E_{inc}^{rel\prime}$) for 11-vertex *nido*-sila-, germana-, stanna-, arsa-, stiba-, thia-, selena- and tellura(carba)boranes and -borates, phosphathiaboranes and -borates and selenathiaboranes produced by E_{inc}', which are more approximate but easier to determine and are accurate enough for the interpretation of general trends which we wish to investigate in the present study.

The numbering scheme for the 11-vertex *nido*-cluster is shown in Fig. 1. The apical position is numbered as 1. The vertices next to the apex (middle belt) are given numbers 2–6, while the vertices of the open face are numbered from 7 to 11 where 7 is connected to 2 and 3. There are six cage vertices with connections to five other cluster atoms, $k_c = 5$ and five peripheral vertices with $k_p = 4$, where, c and p denote cage and peripheral vertices, respectively. In the literature, different numbering patterns have been used for mixed heteroboranes.

Computational details

For all hetero(carba)boranes and -borates except stanna, stiba and tellura(carba)boranes and -borates, geometries were consecutively optimized at B3LYP/3-21G and B3LYP/6-31G(d) using the *Gaussian 98* program [60]. The structures presented in this paper are local minima at B3LYP/6-31G(d). Single point energies were computed at B3LYP/6-311+G(d,p). Zero point vibrational energies from B3LYP/6-31G(d) frequency calculations were included to derive the relative energies for all the isomers.

For stanna, stiba and telluraboranes, geometries were optimized at the B3LYP/LANL2DZ level with additional d-polarization functions [61] for Sn, Sb, Te, B and C atoms ($\zeta = 0.183$, 0.211, 0.237, 0.388, 0.600, respectively). Single point energies were determined at B3LYP/SDD together with p-polarization function for H ($\zeta = 1.000$) and d-polarization function for Sn, Sb, Te, B and C [61] along with an sp set of diffuse functions for Sn, Sb, Te ($\zeta = 0.0231$, 0.0259, 0.0306, respectively) [62] as well as for B and C ($\zeta = 0.0315$ and 0.0438, respectively) [63].

Results and discussion

Structural features for hetero(carba)boranes and -borates

Different structural features for hetero(carba)boranes and -borates are shown in Fig. 2 and their energy penalties are listed in Table 1. Energy penalties for carbon in Table 1 are statistically fitted values taken from our previous work [57, 59]. For all other heteroatoms, the energy penalties are estimated as the energy difference of two structures that differ with respect to one structural feature only.

$Het_{5k}(1)$ and $Het_{5k}(2)$

A heteroatom at a 5k position (1–6) rather than a 4k position (7–11) is indicated by the structural feature Het_{5k} [57]. The apical position (number 1) differs from positions 2–6: the former has only 5k neighbors, the latter has two 4k and three 5k neighbors. Hence, higher energy penalties are observed for position 1, i.e., $Het_{5k}(1)$, as compared to positions 2 through 6, i.e.,

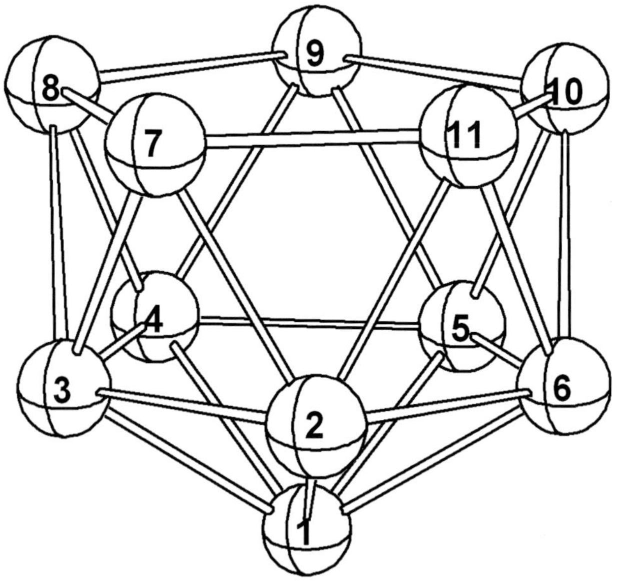

Fig. 1 Numbering scheme for the 11-vertex *nido*-cluster

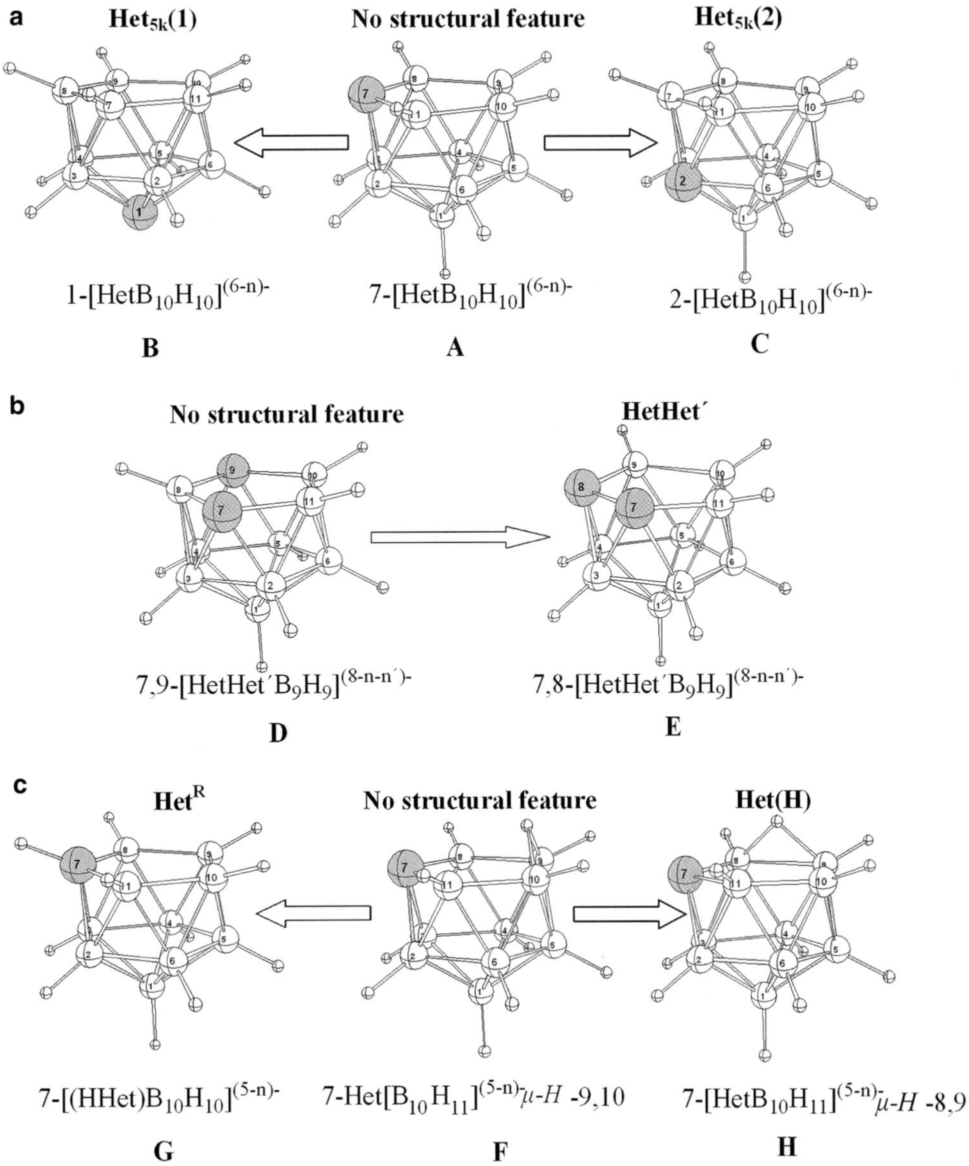

Fig. 2 **a** A heteroatom (Het) at a 5 k apical position (vertex number 1, structure *B*) or in the middle belt (positions 2 through 6, structure *C*) rather than at the open face (positions 7 through 11, structure *A*) represent the structural features $Het_{5k}(1)$ and $Het_{5k}(2)$, respectively. **b** Heteroatom adjacent (*E*) rather than heteroatom apart isomer (*D*) represent the structural feature HetHet′, where Het and Het′ may be equal or different heteroatoms. *n* and *n′* are the number of electrons donated by two heteroatoms (Het and Het′) **c** μ-*H*-8,9 (hydrogen bridge adjacent to heteroatom, *H*) rather than μ-*H*-9,10 (hydrogen bridge far away from heteroatom, *F*) in *nido*-7-$[HetB_{10}H_{11}]^{(4-n)-}$, represent the structural feature Het(H). Hydrogen as an exo-substituent (*G*) rather than bridged between positions 9 and 10 (*F*) produces the structural feature Het^R

$Het_{5k}(2)$ [57]. Estimated $Het_{5k}(1)$ energy penalties for a given heteroatom were obtained by comparing the 7- and 1-isomers of $[HetB_{10}H_{10}]^{(6-n)-}$ and that of $Het_{5k}(2)$ by comparing 7- and 2-isomers of $[HetB_{10}H_{10}]^{(6-n)-}$ (Fig. 2a), where Het = H–C, H–Si, N, H–N, P or H–P etc. and n = number of electrons donated by a given hetero group. $E_{inc}'[Het_{5k}(1)]$ and $E_{inc}'[Het_{5k}(2)]$ for different heteroatoms are listed in Table 1. For the carbon atom at a 5k position in heterocarbaboranes, the statistically fitted energy penalty of 28.0 kcal mol^{-1} obtained originally from 11-vertex *nido*-carboranes will be used [57].

HetHet′

Heteroatom-apart isomers are generally more favorable than heteroatom-adjacent isomers in heteroboranes and -borates [6, 54–59]. The structural feature HetHet′ gives the amount of destabilization caused by two adjacent heteroatoms. For example $7,8\text{-}[C_2B_8H_{10}]^{2-}$ with two adjacent carbon atoms (CC) is 16.3 kcal mol^{-1} less stable than carbon apart 7,9-isomer [57, 59]. The estimated energy penalties for HetHet′ were obtained by comparing the 7,8- and 7,9-isomers of $[HetHet'B_9H_9]^{(8-n-n')-}$ (Fig. 2b), where Het or Het′ may be equal or different heteroatoms and n and n' are the number of electrons donated by Het and Het′. When Het and Het′ are three-electron-donating heteroatoms ($\sum n = 6$), the structures to be compared are dianions, but they are neutral and monoanionic for two four-electron-donating heteroatoms ($n + n' = 8$) and one three and one four-electron-donating heteroatom ($n + n' = 7$), respectively. HetHet′ energy penalties for two adjacent carbon atoms, CC [57], and two adjacent phosphorus atoms, PP [59], are 16.0 and 10.7 kcal mol^{-1}, respectively. HetHet′ energy penalties for Het′ = Het and for Het′ = C are listed in Table 1. The energy penalties for a heteroatom adjacent to a bare phosphorus atom

Table 1 Relative trends of energy penalties [kcal mol^{-1}] for different features in 11-vertex *nido*-hetero(carba)boranes and -borates

Legend			
Het[a] χ[b] r [pm][c] **HetHet**[d] $Het_{5k}(1)$[f] **HetC**[e] $Het_{5k}(2)$[g] Het(H)[h]			

Group 14	Group 15 (bare)	Group 15 (Exo-Substituted)	Group 16
C 2.55[i] 77 **16.0** 28.0 **16.0** 28.0 2.2	**N** 3.04 75 **40.7** 44.3 **23.4** 41.1 0.5	**N^R** 3.04 75 **63.6**[k] 65.6 **36.9** 49.9 6.7	**O** 3.44[j] 73 --- --- --- --- ---
Si 1.90 111 **8.7** 45.0 **8.5** 33.6 4.9	**P** 2.19 106 **12.0** 31.5 **14.7** 27.8 4.3	**P^R** 2.19 106 **36.9** 56.7 **20.1** 43.1 6.8	**S** 2.58 102 **45.0**[l] 52.2 **31.2** 43.8 6.2
Ge 2.01 122 **4.2** 54.2 **7.7** 44.2 4.2	**As** 2.18 119 **6.9** 32.3 **16.0** 28.5 3.8	**As^R** 2.18 119 **26.8** 79.7 **17.3** --- 3.4	**Se** 2.55 116 **35.1** 48.2 **30.3** 40.7 6.1
Sn 1.96 141 **3.1** 69.7 **2.4** --- -1.7	**Sb** 2.05 138 **3.8** 31.8 **15.7** 29.0 4.5	**Sb^R** 2.05 138 **15.8** 92.6 **12.5** 26.2 1.2	**Te** 2.1 135 **29.3** 45.0 **28.6** 34.8 6.3

[a]Heteroatom
[b]Electronegativity values, see Pauling, L. *The Nature of the Chemical Bond. Cornell University Press*: Ithaca, New York, 1960
[c]Covalent radii in pico meter, see Huheey, J.E.; Keiter, E.A.; Keiter, R.L. Inorganic chemistry: Principles of structure and reactivity, 4th edition, HarperCollins, New York, USA, 1993
[d]Energy penalty for two identical adjacent heteroatoms in the 11-vertex *nido*-cluster
[e]Energy penalty for a heteroatom adjacent to a carbon atom in the 11-vertex *nido*-cluster
[f]$Het_{5k}(1)$ is the structural feature for a heteroatom at a 5k apical position (vertex number 1) rather than the ideal 4k open face positions
[g]$Het_{5k}(2)$ is the structural feature for a heteroatom at vertices 2 through 6 rather than at the ideal 4k open face positions
[h]Structural feature Het(H) denotes the amount of destabilization caused by a heteroatom adjacent to a bridged hydrogen atom
[i]Statistically fitted values taken from ref. 31. For all other heteroatoms, energy penalties are estimated by comparing two suitable reference structures which differ with respect to one structural feature
[j]Initial starting 11-vertex *nido*-oxaborane geometries did not survive geometry optimizations due to the expected very high energy penalties of the oxygen atom
[k]The N^RN^R energy penalty could not be accurately obtained as the structure rearranged. The rough energy penalty derived by fixing N7-B2 and N8-B2 distances to be 1.775 Å was even higher (76.5 kcal mol^{-1})
[l]The energy penalty for SS (45 kcal mol^{-1}) also needed to be derived by fixing the S(7)-S(8) bond distance to be 2.34 Å

(HetP) and to an exo-substituted phosphorus atom ($HetP^R$) are listed in Table 2.

Very similar energy penalties were derived for CC (i.e., two adjacent carbon atoms) in carboranes (16.0 kcal mol^{-1}) [57], phosphacarbaboranes (18.3 kcal mol^{-1}) [59], exo-substituted azacarbaboranes (15.4 kcal mol^{-1}) [59] and thiacarbaboranes (17.7 kcal mol^{-1}). Hence, we use an average value of 17.0 kcal mol^{-1} for E_{inc}[CC] in all heterocarbaboranes considered in this work.

Table 2 Energy penalties [kcal mol^{-1}] for $HetP^R$ and HetP together with covalent radius of heteroatom (Het)

HetHet′	R_{Het} [pm]	E_{inc}' [kcal mol^{-1}]
NP	71	18.8
CP	77	15.1
PP	93	10.7
N^RP^R	71	42.5
P^RP^R	93	36.9
SP^R	104	38.8
SeP^R	117	35.8

Het(H)

This structural feature presents the amount of destabilization caused by a heteroatom (Het) adjacent to a hydrogen bridge. Comparing *nido*-7-$[HetB_{10}H_{11}]^{(5-n)-}$ isomers, (n = number of electrons donated by Het) with μ-*H*-8,9 and μ-*H*-9,10 hydrogen positions, directly gives an estimated energy penalty for the structural feature Het(H) (Fig. 2c). This structural feature has a relatively small destabilizing effect. For example, the energy penalty for C(H) was determined to be 2.2 kcal mol^{-1} for carboranes [57]. The energy penalties of other heteroatoms adjacent to a hydrogen bridge are listed in Table 1. The largest Het(H) energy penalty (9.4 kcal mol^{-1}) is observed for the four-electron-donating P^R heterogroup, while tin has the smallest (even negative) energy penalty E_{inc}'[Sn(H)] = −1.7 kcal mol^{-1}. It is the only negative energy penalty observed for any heteroatom structural feature in 11-vertex *nido*-heteroboranes.

Het^R

This structural feature allows to compare bare (three-electron donating) and exo-substituted (four-electron donating) group 14 heteroatoms. *nido*-7-$[HetB_{10}H_{11}]^{2-}$ (μ-*H*-9,10) and *nido*-7-$[(HHet)B_{10}H_{10}]^{2-}$ (Fig. 2c) give a direct estimate of the energy penalty of Het^R for group 15 heteroatoms. Generally, three-electron-donating nitrogen and phosphorus atoms (N and P) have smaller energy penalties as compared to four electron donating exo-substituted nitrogen and phosphorus (N^R and P^R) atoms [59]. The same is true for bare arsenic (As) and antimony (Sb) atoms in the 11-vertex *nido*-cluster which have generally smaller energy penalties as compared to exo-substituted arsenic (As^R) and antimony (Sb^R) atoms (see Table 1).

Energy penalties as periodic properties of heteroatoms in 11-vertex *nido*-clusters

In this section, the general trends of HetHet′, $Het_{5k}(1)$ and $Het_{5k}(2)$ energy penalties will be discussed.

HetHet and HetC energy penalties decrease along group 14 (C → Sn), 15 (N → Sb) and 16 (S → Te) and increase along the periods (C → N, Si → S, Ge → Se, Sn → Te, see Table 1). The magnitude of energy penalties depends largely upon the extent of electron localization, which is determined primarily by the number of electrons donated by a heteroatom and secondarily by the electronegativity of the heteroatom. All the heteroatoms in Table 1 formally donate more than two electrons (two electrons are donated by a BH vertex) to the total of 26 skeletal electrons required in an 11-vertex *nido*-cluster and hence cause stronger electron localization as compared to a BH vertex. Two adjacent heteroatoms result in a larger degree of electron localization on two adjacent vertices and hence a positive HetHet energy penalty. This HetHet energy penalty is more positive for three-electron-donating group 15 heteroatoms as compared to the three-electron-donating group 14 heteroatoms. This is due to the larger electronegativity of three-electron-donating group 15 members. Four-electron-donating group 15 members have even higher electron localization due to four rather than three electrons localized at one vertex. Group 16 heteroatoms have even higher energy penalties as compared to group 15 heteroatoms due to larger electronegativity of the group 15 heteroatoms. It is interesting to note that neighboring NH groups have such a large destabilizing effect that the energy penalty could only be estimated by fixing the N(7)-B(2) and N(8)-B2 distances as the cluster shape was destroyed upon free geometry optimization [59]. Considering the general trends, the energy penalties for oxygen should be the largest but none of the five structural features for 11-vertex *nido*-oxaboranes could be determined as none of the oxaborane starting geometries optimized to a *nido*-11-vertex cluster geometry. Among the heteroatoms in Table 1, oxygen is the only one for which no experimentally known 11-vertex *nido*-heteroborane exists. The smallest HetHet energy penalty (3.1 kcal mol^{-1}) is found for tin (on the left bottom of Table 1).

Geometric consequences also seem to be important. Incorporation of one large heteroatom requires geometric distortion of the cluster. Incorporating another large heteroatom next to the first enhances the geometric distortion but to a lesser extent as compared to placing it at a yet undistorted site. Although this effect is overruled by the opposing electronic effects, it considerably reduces the energy penalties for two adjacent larger heteroatoms. When there is a significant electronegativity difference between boron and the heteroatoms, the electronic effect dominates. However, when the electronegativity of the heteroatom is very close to that of boron, the relative position of hetero-groups does not influence the electronic situation much and the geometric consequences are important.

Figure 3 shows such general trends for HetHet′ and HetC energy penalties, which are indirectly proportional to the covalent radii (directly proportional to electronegativity) within one group. Table 2 also shows very similar effects for $HetP^R$ and HetP energy penalties, where one heteroatom is a phosphorus atom.

Energy penalties for $Het_{5k}(1)$ and $Het_{5k}(2)$ increase down group 14 but decrease down group 16. For both three- as well as four-electron-donating heteroatoms in group 15, however, they show mixed trends (Fig. 4).

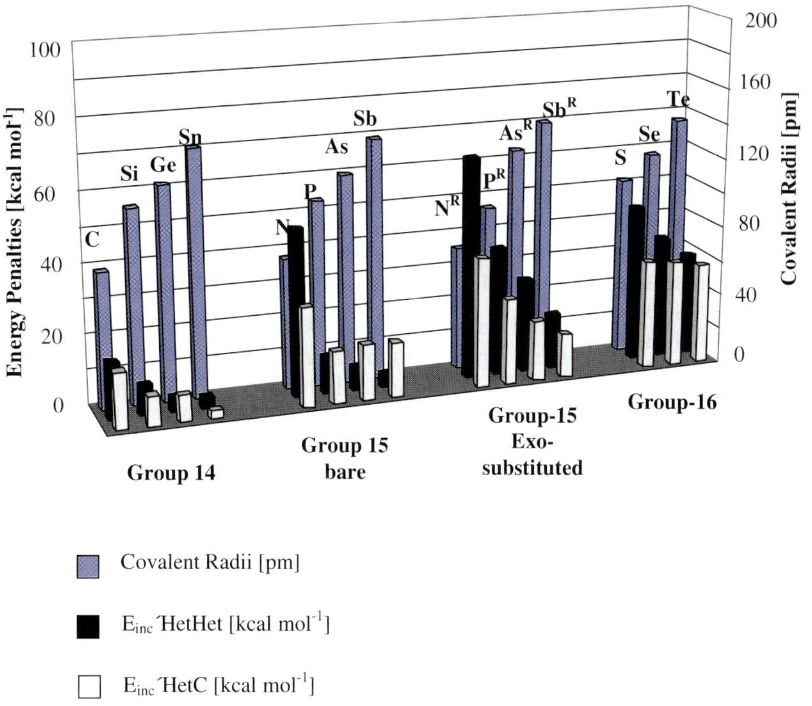

Fig. 3 Covalent radii, HetHet and HetC energy penalties for group 14, group 15 and group 16 heteroatoms. HetHet and HetC energy penalties for heteroatoms increase with decrease in covalent radii

and -borates, and selenathiaboranes and -borates. $\Delta E'$ is the difference between $E_{inc}^{rel\prime}$ and E_{calc}.

Thia(carba)boranes and -borates

Twenty-five isomers of thia(carba)boranes and -borates from *nido*-$SB_{10}H_{12}$ to *nido*-$SC_2B_8H_{10}$ are considered in this study. The estimated energy penalties for $S_{5k}(1)$, $S_{5k}(2)$, SS, SC, CC and S(H) were obtained as explained in the Structural features for hetero(carba)boranes and -borates section. A total of nine 11-vertex *nido*-thia(carba)borane and -borate clusters is experimentally known (labeled by "a" in Table 3, also see Fig. 5). Metal complexes of *nido*-$[SB_{10}H_{10}]^{2-}$ (**CA**) were also reported [64–67]. Two experimentally unknown $SC_2B_8H_{10}$ isomers, **GC** and **GD** (see Table 3) are predicted as strong candidates for synthesis because of their competitive thermodynamic stabilities.

The experimentally known [3, 34, 68] most stable *nido*-$SB_{10}H_{12}$ isomer, i.e., *nido*-7-$SB_{10}H_{12}$ (**AA**) has a sulfur atom at the open face with two bridged hydrogen atoms adjacent to the sulfur atom (structural feature S(H), twice). Both $E_{inc}^{rel\prime}$ and E_{calc} have very similar relative energy values for **AA** (*nido*-2-$SB_{10}H_{12}$), **AB** (*nido*-2-$SB_{10}H_{12}$) and **AC** (*nido*-1-$SB_{10}H_{12}$) (Table 3).

One extra hydrogen atom in *nido*-7-$[SB_{10}H_{11}]^-$ (**BA**) [3] bridges positions 9 and 10, resulting in no disfavoring structural feature but is adjacent to the sulfur atom in isomer **BB**, resulting in E_{inc}'[S(H)] = 6.2 kcal mol^{-1}. **BC**, i.e., *nido*-2-$[SB_{10}H_{11}]^-$ has a sulfur atom at position number 2 ($E_{inc}'[S_{5k}(2)]$ = 43.8 kcal mol^{-1}) and hence the structure is higher in energy than both **BA** and **BB**.

The absence of hydrogen bridges in *nido*-$[SB_{10}H_{10}]^{2-}$ results in only three possible isomers, i.e., *nido*-7-$[SB_{10}H_{10}]^{2-}$ (**CA**), *nido*-2-$[SB_{10}H_{10}]^{2-}$ (**CB**) and *nido*-1-$[SB_{10}H_{10}]^{2-}$ (**CC**), used to derive $E_{inc}'[S_{5k}(2)]$ = 43.8 kcal mol^{-1} and $E_{inc}'[S_{5k}(1)]$ = 52.2 kcal mol^{-1}.

The experimentally known [69] *nido*-7,9-$S_2B_9H_9$ (**DA**) is the most stable isomer as it lacks any structural feature. None of the dithiaborane starting geometries with two adjacent sulfur atoms optimized successfully but converged to rearranged structures. However, a rough estimate for the SS feature was obtained by fixing the S(7)-S(8) distance in 7,8-$S_2B_9H_9$ to be 2.34 Å (45.5 kcal mol^{-1}). Obviously, the SS feature, like N^RN^R [59], is incompatible with the *nido*-11-vertex cluster due to too large destabilization.

nido-7,9-SCB_9H_{11} with μ-*H*-10,11 (**EA**) [70], the most stable SCB_9H_{11} isomer, has non-adjacent carbon and sulfur atoms. Isomers **EB** through **ED** are at least 25 kcal mol^{-1} less stable than **EA**. A similar profound preference is found for the heteroatom apart *nido*-7,9-isomer (**FA**) [70] among $[SCB_9H_{10}]^-$ structures.

The experimentally known *nido*-7,9,10-$SC_2B_8H_{10}$ (**GA**) [3] is the most stable of the seven computed isomers. *nido*-7,8,9-$SC_2B_8H_{10}$ (**GE**) [3] and *nido*-7,8,10-$SC_2B_8H_{10}$ (**GB**) [3] with E_{calc} = 32.9 and 14.2 kcal mol^{-1}, respectively are also experimentally known. 8,2,10- (**GC**) and 7,1,9-$SC_2B_8H_{10}$ (**GD**) are thermodynamically more stable than 7,8,9-$SC_2B_8H_{10}$ (**GE**) [3], but are still experimentally unknown.

Phosphathiaboranes and -borates

The relative stabilities as determined from DFT computations and from structural increments for a few phosphathiaboranes are compared in Table 4. $[PSB_9H_9]^-$ structures lack extra hydrogen atoms and possess bare-phosphorus atom/s only. For *nido*-PSB_9H_{10}, however, both bare and exo-substituted phosphorus atoms are considered. The energy penalties derived for a phosphorus atom in phospha(carba)boranes and -borates [59] and for a sulfur atom in thia(carba)boranes and -borates (this paper) along with energy penalties for PS (derived by comparing *nido*-7,9-$[PSB_9H_9]^-$ with *nido*-7,8-$[PSB_9H_9]^-$) and P^RS (derived by comparing *nido*-7,9-(PH)SB_9H_9 with *nido*-7,8-(PH)SB_9H_9 can be used to estimate the relative stabilities of phosphathiaboranes. The estimated relative energies of four *nido*-PSB_9H_{10} isomers (i.e., **HA**–**HD** that differ in more than one feature) were found to be in good agreement with the relative energies computed at B3LYP/6-311+G(d,p)//B3LYP/6-31G(d) (see Table 4, **HA**–**HD**). 7,9-PSB_9H_{10} μ-*H*-10,11 (**HA**) with the structural features P(H) and S(H) has the least $\sum E_{inc}'$, $E_{inc}^{rel\prime}$ and E_{calc} values but is still experimentally unknown. *nido*-7,9-PSB_9H_{10} with an exo-substituted phosphorus atom (**HB**) is computed to be 3.4 kcal mol^{-1} higher in

Table 4 Estimated energy penalties (E_{inc}'), estimated relative energies ($E_{inc}^{rel\prime}$) for phosphathiaboranes. DFT computed relative energies are also reported for **HA** to **HD**. All values are in kcal mol^{-1}

	Compound	μ-*H*-	P(H) 2.2	P^R 13.3	S(H)′ 6.1	PS′ 21.4	P^RS' 38.8	$\sum E_{inc}'$	$E_{inc}^{rel\prime}$	E_{calc}	$\Delta E'$
HA[a]	7,9-PSB_9H_{10}	10,11	1		1			8.3	0.0	0.0	0.0
HB[b]	7,9-(HP)SB_9H_{10}			1				13.3	5.0	3.4	1.6
HC	7,8-PSB_9H_{10}	10,11	1			1		23.6	15.3	13.8	1.5
HD	7,8-(HP)SB_9H_{10}			1			1	52.1	43.8	42.2	1.6
IA	7,9-$PSB_9H_9^-$							0.0	0.0	0.0	0.0
IB	7,8-$PSB_9H_9^-$					1		21.4	21.4	21.4	0.0

[a]Strong candidate for synthesis
[b]7-Ph–**HB**, i.e., 7-Ph derivative of 7,9-PSB_9H_{10} is experimentally known

Table 5 Estimated energy penalties (E_{inc}'), estimated relative energies ($E_{inc}^{rel}{}'$) for selenaboranes and -borates. DFT computed relative energies are also reported for some structures. All values are in kcal mol^{-1}

	Compound	μ-*H*-	$Se_{5k}(1)'$ 48.2	$Se_{5k}(2)'$ 40.7	Se(H)′ 6.1	SeSe′ 35.1	$\sum E_{inc}'$	$E_{inc}^{rel}{}'$	E_{calc}	$\Delta E'$
JA[a]	7-$SeB_{10}H_{12}$	8,9; 10,11			2		12.2	0.0	0.0	0.0
JB	2-$SeB_{10}H_{12}$	7,8; 9,10		1	1		46.8	34.6	39.5	-4.9
KA[a]	7-$SeB_{10}H_{11}^-$	9,10					0.0	0.0	0.0	0.0
KB	7-$SeB_{10}H_{11}^-$	8,9			1		6.1	6.1	6.1	0.0
KC	1-$SeB_{10}H_{11}^-$	7,8	1				48.2	48.2	52.6	-4.4
LA[b]	7-$SeB_{10}H_{10}^{2-}$						0.0	0.0	0.0	0.0
LB	2-$SeB_{10}H_{10}^{2-}$			1			40.7	40.7	40.7	0.0
MA	7,9-$Se_2B_9H_9$						0.0	0.0	0.0	0.0
MB[a]	7,8-$Se_2B_9H_9$					1	35.1	35.1	35.1	0.0

[a]Experimentally known isomers
[b]Cyclopentadienyl metal derivatives are experimentally known

Table 6 Estimated energy penalties (E_{inc}'), estimated relative energies ($E_{inc}^{rel}{}'$) for selenacarbaboranes and -borates. DFT computed relative energies are also reported for some structures. All values are in kcal mol^{-1}

	Compound	μ-*H*-	C_{5k} 28.0	C(H) 2.1	CC 17.0	$Se_{5k}(1)'$ 48.2	$Se_{5k}(2)'$ 40.7	Se(H)′ 6.1	SeC′ 30.3	$\sum E_{inc}'$	$E_{inc}^{rel}{}'$	E_{calc}	$\Delta E'$
NA[a]	7,9-$SeCB_9H_{11}$	10,11		1				1		8.2	0.0	0.0	0.0
NB[b]	7,8-$SeCB_9H_{11}$	9,10		1					1	32.4	24.2	24.8	-0.6
NC	7,8-$SeCB_9H_{11}$	10,11						1	1	36.4	28.2	26.5	1.7
ND	1,7-$SeCB_9H_{11}$	9,10		1			1			42.8	34.6	32.8	1.8
NE	1,7-$SeCB_9H_{11}$	8,9		1		1				50.3	42.1	46.7	-4.6
NF	2,4-$SeCB_9H_{11}$	9,10	1	1			1			70.8	62.6	65.0	-2.4
OA	7,9-$SeCB_9H_{10}^-$									0.0	0.0	0.0	0.0
OB	7,8-$SeCB_9H_{10}^-$								1	30.3	30.3	30.3	0.0
OC	7,1-$SeCB_9H_{10}^-$		1							28.0	28.0	32.6	-4.6
PA[c]	7,9,10-$SeC_2B_8H_{10}$				1					17.0	0.0	0.0	0.0
PB	7,8,10-$SeC_2B_8H_{10}$								1	30.3	13.3	2.5	9.8
PC	7,1,9-$SeC_2B_8H_{10}$		1							28.0	11.0	7.6	3.4
PD	7,8,9-$SeC_2B_8H_{10}$				1				1	47.3	30.3	22.5	8.8
PE	7,8,11-$SeC_2B_8H_{10}$								2	60.6	42.6	37.4	5.2

[a]Strong candidate
[b]7-Cycloheanamine derivative is experimentally known
[c]Experimentally known isomer

energy than the former and its phenyl derivative i.e., *nido*-7-Ph-7,9-PSB_9H_9 was experimentally characterized [3].

Selena(carba)boranes and -borates

Estimated energy penalties were used to give the relative stability order of 25 selena(carba)boranes and -borates (Tables 5 and 6). The relative stability order is correctly reproduced in most cases, yet ΔE′ (the difference of $E_{inc}^{rel}{}'$ and E_{calc}) is larger for $SeC_2B_8H_{10}$ isomers (up to 9.8 kcal mol^{-1} for **PB**).

The most stable $SeB_{10}H_{12}$ isomer i.e., *nido*-7-$SeB_{10}H_{12}$ (**JA**) [39] has the selenium atom at vertex number seven with hydrogens bridging between 8/9 and 10/11 positions (structural feature Se(H) twice). The increment system suggests the deprotonated species, i.e., *nido*-7-[$SeB_{10}H_{11}$]$^-$ [35–41], with a hydrogen bridging positions 9/10 (**KA**) rather than positions 8/9 (**KB**) to be the most stable as in the case of exo-substituted *nido*-7-[(PH)$B_{10}H_{12}$]$^-$ [59]. *nido*-7-[$SeB_{10}H_{10}$]$^{2-}$ (**LA**) was reported as a ligand in complexes with different metal fragments [35–38, 42, 43, 71, 72]. The geometry of *nido*-7,8-$Se_2B_9H_9$ (**MB**) [73–76] unlike that of *nido*-7,8-$S_2B_9H_9$ could successfully be optimized and is 35.1 kcal mol^{-1} higher in energy than the experimentally still unknown but energetically more favorable 7,9-isomer (**MA**). Similarly, the heteroatom apart *nido*-7,9-$SeCB_9H_{11}$ (**NA**), the most stable $SeCB_9H_{11}$ isomer, is

Table 7 Estimated energy penalties (E_{inc}'), estimated relative energies ($E_{inc}^{rel}{}'$) for selenathiaboranes. All values are in kcal mol^{-1}

	Compound	$E_{inc}^{rel}{}'$	E_{calc}	ΔE	Structural feature
	7,9-$SeSB_9H_9$	0.0	0.0	0.0	None
QB	7,8-$SeSB_9H_9$	40.2	40.2	0.0	SSe′
QC	2,9-$SeSB_9H_9$	40.7	36.0	4.7	$Se_{5k}(2)'$
QD	9,2-$SeSB_9H_9$	43.8	38.3	5.5	$S_{5k}(2)'$
QE	1,7-$SeSB_9H_9$	48.1	51.9	−3.8	$Se_{5k}(1)'$
QF	7,1-$SeSB_9H_9$	52.2	54.7	−2.5	$S_{5k}(1)'$

Table 8 Estimated relative energies (kcal mol^{-1}) of 7,9,10-, 7,8,10-, 7,8,9- and 7,8,11-isomers in $[HetC_2B_8H_{10}]^{(4-n)-}$ [a,b]

Heteroatom (cluster charge)	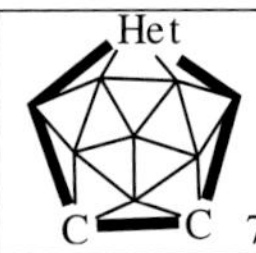 7,9,10-[c]	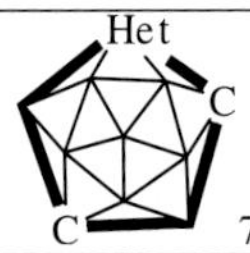 7,8,10-[d]	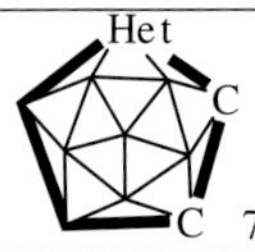 7,8,9-[e]	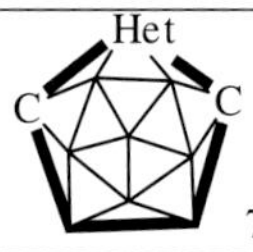 7,8,11-[f]
Het = H-Si (-1)	CC = 17.0 E_{rel} = 7.5	SiC = 8.5 E_{rel} = 0.0	SiC + CC = 25.5 E_{rel} = 17.0	2*SiC = 17.0 E_{rel} = 8.5
Het = H-Ge (-1)	CC = 17.0 E_{rel} = 9.3	GeC = 7.7 E_{rel} = 0.0	GeC + CC = 24.7 E_{rel} = 17.0	2*GeC = 15.4 E_{rel} = 7.7
Het = H-Sn (-1)	CC = 17.0 E_{rel} = 14.6	SnC = 2.4 E_{rel} = 0.0	SnC + CC = 19.4 E_{rel} = 17.0	2*SnC = 4.8 E_{rel} = 2.4
Het = N (-1)	CC = 17.0 E_{rel} = 0.0 (0.0)	NC = 28.4 E_{rel} = 11.4 (6.1)	NC + CC = 40.0 E_{rel} = 23.0	2*NC = 56.8 E_{rel} = 39.8
Het = P (-1)	CC = 17.0 E_{rel} = 1.9 (3.0)	PC = 15.1 E_{rel} = 0.0 (0.0)	PC + CC = 32.1 E_{rel} = 17.0 (17.9)	2*PC = 31.2 E_{rel} = 15.1 (18.1)
Het = As (-1)	CC = 17.0 E_{rel} = 1.0	AsC = 16.0 E_{rel} = 0.0	AsC + CC = 33.0 E_{rel} = 17.0	2*AsC = 32.0 E_{rel} = 16.0
Het = Sb (-1)	CC = 17.0 E_{rel} = 1.3	SbC = 15.7 E_{rel} = 0.0	SbC + CC = 32.7 E_{rel} = 17.0	2*SbC = 31.4 E_{rel} = 15.7
Het = H-N (0)	CC = 17.0 E_{rel} = 0.0 (0.0)	N^RC = 37.2 E_{rel} = 20.2 (18.8)	N^RC + CC = 44.2 E_{rel} = 37.2 (41.3)	2*N^RC = 74.4 E_{rel} = 57.4 (58.8)
Het = H-P (0)	CC = 17.0 E_{rel} = 0.0 (0.0)	P^RC = 24.3 E_{rel} = 7.3 (5.2)	P^RC + CC = 43.6 E_{rel} = 23.6 (24.9)	2*P^RC = 48.6 E_{rel} = 31.6 (29.5)
Het = H-As (0)	CC = 17.0 E_{rel} = 0.0	As^RC = 17.3 E_{rel} = 0.3	As^RC + CC = 34.3 E_{rel} = 17.3	2*As^RC = 34.6 E_{rel} = 17.6
Het = H-Sb (0)	CC = 17.0 E_{rel} = 4.5	Sb^RC = 12.5 E_{rel} = 0.0	Sb^RC + CC = 29.5 E_{rel} = 17.0	2*Sb^RC = 25.0 E_{rel} = 12.5
Het = S (0)	CC = 17.0 E_{rel} = 0 (0.0)	SC = 32.0 E_{rel} = 15.0 (13.1)	SC + CC = 49.0 E_{rel} = 32.0 (32.9)	2*SC = 64.0 E_{rel} = 47.0 (48.8)
Het = Se (0)	CC = 17.0 E_{rel} = 0 (0.0)	SeC = 30.3 E_{rel} = 12.7 (2.5)	SeC + CC = 47.3 E_{rel} = 30.3 (22.5)	2*SeC = 60.6 E_{rel} = 43.6 (37.4)
Het = Te (0)	CC = 17.0 E_{rel} = 0	TeC = 28.6 E_{rel} = 11.6	TeC + CC = 45.6 E_{rel} = 28.6	2*TeC = 57.2 E_{rel} = 40.2

[a]Het may be a three- or four-electron-donating heteroatom. *n* corresponds to the number of electrons donated by a given heteroatom
[b]B3LYP/6-311 + G(d,p)//B3LYP/6-31G(d) + ZPE computed relative stabilities of various $[HetC_2B_8H_{10}]^{(4-n)-}$ isomers are listed in parenthesis for various heteroatoms. These values are usually very close to the values predicted by estimated energy penalties
[c]7,9,10-$[NC_2B_8H_{10}]^-$, 7,9,10-$(HN)C_2B_8H_{10}$, 7,9,10-$SC_2B_8H_{10}$, 7,9,10-$SeC_2B_8H_{10}$ are experimentally known
[d]7,8,10-$SC_2B_8H_{10}$ is experimentally known
[e]7,8,9-$[NC_2B_8H_{10}]^-$ and 7-Me and 7-Ph derivatives of 7,8,9-$(HP)C_2B_8H_{10}$ are experimentally known
[f]7-Ph derivatives of 7,8,11-$(HP)C_2B_8H_{10}$ is experimentally known

still experimentally unknown although the 7-cyclohexanamine derivative of the 7,8-isomer (**NB**) is experimentally known [77]. *nido*-7,9,10-$SeC_2B_8H_{10}$ (**PA**) is experimentally known [73] and other computed $SeC_2B_8H_{10}$ structures (**PB–PE**) are thermodynamically less stable (Table 6).

Selenathiaboranes

The energy penalty (40.2 kcal mol^{-1}) for the structural feature SSe was obtained as the energy difference of 7,8- and 7,9-$SeSB_9H_9$. The latter is more stable and is experimentally known [69]. Relative energies of five $SeSB_9H_9$ isomers are given in Table 7.

Estimated energy penalties (E_{inc}') and corresponding estimated relative stabilities ($E_{inc}^{rel}{}'$) for other 11-vertex nido-hetero(carba)boranes and -borates

Estimated energy penalties for sila-, germa-, stanna-, bare and exo-substituted arsa- and stiba(carba)boranes and -borates are reported in Table 1, which can be used to produce the $E_{inc}^{rel}{}'$ for the 11-vertex *nido*-hetero(carba)boranes and -borates with H–Si, H–Ge, H–Sn, As, H–As, Sb and H–Sb heterogroups, respectively.

Prediction of thermodynamically most stable mixed heteroboranes and -borates with three open face heteroatoms

Energy penalties for the HetHet′ structural features describe the relative energies of open-face heteroboranes with two equal heteroatoms, for example, $[C_2B_9H_{11}]^{2-}$ [57], $P_2B_9H_{11}$ [59], $Se_2B_9H_9$ or that of heteroboranes with two different heteroatoms, e.g., 7,8- and 7,9-isomers of $[PSB_9H_9]^-$ and PSB_9H_{10}, $SeSB_9H_9$ etc. However, it is complex to predict the thermodynamically most stable isomer in mixed heteroboranes with three open-face heteroatoms, e.g., $[P_2CB_8H_9]^-$ [78], $[PC_2B_8H_{10}]^-$ [79, 80], $SC_2B_8H_{10}$ [3], $SeC_2B_8H_{10}$ [73], $NC_2B_8H_{11}$ [81], $[NC_2B_8H_{10}]^-$ [81]. Here we present only $[HetC_2B_8H_{10}]^{(4-n)-}$ examples, (where n = number of electrons donated by a heterogroup, and Het may be a three-electron-donating heteroatom/group, i.e., H–C, H–Si, H–Ge, H–Sn, N, P, As, Sb, or a four-electron-donating heteroatom/group, i.e., H–N, H–P, H–As, H–Sb, S, Se, Te (Table 8). All four possibilities for $[HetC_2B_8H_{10}]^{(4-n)-}$ structures with open face heteroatoms, i.e., 7,9,10-, 7,8,10-, 7,8,9- and 7,8,11-$[HetC_2B_8H_{10}]^{(4-n)-}$ will be discussed.

Both 7,9,10- and 7,8,10-isomers of $[HetC_2B_8H_{10}]^-$ have one structural feature each, i.e., CC and HetC, respectively. However, 7,8,9- and 7,8,11-isomers of $[HetC_2B_8H_{10}]^-$ have two structural features each, i.e., HetC + CC and 2·HetC, respectively. For group 14 heteroatoms, i.e., H–Si, H–Ge and H–Sn, the HetC, i.e., SiC, GeC and SnC energy penalties are smaller than that of CC and therefore 7,8,10-isomers (i.e., isomers with the HetC structural feature) are more stable. The 7,8,11-isomers with twice the structural feature HetC for three-electron-donating group 14 heteroatoms is not a too high energy option. HetC is very small for group 14 heteroatoms and therefore the 7,8,11-isomers of $SnC_2B_8H_{10}$ is only 2.4 kcal mol^{-1} higher in energy than the 7,8,10-isomer (see Table 8). In the case of three-electron-donating bare nitrogen atom (N), however, the NC structural feature has a larger disfavoring effect than CC, and therefore the 7,9,10-isomer (with structural feature CC) is more stable than the 7,8,10-isomer (with structural feature NC). But for other three-electron-donating group 15 heteroatoms, i.e., P, As, Sb, HetC has less disfavoring effect than CC and therefore the 7,8,10-isomer is more favorable for $[PC_2B_8H_{10}]^-$, $[AsC_2B_8H_{10}]^-$, $[SbC_2B_8H_{10}]^-$.

The estimated relative stabilities for $HetC_2B_8H_{10}$ structures for four-electron-donating heteroatoms are listed in Table 8. H–N and H–P have HetC energy penalties ($E_{inc}'[N^RC]$ = 36.0 kcal mol^{-1} and $E_{inc}'[P^RC]$ = 23.6 kcal mol^{-1}) much larger than CC ($E_{inc}'[CC]$ = 17.0 kcal mol^{-1}) and hence 7,9,10-isomers with structural feature CC are more favorable than the 7,8,10-isomers. For H–As, however, 7,8,10-$AsC_2B_8H_{11}$ (with structural feature As^RC ($E_{inc}'[As^RC]$ = 17.3 kcal mol^{-1}) and 7,9,10-$AsC_2B_8H_{11}$ with the structural feature CC ($E_{inc}'[CC]$ = 17.0 kcal mol^{-1}) are very similar in energy. Since HetHet′ energy penalties decrease down the group, the HetC energy penalty (Sb^RC) for a four-electron-donating antimony atom (Sb^RC) is 4.8 kcal mol^{-1} less than that of As^RC and therefore the 7,8,10-isomer is more stable for $(HSb)C_2B_8H_{10}$ as compared to the 7,9,10-isomer (7,9,10-isomer has structural feature CC and $E_{inc}'[CC] > E_{inc}'[Sb^RC]$.

HetC energy penalties for all four-electron-donating group 16 heteroatoms are much higher than CC and therefore 7,9,10-$HetC_2B_8H_{10}$ isomers are thermodynamically more stable than 7,8,10-isomers. 7,8,9- and 7,8,11-isomers have more than one structural feature, i.e., HetHet + HetC and 2·HetC, respectively, and therefore have even larger disfavoring effects for four-electron-donating heteroatoms.

Conclusions

Estimated energy penalties present a convenient method to predict the relative stabilities of 11-vertex *nido*-heteroboranes and -borates. Energy penalties for adjacent heteroatoms increase along the period and decrease down the group. Four-electron-donating heteroatoms generally have larger energy penalties than those of three-electron-donating heteroatoms. Larger heteroatoms usually have larger $Het_{5k}(1)$ and $Het_{5k}(2)$ energy penalties and smaller HetHet′ energy penalties, indicating that they prefer open-face vertices and that the destabilizing effect of adjacent heteroatoms is smaller for larger heteroatoms. Most stable mixed heteroboranes with more than two open-face heteroatoms have different but easily predictable heteroatom positions in the thermodynamically most stable 11-vertex *nido*-heteroborane isomers. Energy penalties are likely to have periodic trends in other polyborane clusters.

Supplementary material

Cartesian coordinates of the optimized geometries of 11-vertex *nido*-heterocarbaboranes and -borates considered in this paper are listed in Appendices I through V.

Acknowledgement Financial support by DFG is gratefully acknowledged.

References

1. Štibr B (2002) Collec Czech Chem Commun and references therein 67:843–868
2. Haubold W, Keller W, Sawitzki G (1988) Angew Chem Int Ed Engl 27:925–926
3. Shedlow AM, Sneddon LG (1998) Inorg Chem 37:5269–5277
4. Todd LJ, Little JL, Silverstein HT (1969) Inorg Chem 8:1698–1703
5. Little JL, Whitesell MA, Chapman RW, Kester JG, Huffman JC, Todd LJ (1993) Inorg Chem 32:3369–3372
6. Williams RE (1992) Chem Rev and references therein 92:177–207
7. Fox MA, Goeta AE, Hughes AK, Johnson AL (2002) J Chem Soc Dalton Trans 9:2009–2019
8. Fox MA, Greatrex R, Nikrahi A, Brain PT, Picton MJ, Rankin DWH, Robertson HE, Bühl M, Li L, Beaudet RA (1998) Inorg Chem 37:2166–2176
9. Dirk W, Paetzold P, Radacki K (2001) Z Anorg Allg Chem 627:2615–2618
10. Dopke JA, Bridges AN, Schmidt MR, Gaines DF (1996) Inorg Chem 35:7186–7187
11. Wesemann L, Englert U, Seyferth D (1995) Angew Chem 107:2345–2436
12. Wesemann L, Englert U, Seyferth D (1995) Angew Chem Int Ed Engl 34:2236–2238
13. Dopke JA, Powel DR, Hayashi RK, Gaines DF (1998) Inorg Chem 37:4160–4161
14. Loffredo RE, Norman AD (1971) J Am Chem Soc 93:5587–5588
15. Wesemann L, Trinkaus M, Ruck M (1999) Angew Chem Int Ed 38:2375–2377
16. Wesemann L, Ramjoie Y, Trinkaus M, Spaniol TP (1998) Eur J Inorg Chem 9:1263–1268
17. Wesemann L, Ramjoie Y, Trinkaus M, Ganter B (1997) Inorg Chem 36:5192–5197
18. Greenwood NN, Youll B (1975) J Chem Soc Dalton Trans 2:158–162
19. Dupont TJ, Loffredo RE, Haltiwanger RC, Turner CA, Norman AD, (1978) Inorg Chem 17:2062–2067
20. Loffredo RE, Dupont TJ, Haltiwanger RC, Norman AD (1977) J Chem Soc Chem Commun 4:121–122
21. Loffredo RE, Drullinger LF, Slater JA, Turner CA, Norman AD (1976) Inorg Chem 15:478–480
22. Todd LJ, Burke AR, Garber AR, Silverstein HT, Storhoff BN (1970) Inorg Chem 9:2175–2179
23. Bould J, Kennedy JD, Ferguson G, Tony DF, O'Riordan GM, Spalding TR (2003) Dalton Trans 23:4557–4564
24. O'Connell D, Patterson JC, Spalding TR, Ferguson G, Gallagher JF, Li Y, Kennedy JD, Macias R, Thornton-Pett M, Holub J (1996) J Chem Soc Dalton Trans 15:3323–3333
25. Fontaine XLR, Kennedy JD, McGrath M, Spalding TR (1991) Magn Reson Chem 29:711–720
26. Wright WF, Garber AR, Todd LJ (1978) J Magn Reson 30:595–602
27. Little JL, Pao SS (1978) Inorg Chem 17:584–587
28. Little JL, Pao SS, Sugathan KK (1974) Inorg Chem 13:1752–1756
29. Hanusa TP, Roig de Parisi N, Kester JG, Arafat A, Todd LJ (1987) Inorg Chem 26:4100–4102
30. Little JL (1979) Inorg Chem 18:1598–1600
31. Yamamoto T, Todd LJ (1974) J Organomet Chem 67:75–80
32. Colquhoun HM, Greenhough TJ, Wallbridge MGH (1979) J Chem Research 7:248
33. Valnot JY (1978) Synthesis 8:590–592
34. Pretzer WR, Rudolph RW (1976) J Am Chem Soc 98:1441–1447
35. Ferguson G, Gallagher JF, McGrath M, Sheehan JP, Spalding TR, Kennedy JD (1993) J Chem Soc Dalton Trans 1:27–34
36. Ferguson G, Parvez M, MacCurtain JA, Dhubhghaill ON, Spalding TR, Reed D (1987) J Chem Soc Dalton Trans 4:699–704
37. Little JL, Friesen GD, Todd LJ (1977) Inorg Chem 16:869–872
38. Faridoon, Dhubhghaill ON, Spalding TR, Ferguson G, Kaitner B, Fontaine XLR, Kennedy JD (1989) J Chem Soc Dalton Trans 9:1657–1668
39. Ferguson G, Hampden-Smith MJ, Dhubhghaill ON, Spalding TR (1988) Polyhedron 7:187–193
40. Barriola AM, Hanusa TP, Todd LJ (1980) Inorg Chem 19:2801–2802
41. Reed D, Ferguson G, Ruhl BL, Dhubhghaill ON, Spalding TR (1988) Polyhedron 7:17–23
42. Faridoon S, Trevor R, Ferguson G, Kennedy JD, Fontaine XLR (1989) J Chem Soc Chem Commun 14:906–908
43. Faridoon O, Dhubhghaill ON, Spalding TR, Ferguson G, Kaitner B, Fontaine XLR, Kennedy JD (1988) J Chem Soc Dalton Trans 11:2739–2745
44. Thornton-Pett M, Kennedy JD, Spalding F, Spalding TR (1995) Act Cryst C51:840–843
45. Ferguson G, O'Connell D, Spalding TR (1994) Act Cryst C50:1432–1434
46. Ferguson G, Gallagher JF, Sheehan JP, Spalding TR, Kennedy JD, Macias R (1993) J Chem Soc Dalton Trans 20:3147–3148
47. Sheehan JP, Spalding TR, Ferguson G, Gallagher JF, Kaitner B, Kennedy JD (1993) J Chem Soc Dalton Trans 1:35–42
48. Faridoon, McGrath M, Spalding TR, Fontaine XLR, Kennedy JD, Thornton-Pett M (1990) J Chem Soc Dalton Trans 6:1819–1829
49. Ferguson G, Lough AJ, Faridoon, McGrath MN, Spalding TR, Kennedy JD, Fontaine XLR (1990) J Chem Soc Dalton Trans 6:1831–1839
50. Ferguson G, Gallagher JF, Sheehan JP, Spalding TR (1998) J Organomet Chem 550:477–480
51. Ferguson G, Kennedy JD, Fontaine XLR, Faridoon, Spalding TR (1988) J Chem Soc Dalton Trans 10:2555–2564
52. Mceneaney PA, Spalding TR, Ferguson G (1997) J Chem Soc Dalton Trans 2:145–147
53. Ferguson G, Kennedy JD, Fontaine XLR, Faridoon, Spalding TR (1988) J Chem Soc Dalton Trans 10:2555–2564
54. Williams RE (1965) J Am Chem Soc 87:3513–3515
55. Williams RE (1970) Carboranes. In: Brotherton RJ, Steinberg H (eds) Progress in Boron Chemistry, Vol 2, Chapter 2. Pergamon Press, Oxford, England, p 57
56. Hofmann M, Fox MA, Greatrex R, Schleyer PvR, Williams RE (2001) Inorg Chem 40:1790–1801
57. Kiani FA, Hofmann M (2004) Inorg Chem 43:8561–8571
58. Kiani FA, Hofmann M (2005) Eur J Inorg Chem 12:2545–2553
59. Kiani FA, Hofmann M (2005) Inorg Chem 44:3746–3754
60. Frisch MJ, Trucks GW, Schlegel HB, Scuseria GE, Robb MA, Cheeseman JR, Zakrzewski VG, Montgomery JA, Stratmann RE, Burant JC, Dapprich S, Millam JM, Daniels AD, Kudin KN, Strain MC, Farkas O, Tomasi J, Barone V, Cossi M, Cammi R, Mennucci B, Pomelli C, Adamo C, Clifford S, Ochterski J, Petersson GA, Ayala PY, Cui Q, Morokuma K, Malick DK, Rabuck AD, Raghavachari K, Foresman JB, Cioslowski J, Ortiz JV, Stefanov BB, Liu G, Liashenko A, Piskorz P, Komaromi I, Gomperts R, Martin, RL, Fox DJ, Keith T, Al-Laham MA, Peng CY, Nanayakkara A, Gonzalez C, Challacombe M, Gill PMW, Johnson BG, Chen W, Wong MW, Andres JL, Head-Gordon M, Replogle ES, Pople JA (1998) Gaussian 98. Gaussian Inc, Pittsburgh, PA
61. Huzinaga S, Andzelm J (1984) Gaussian basis sets for molecular calculations. Elsevier, Amsterdam, pp 23–25
62. These values were optimized for the atomic ground state anion, using ROHF with a flexible ECP basis set, by Ted Packwood at NDSU. Diffuse functions for these and other heteroatoms may be found at the website: http://phoenix.liu.edu/~nmatsuna/gamess/refs/basis.refs.html

63. Clark T, Chandrasekhar J, Spitznagel GW, Schleyer PvR (1983) J Comput Chem 4:294–301
64. Kang SO, Carroll PJ, Sneddon LG (1989) Inorg Chem 28:961–964
65. Kang SO, Carroll PJ, Sneddon LG (1988) Organometallics 7:772–776
66. Zimmerman GJ, Sneddon LG (1981) J Am Chem Soc 103:1102–1111
67. Thompson DA, Rudolph RW (1976) J Chem Soc Chem Commun 19:770–771
68. Kang SO, Sneddon LG (1988) Inorg Chem 27:3298–3300
69. Friesen GD, Barriola A, Daluga P, Ragatz P, Huffman JC, Todd LJ (1980) Inorg Chem 19:458–462
70. Holub J, Kennedy J D, Jelínek T, Štíbr B (1994) Inorg Chem 8:1317–1323
71. Ferguson G, Faridoon, Spalding TR (1988) Acta Cryst C44:1368–1371
72. Ferguson G, Ruhl BL, Ni DO, Spalding TR (1987) Acta Cryst C43:1250–1253
73. Base K, Štibr B (1977) Chem Ind 22:919–920
74. Friesen GD, Barriola A, Todd LJ (1978) Chem Ind 16:631
75. Base K (1983) Collect Czech Chem Commun 48:2593–2603
76. Schultz RV, Huffman JC, Todd LJ (1979) Inorg Chem 18:2883–2886
77. Arafat A, Friesen GD, Todd LJ (1983) Inorg Chem 22:3721–3724
78. Bakardjiev M, Holub J, Štíbr B, Hnyk D, Wrackmeyer B (2005) Inorg Chem DOI 10.1021/ic048232c
79. Štíbr B, Holub J, Bakardjiev M, Hnyk D, Tok OL, Milius W, Wrackmeyer B (2002) Eur J Inorg Chem 9:2320–2326
80. Holub J, Ormsby DL, Kennedy J D, Greatrex R, Štibr B (2000) Inorg Chem Commun 3:178–181
81. Plešek J, Štíbr B, Hnyk D, Jelínek T, Heřmánek S, Kennedy JD, Hofmann M, Schleyer PvR (1998) Inorg Chem 37:3902–3909

number of complex materials-development problems including polymers, electronic materials and catalysts.

Some of the promises not delivered yet by HTE are discovery of knowledge and correlation of the function of a solid with its structure. Here things become difficult. While most drugs are clearly identified by their molecular structure, materials are complex entities that do not lend themselves to the idea of structure–activity relationships. Heterogeneous catalysts are typical complex materials and were chosen as models in this study. Today's state of theory does not allow us to predict the catalytic activity of new compositions. Catalyst optimization is therefore a tedious process of materials preparation followed by individual testing. Correlating activity with useful descriptors, such as composition, would open the complex field of materials to theoretical modeling and predictions. In an interesting study, the groups of Schüth and Mirodatos investigated 467 catalysts and their catalytic performance for the oxidation of propene with oxygen. The catalysts were of different origin and prepared by various methods [7]. The authors searched for general descriptors that correlate with the catalytic activity described by the dataset [8]. Because of the difficulty in assigning appropriate structure parameters to materials, a complex strategy was developed. In the dataset, parameters such as mean electron affinity, mean electronegativity or mean molar mass of all elements in the catalysts were identified as descriptors, while chemical composition does not correlate with the function of the materials [7]. This was to be expected, since catalytic activity is very sensitive to the preparation method of the catalyst. The function of a catalyst is not defined by composition, but by the exact synthesis protocol and the synthesis procedure, such as co-precipitation, impregnation, sol–gel-synthesis or powder synthesis and the precursors used. Changing the sequence of additions and allowing different drying times and calcination temperatures will result in a variety of different materials of identical composition, but different catalytic activity and selectivity. Such a dataset cannot be used in the search for the optimal composition of a catalyst. When optimization of chemical composition becomes the issue of HTE, great care has to be taken that the synthesis of the materials of varying composition is carried out under identical synthetic conditions. This implies synthesis recipes that tolerate a broad variation in chemical composition. We have found that this can be achieved very generally with specially optimized sol–gel-procedures. With such a set of materials, catalytic activity, measured under identical reaction conditions, should correlate with chemical composition and can be called QCAR (quantitative composition–activity relationship) [9]. If such correlations can be found, modeling catalytic activity becomes possible, as would be the prediction of optimal compositions. One essential aspect for such an approach is that the dependence of the material's function (catalytic activity or other desired properties) on chemical composition is continuous and not erratic.

QCAR modeling applied to heterogeneous catalysts

The main goal of this work is the modeling of the function of materials dependent on their composition. As model materials a discrete set of catalysts has been prepared. The complete composition space of the mixed oxides of Ni, Cr, Mo, Co and Mn has been synthesized (spacing steps of 10%, 1,001 catalyst samples) based on a compositionally tolerant sol–gel-recipe developed for this purpose. The materials have been prepared with the help of a pipetting robot in arrays of HPLC-vials through the control of the software Plattenbau [10]. The thus prepared sols were allowed to gel, then calcined at 300°C and manually transferred into the wells of a 207-well slate plate. Prior to the measurement, the reactor was heated to 400°C in synthetic air for 3 h. Catalytic activity for the oxidation of propene to acroleine was monitored at 350°C (28.6 vol.% propene in synthetic air at ambient pressure) on a high-throughput reactor by sequential rapid analysis with a micro GC [11]. This dataset not only contains the activity and selectivity of the mixed oxides composed of five metals, but also all lower mixed oxides (pure oxides, binary, ternary and quaternary mixed oxides).

Parameter space

Mixed metal oxides are potential heterogeneous catalysts, composed of n different metals in various contents and oxidation states. Since the oxidation states, and therefore the oxygen content of the materials, may vary with preparation conditions, pretreatment or during catalysis, it is not specified. The total metal content (defined as 100%) is uniquely determined if the contents of $n-1$ metals are specified. From a mathematical point of view, a catalyst of n metals can be regarded as a point from an n-dimensional space where each coordinate axis corresponds to one element. Figure 1 illustrates this for ternary catalysts. Ternary mixtures can be regarded as points lying on the diagonal of the unit cube in the Cartesian coordinate system if the composition is interpreted as Cartesian coordinates within the range from 0 to 1 (0–100%).

Because of the redundancy of the nth dimension, the points can be transformed by an adequate coordinate transform into a $(n-1)$th dimensional representation. Figure 2 shows the three-dimensional case.

Thus, a ternary composition spread can be illustrated in two dimensions. Analogously, catalysts consisting of four different metals can be illustrated in three dimensions: they form a tetrahedron as illustrated in Fig. 3. Clearly visible is the rather steady catalytic activity. A clear region of higher activity is in the tetrahedron towards higher Co and Cr contents. This also verifies that the catalytic activity changes smoothly with composition and may provide a good basis for modeling.

From Figs. 1 and 2 it becomes clear what the mathematical model should achieve. Modeling the three dimen-

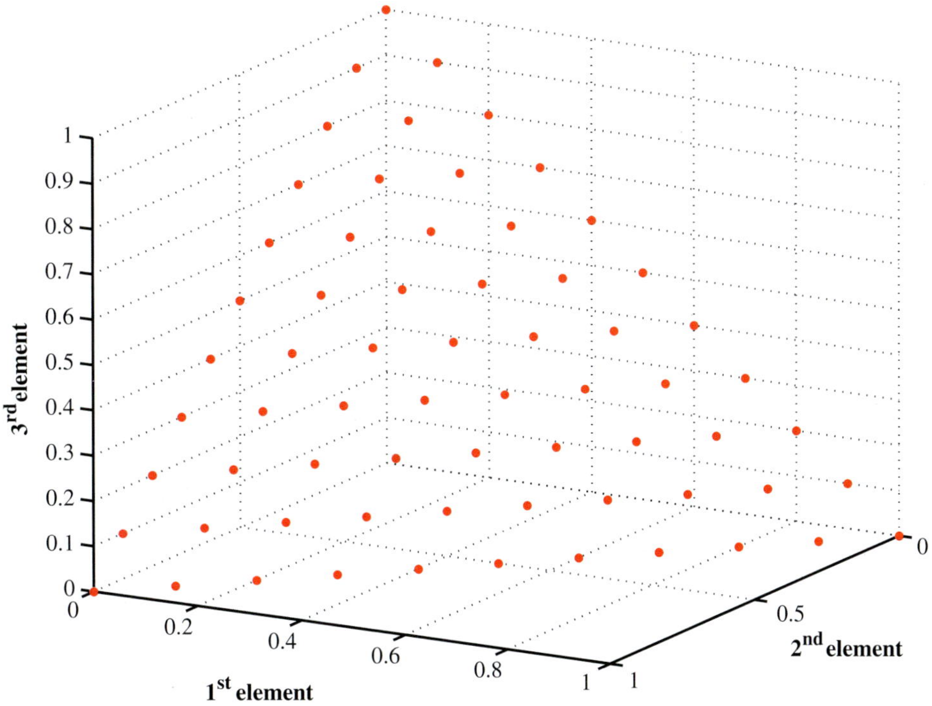

Fig. 1 Ternary composition spread interpreted as Cartesian coordinates in three dimensions

sional case, the result is a triangular surface providing catalytic activity for any composition within the parameter space. In more than two dimensions, the result will be a hyperplane describing the data. The goal of interpolation or approximation is to reconstruct an underlying function (e.g. surface) that may be evaluated at any desired set of positions. This serves to propagate the information associated with the discrete data smoothly onto all positions in the domain. The field of mathematical approximation techniques is very large and the approximation in itself can be done in a variety of ways such that no unique solution exists. Using different approaches, the approximating surfaces differ in shape and smoothness according to the model chosen. As many real-world applications, this problem ranks among the so-called *ill-posed* problems. The following figure stresses the spectrum of possible interpolations and gives an idea what ill-posed means in this context.

Figure 4 illustrates in a simple way what the difficulties of interpolation and approximation problems are. All four curves interpolate or approximate the data points but possess different shapes. As with catalyst data the underlying composition–activity relationship is an unknown function, the goodness of fit cannot be calculated directly (e.g. least-squares distance, mean squared error etc.) but one has to think about a criterion that handles the problem structure.

For the interpolation of catalyst data, we assume that samples of the same composition should show nearly the same catalytic behavior within the range of synthesis and measurement errors. Samples of similar compositions are assumed to cause a similar catalytic activity such that the approximating curve should not exhibit jumps or sharp bends but is assumed to be of rather smooth shape.

Due to synthesis and measurement errors, the catalyst data cannot be considered as absolute values but as values plus a certain amount of noise and also outliers. Therefore, the use of approximation techniques instead of interpolation approaches to model the catalytic surfaces appears to be more suitable.

Data

For our modeling purpose we chose a data set consisting of 66 ternary heterogenous catalysts containing Cr, Mo and Ni in 10%-wise variation without replication. The activity values for the oxidation of propene can be seen in Fig. 2. Among these catalysts the main activity occurred for mixtures containing Mo and Ni in nearly equal contents and little Cr. The data point containing only Cr and Mo as well as the point having the same Cr content but containing more Ni show higher activity indicative of local maxima. Both points were suspected to be outliers due to measurement errors since there are no more active samples within their direct neighborhoods. To prove this, 50 replicates including 2–3 repetitions were prepared and analyzed together with more local samples (5 and 10% spacing) in the direct neighborhood. The data show no significant differences between the data points and their direct neighborhood, which identifies the two original measurement points in the left middle and the middle of the bottom of Fig. 2b as outliers. In the following sections the results of three different models are discussed and their behavior in coping with possible outliers is stressed.

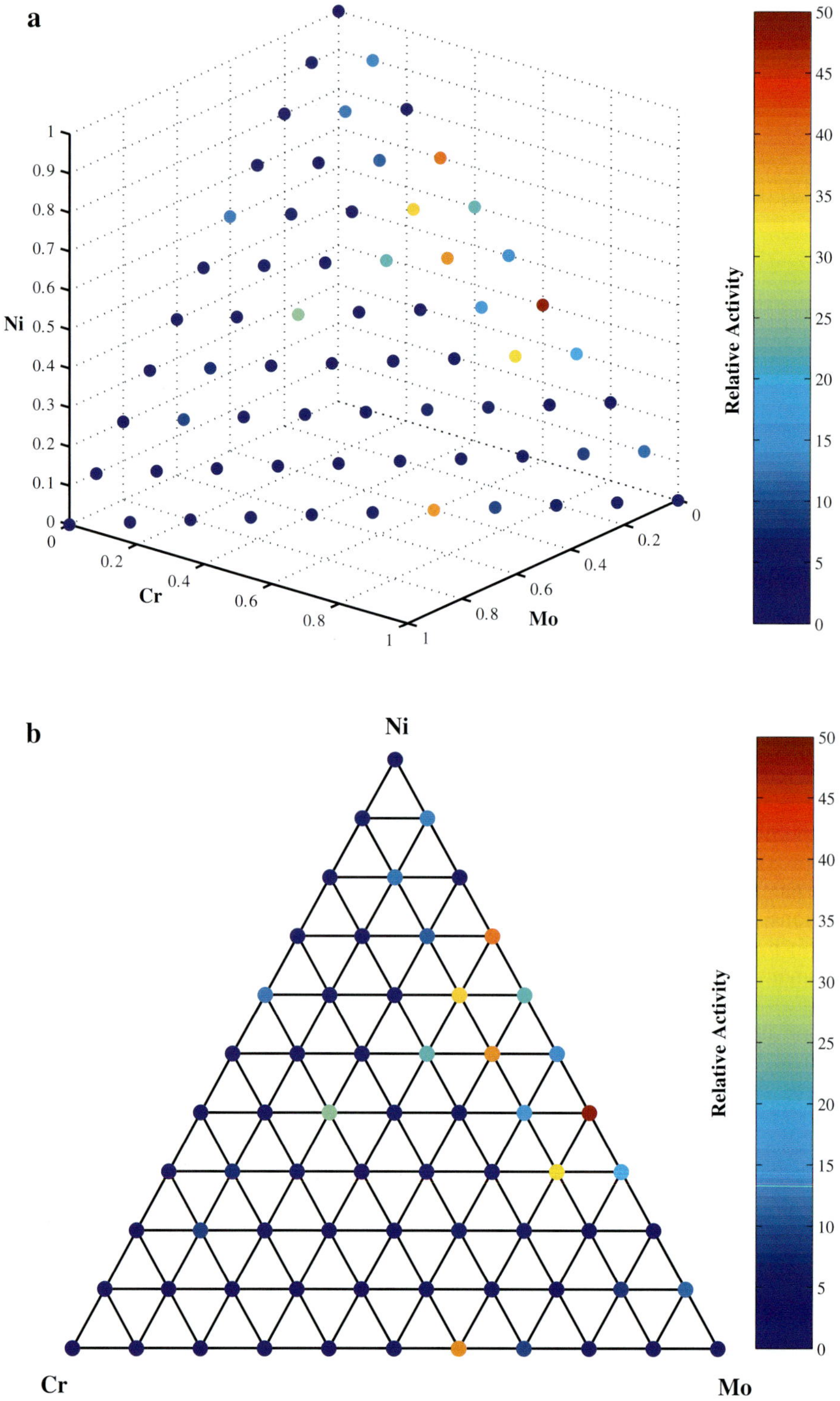

Fig. 2 **a** Ternary composition spread illustrated in three dimensions. The color corresponds to the measured activity at 350°C. **b** Ternary composition spread projected onto two dimensions. The color corresponds to the measured activity

For our approach, we have chosen variations of three parameters (ternary composition spread) for demonstration, because these can be visualized easily by projection onto a triangular plane (see above). Parameter sets of higher order are clearly of interest and a tetrahedral presentation (quaternary composition spread) has also been shown above. A more attractive visualization approach is presented below based on the modeling of the parameter space of quaternary mixed oxide catalysts.

Support vector machines (SVM)

The use of support vector machines was first introduced by V. Vapnik and his coworkers in 1992 [12]. Similar

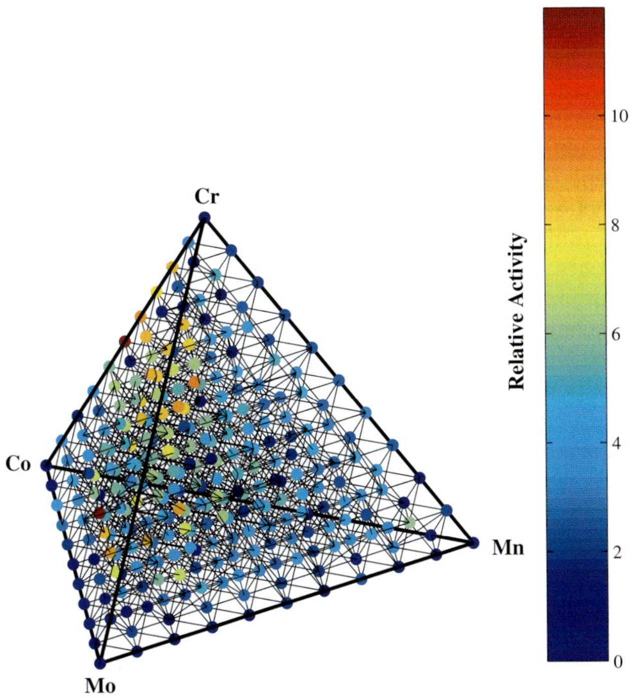

Fig. 3 Quaternary composition spread in three dimensions. The color corresponds to the measured activity

approaches have been known in machine learning theory since 1960, but it was not until 1992 that all of the relevant features were put together to define the so-called "maximal margin classifier", the basic Support Vector Machine. SVM can be used to solve classification problems but also regression problems and are often employed for feature extraction. For a detailed introduction to the theory of SVM and the mathematical background we refer to [13] and [14].

SVM have been applied to our catalytic data to perform regression, i.e. by means of SVM regression. The activities of non-synthesized catalyst samples have been estimated and checked whether calculated activities correlated with

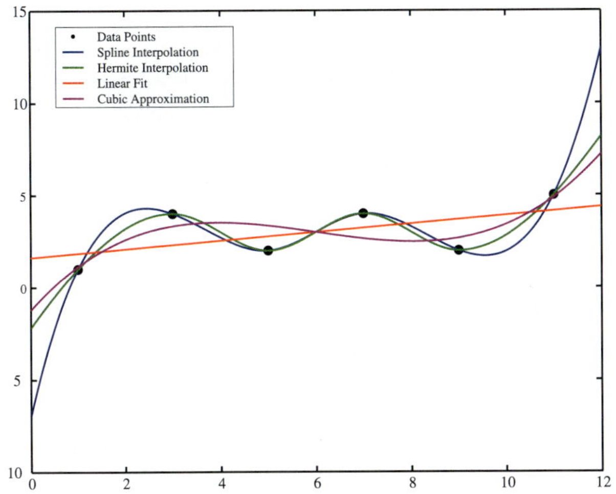

Fig. 4 Possibilities to interpolate and approximate given data points

measured activities (cf. [15]). The local fitting procedure of SVM seems to be an advantage, since there is no reason to assume that catalytic activity across the complete parameter space can be described by a simple function.

Figure 5 illustrates the application of support vector regression to the dataset introduced in Fig. 2. Among the 66 heterogenous catalysts selected, Cr, Mo and Ni composition varies from 0–100% with an increment of 10%.

Compared to Fig. 2 it can be seen that the most active regions are well identified by the model. Variation of the resolution of the method can be used to smooth the activity surface. In Fig. 5 the isolated points of activity have been confirmed as outliers (s.a.).

In Reference [15], another ternary data set has been tested and modeled using SVM. After the calculations, several new samples were synthesized and tested by IR-thermography, gas chromatography or mass spectroscopy for certain reactions. Good agreement between predicted and experimental values was found, confirming the validity of the approach.

Data approximation with multilevel b-splines

As mentioned above, data approximation refers to the problem of fitting a smooth surface through a scattered or uniform distribution of data. This subject is of practical importance in many areas of science and engineering such as physics, geology, chemistry, meteorology, mining etc. All these fields require data approximation or interpolation techniques to determine values at arbitrary positions. This is exactly what will be done with catalyst data and therefore some techniques from this mathematical field are also applied. Publications dedicated to the theory of scattered data interpolation are for example [16, 17], which give good surveys on this topic. For modeling the catalyst data, a multilevel B-spline approach is followed according to [18].

The basic idea of this approximation technique can be seen as follows.

For a three-dimensional application, consider a dataset $P=\{(x_c, y_c, z_c)\}$ in 3D-space where (x_c, y_c) is a point in the xy-plane and z_c is considered as the corresponding activity value. Let $\Omega = \{(x,y)|0 \leq x < m, 0 \leq y < n\}$ be a rectangular domain in the xy-plane. To approximate the data points, an approximation function f is formulated as a uniform bicubic B-spline function, which is defined by a control lattice Φ overlaid on domain Ω. Without loss of generality, it is assumed that Φ is an $(m+3) \times (n+3)$ lattice which spans the integer grid in Ω, cf. Fig. 6.

Let φ_{ij} be the value of the ij-th control point on lattice Φ, located at (i,j) for $i=-1, 0, \ldots, m+1$ and $j=-1, 0, \ldots, n+1$. The approximation function is defined in terms of these control points by

$$f(x,y) = \sum_{k=0}^{3} \sum_{l=0}^{3} B_k(s) B_l(t) \phi_{(i+k)(j+l)} \tag{1}$$

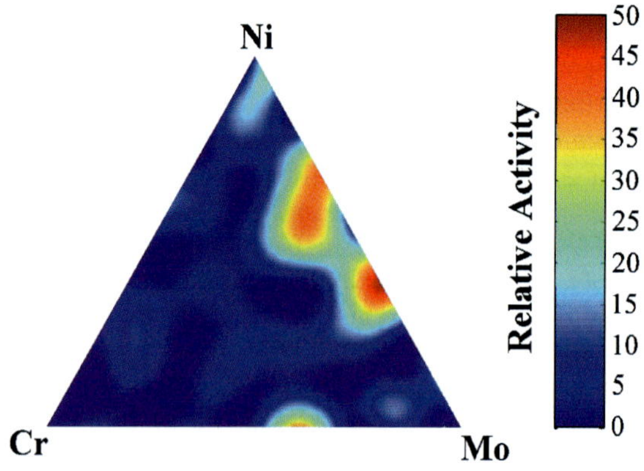

Fig. 5 Estimated activities by support vector machines

where $i = \lfloor x \rfloor - 1$, $j = \lfloor y \rfloor - 1$, $s = x - \lfloor x \rfloor$ and $t = y - \lfloor y \rfloor$. B_k and B_l are uniform cubic B-spline basis functions defined as:

$$B_0(t) = \frac{(1-t)^3}{6}$$
$$B_1(t) = \frac{(3t^3 - 6t^2 + 4)}{6}$$
$$B_2(t) = \frac{(-3t^3 + 3t^2 + 3t - 1)}{6}$$
$$B_3(t) = \frac{t^3}{6}$$

where $0 \leq t < 1$. They serve to weight the contribution of each control point to $f(x, y)$ based on its distance to (x, y). With this formulation, the problem of deriving function f is reduced to solving for the control points in Φ that best approximate the data points in the least square sense. According to Eq. 1 each point in P influences 4×4 neighboring control points in Φ and analogously, 4×4 control points of the lattice Φ are needed to determine the approximated value at any location in Ω.

It is clear that the resulting surface depends on the fineness of lattice Φ. As Φ becomes finer, the influence of a data point is limited to smaller neighborhoods. This enables P to be more closely approximated, although f will tend to contain local peaks near the data points (danger of local biases). Figure 7 illustrates this behavior.

Using this approach, there is a trade-off between the shape smoothness and accuracy of the approximation function that has to be handled. This is solved by applying a multilevel B-spline approach with a hierarchy of control lattices to generate a sequence of functions f_k whose sum approaches the desired approximation function. In the sequence, a function from a coarse lattice provides a rough approximation, which is further refined in accuracy by functions derived from fine lattices. When Φ is sufficiently fine compared to the data distribution, f interpolates the data.

The application of this approach to catalyst data is exemplified in Fig. 8, where the same ternary composition spread of catalysts used for SVM above is approximated. Comparing the result to Fig. 2 (discrete data points) reveals that the structure of the data set is well represented. In contrast to Fig. 5 (SVM regression) this approach leads to maxima that tend to be more local than those of SVM.

Simulated Gauss functions have been used to compare the approximation accuracy of SVM to B-splines [15]. The SVM regression yielded a smaller mean squared error (MSE) in this test, although we cannot conclude that in general SVM regression is more suitable to approximate catalyst data since the real function f is not known as it was in [15]. The B-spline approximation is presently limited by its application to a maximum of three dimensions. Challenges lie in the extension of this method to datasets that range in spaces with more than three dimensions The formalism for applications to spaces of higher order has not yet been developed. Another possibility could be the use of a polynomial-response surface with interaction terms. For our purpose this is not reasonable due to a lack of data. Furthermore, these techniques cause responses outside the sampled space yielding +/− infinity instead of zero. Therefore, we have chosen to try Kriging as another approach.

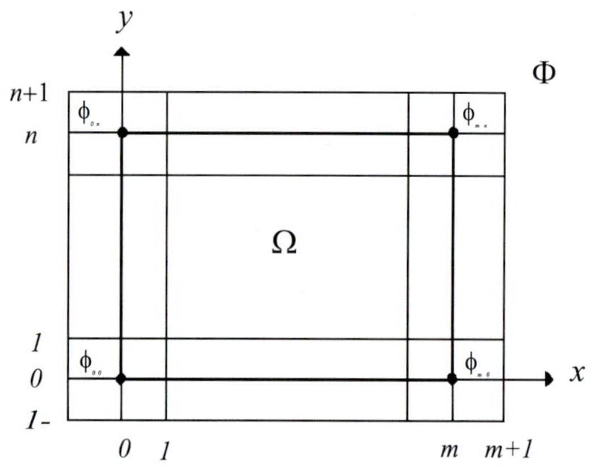

Fig. 6 The configuration of control lattice Φ over Ω (cf. [18])

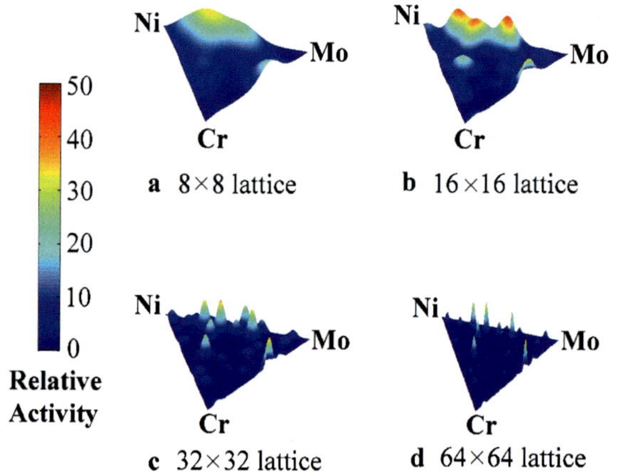

Fig. 7 B-spline approximation using different control lattice resolutions

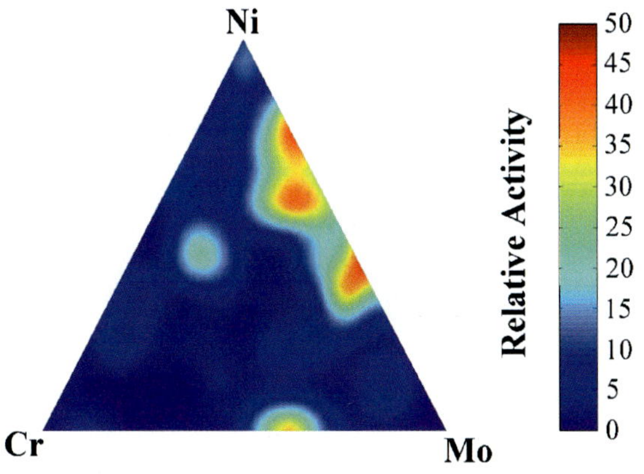

Fig. 8 Estimated activities by a multilevel B-spline approximation

Kriging

In this section the technique of Kriging is discussed to solve the approximation problem introduced above. The term Kriging is known in the field of geostatistics or spatial statistics and describes modeling by random processes. Kriging was named after a South African mining engineer, D. Krige, who first popularized stochastic methods for spatial predictions, especially for mining. Although originally developed for mining problems, Kriging can also be applied to many other fields and applications with similar problem structure. A brief overview of Kriging is given in this section. For a comprehensive review the reader is referred to the literature [19–21].

A central problem in geostatistics is the reconstruction of a phenomenon over a domain on the basis of values observed at a limited number of points. The main difference to the other techniques described above is that Kriging starts from a statistical model rather than a model of the approximation function. Formally, Kriging appears to be similar to the prediction problem in time series where the signal is to be predicted out of given values of the past, usually at regular time intervals, for some time in the future. With time series, the signal is analyzed first by computing and modeling the spectrum, and then a predictor is designed. Kriging follows a similar approach but in a spatial setting, where there is no general concept of past and future. Within this section the use of Kriging is applied to model catalyst data to be able to predict the activity of untested samples. In contrast to Reference [9], the results of the Kriging model are discussed in comparison to the other approximation techniques described above.

The basic idea of Kriging can be described as follows:

Using the Kriging approach, catalyst data are modeled as a random process

$$Z(x) = Z_0(x) + e(x)$$

with $x \in D \subset \mathrm{IR}^n$ and D is the experiment's design region. Z_0, some smoothly varying term, is assumed to depend only on the position x and e is a realization of a random process with zero mean and covariance C:

$$C(e) = C(x_i, x_j).$$

Let $S=\{x_1,...,x_N\}$ be the set of all experimental samples, i.e. S contains all measured catalysts. Usually, the covariance is assumed to depend only on distance h of two samples, i.e.,

$$C(x_i, x_j) = C(h_{ij})$$

For modeling catalyst data, mean and covariance are unknown and have to be estimated out of the data. The prediction of unmeasured samples is done by calculating a *best linear unbiased estimator* (BLUE) for each location. This denotation can be explained as follows:

- Best: estimator of smallest variance among all other estimators.
- Linear: according to its construction, linear combination of all observations.
- Unbiased: no bias (i.e. the expected value of the random error is zero).

Although Kriging is generally considered "unbiased", this holds only as long as a stochastic process is studied and the correct covariance is used.

Another assumption is made concerning the measured values: data points that are quite close together are assumed to be somehow "linked" to each other and this spatial influence decreases with increasing distance of points until it ebbs away. To analyze this spatial behavior, the covariance between two samples is to be considered. Instead of the covariance, the sample variogram can also be considered to model the spatial behavior of the dataset. The sample variogram is calculated by:

$$\gamma(h) = \frac{1}{2N(h)} \sum_{i=1}^{N(h)} (Z(x_i) - Z(x_i + h))^2$$

where $N(h)$ is the number of data points with distance h. Figure 9 illustrates a typical sample variogram and explains its common descriptors *range, sill* and *nugget effect.*

The range can be considered as the distance where covariance disappears and randomness dominates. This means that points that lie more than the range apart from each other are considered to be uncorrelated. The nugget effect describes a discontinuity at the origin. For example, in gold deposits, gold commonly occurs as nuggets of pure metal that might be much smaller than the size of a sample. This results in a strong grade variability in the sample, even when physically very close. In that case, the variogram shows a discontinuity at the origin. In practice, the nugget effect might be due to some kind of microstructure, namely a component of the phenomenon with a range shorter than the sampling distance or due to a structure with a range shorter than the smallest inter-point distance. However, measurement and synthesis errors also cause a nugget effect.

To calculate the Kriging estimators, the discrete variogram points must be modeled by a variogram function. There exist many approaches and it depends on the problem structure to choose the right variogram function. We refer to [19–21] for the description of different variogram models. By means of the sample variogram, the Kriging system can be set up and solved for the Kriging weights w_i that assign a weight to each sample value x_i.

For the sake of simplicity, the derivation of the Kriging system is not discussed in detail here but we refer to [19–21] for extensive studies.

The Kriging estimator at location x^*, $Z(x^*)$ can be derived as a linear combination of all data points together with their calculated weights, i.e.,

$$Z(x*) = \sum_{i=1}^{N} w_i x_i$$

where N denotes the number of given data points. Figure 10 illustrates the result of a Kriging approximation applied to the catalyst data described above. Compared to Figs. 2, 5 and 6 it can be concluded, that all three methods provide comparable models of the activity–composition relationship.

Determining the nugget effect of a Kriging model must be done thoroughly because this parameter controls the extent of approximation. Models with zero nugget effect yield an interpolation of data and the larger the nugget effect the smoother the fitted surface becomes. This enables researchers to adapt the smoothness of the desired surface in a very comfortable way. Another advantage of Kriging lies in its excellent ability also to approximate data in more than three dimensions. The computational effort of Kriging directly corresponds to the number of dimensions. For example, datasets containing four elements lead to Kriging systems with matrix sizes of 287×287 entries, which can be easily solved on a PC. This means Kriging can be used excellently in more than four dimensions provided that there are datasets available. For more than five elements the synthesizing and screening time of such a catalytic library is still too long to establish valuable datasets.

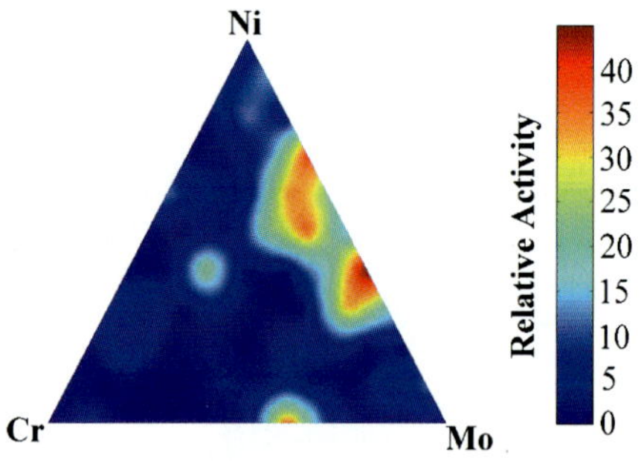

Fig. 10 Approximated surface using the Kriging approach

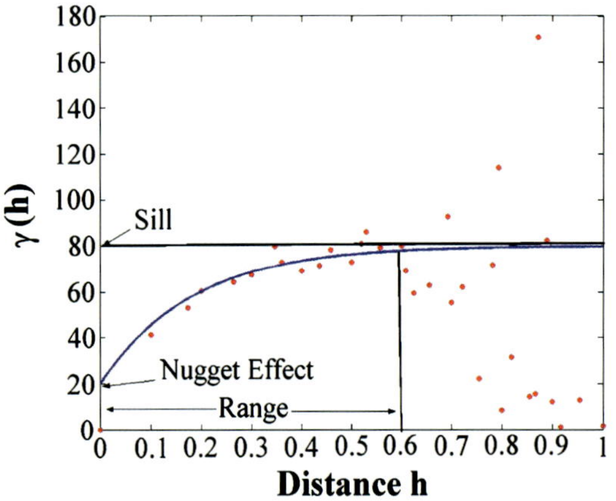

Fig. 9 Sample variogram corresponding to the data set illustrated in Fig. 2 with fitted variogram function

Visualization of a quaternary composition spread

According to Fig. 3, a quaternary composition spread can be illustrated as a tetrahedron. The Kriging approach described above is applied to a dataset in four dimensions containing the elements molybdenum, manganese, cobalt and chromium. Fig. 11 illustrates how the structure of the approximated values within the tetrahedron can be visualized by using slice plots. The catalysts were tested for the oxidation of propene at 300°C. We are mainly interested in catalysts that show high selectivity to acroleine among other products such as acetone, propionaldehyde or 1,5-hexadiene.

From Fig. 11 it can be seen, for example, how the nickel content influences the selectivity of the catalysts. The first plot of Fig. 11 illustrates the selectivity of catalysts without any nickel. In the second plot of Fig. 11 all catalysts containing 10% nickel are shown and it can be seen that there is the region of highest selectivity (colored in red). In the following slices of higher Ni-content (Fig. 11) the selectivity to acroleine decreases. Clearly all four elements contribute to the area of highest selectivity. Very important

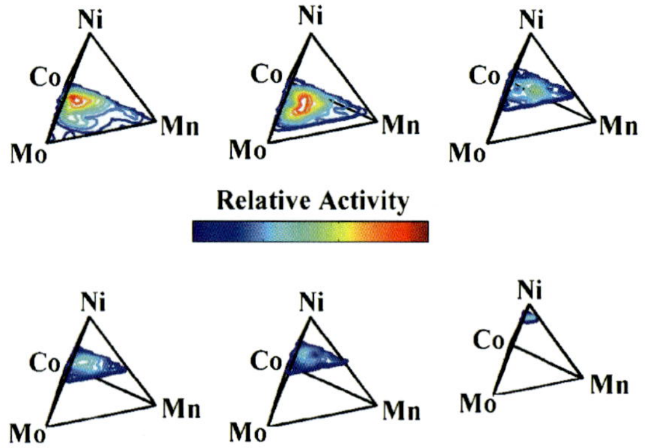

Fig. 11 Visualization of a ternary catalyst system and its approximation using slice plots

here is that the areas of higher activities are supported by the activity of the neighboring points. Points of isolated high activity are most likely outliers, since in all our composition spread studies of the last few years, the surfaces have all been very smooth and continuous. This way of visualization helps to look for trends and redundant elements among quaternary catalysts and enables the researcher to plan the next experiments for a more efficient search of active catalysts. In practice, this visualization can be done more effectively directly on the monitor using the fourth (here Ni) and higher parameters as sliders.

Conclusions and outlook

Through a systematic preparation of mixed oxides covering the whole composition space by a sol–gel preparation procedure tolerant of composition, the catalytic activity has been determined. It has been demonstrated that catalytic activity for propene oxidation to acroleine with air is a function of chemical composition if the materials studied are prepared by the same method. Three different mathematical methods, support vector machines, B-splines and Kriging have been applied to model the datasets. All methods provide comparable results and can therefore be applied for QCAR. While B-splines can, up to now, only be applied to datasets with functional relationships dependent on up to three parameters, SVM require a larger computational effort. Kriging is favored because it can be applied readily to multiple parameter spaces while its computation is still fast enough. Parts of our future work will deal with the evaluation of the predictive power of each method. Furthermore, replications are now included in all experiments.

This QCAR-modeling provides a new entry into prediction of materials properties. It can be used for materials optimization as well as for the design of experiments in extended high-throughput searches for new materials.

References

1. Potyrailo RA, Takeuchi I (2005) Meas Sci & Technol 16:1–4
2. Potyrailo RA, Amis EJ (2003) High-throughput analysis: a tool for combinatorial materials science, 1st edn. Kluwer, New York
3. Takeuchi I, Xiang XD (2003) Combinatorial materials synthesis, 1st edn. Marcel Dekker Inc, New York Basel
4. Maier WF (2003) Appl Catal A: General 254:1–2
5. Koinuma H, Xiang XD, Kawasaki M, Chikyow T (2002) Combinatorial materials science and technology. Elsevier Science B.V.
6. Cawse JN (2003) Experimental design for combinatorial and high throughput materials development. Wiley, Hoboken
7. Klanner C, Farrusseng D, Baumes L, Lengliz M, Mirodatos C, Schüth F (2004) Angew Chem 116:5461–5463
8. Farrusseng D, Klanner C, Baumes L, Lengliz M, Mirodatos C, Schüth F (2005) QSAR & Comb Sci 24:78–93
9. Scheidtmann J, Klär D, Saalfrank JW, Schmidt T, Maier WF (2005) QSAR & Comb Sci 2:203–210
10. Scheidtmann J, Saalfrank JW, Maier WF (2003) Plattenbau - automated synthesis of catalysts and materials libraries. In: Anpo M, Onaka M, Yamashita H (eds) Studies in surface science and catalysis, 145. Elsevier, Tokyo, pp 13–21
11. Weiss PA, Saalfrank JW, Scheidtmann J, Schmidt HW, Maier WF (2003) In: Potyrailo RA, Amis EJ (eds) High-throughput analysis: a tool for combinatorial materials science, 1st edn. Kluwer, New York, pp 125–153
12. Vapnik VN (1995) The nature of statistical learning theory, 2nd edn. Springer, Berlin Heidelberg New York
13. Shawe-Taylor J, Cristianini N (2004) Kernel methods for pattern analysis, 1st edn. Cambridge University Press
14. Hastie T, Tibshirani R, Friedman J (2001) The elements of statistical learning, 1st edn. Springer, Berlin Heidelberg New York
15. Stutz B (2004) Suchstrategien für Optimale Katalysatoren in der Kombinatorischen Chemie. Diplomarbeit, Universität des Saarlandes
16. Franke R, Nielson GM (1991) Scattered data interpolation and applications: a tutorial and survey In: Hagen H, Roller D (eds) Geometric modelling: methods and their application, Springer, Berlin Heidelberg New York
17. Franke R (1982) Math Comp 38:181–200
18. Lee S, Wolberg G, Shin SY (1997) IEEE Transactions on Visualization and Computer Graphics 3:228–244
19. Cressie N (1993) Statistics for spatial data, revised edn. Wiley, New York
20. Rivoirard J (1994) Introduction to disjunctive kriging and nonlinear geostatistics. Oxford University Press, Oxford
21. Chilès JP, Delfiner P (1990) Geostatistics—modelling spatial uncertainty. Wiley, New York, p 1999

J Mol Model (2006) 12: 621–629
DOI 10.1007/s00894-005-0029-3

ORIGINAL PAPER

Alexander Alex · Elke Hänsele · Timothy Clark

The ethylene/metal(0) and ethylene/metal(I) redox system: model ab initio calculations

Received: 19 December 2004 / Accepted: 8 July 2005 / Published online: 9 December 2005

Abstract Ab initio calculations (coupled cluster with single and double excitations; CCSD) have been used to investigate the model redox systems ethylene:M(0) (M = Li, Na, K, Rb, Cs) and ethylene:M(I) (M = Be, Mg, Ca, Sr, Ba, Zn, Cd, Hg). Within C_{2v} symmetry, the ground ($^2A_1$) states correspond to the charge distribution given in the title. The lowest ($^2B_2$) excited states correspond, somewhat counter intuitively, to the ethylene$^{\bullet -}$/M(II) ion pair.These trends can be rationalized on the basis of simple electrostatic and configuration-mixing arguments that lead to two simple equations for predicting the electron-transfer energies for oxidation or reduction of the ethylene. The electron-transfer energies to the $^2B_2$ ion pairs are dominated by the electrostatic ion-pairing energies.

Keywords Electron transfer · Ab initio · CCSD(T) · Ethylene

Introduction

We have shown previously [1–3] that the catalysis of simple closed-shell reactions by Group II metal radical cations can be explained as the result of electron transfer from the metal to the substrate during the reactions. This electron transfer is also known experimentally for transition-metal complexes [4–7] and the resulting geometrical changes in the substrates have been studied extensively [8]. In this paper we focus our attention on electron transfer from different Group II and Group XII metal radical cations to ethylene as a model system in order to understand this type of catalysis better. We will show that the energy required for electron transfer is governed mainly by simple electrostatic interactions and physical properties of the metals such as ionization potentials and ionic radii.

We now report model ab initio molecular orbital calculations on electron-transfer reactions of the ground state ($^2A_1$ in C_{2v} symmetry) complexes $M^{\bullet}:C_2H_4$ (M = Li, Na, K, Rb, Cs) and $M^{\bullet +}:C_2H_4$ (M = Be, Mg, Ca, Sr, Ba, Zn, Cd, Hg). Electron transfer from the metal to ethylene (the reductive process) leads to the first excited ($^2B_2$ in C_{2v} symmetry) $M^{+}:C_2H_4^{\bullet -}$ and $M^{2+}:C_2H_4^{\bullet -}$ states. Transfer of an electron from the ethylene to the metal (the oxidative process) leads to a higher-lying $^2A_1$ excited state, as shown in Scheme 1, where the three relevant molecular orbitals (MOs) and their occupancies in the three different states are shown schematically. For the purposes of this analysis, and for most of our calculations, we have constrained the structures to C_{2v} symmetry. In most cases, the C_{2v} structures are also minima, as shown below.

The bonding combination (Ψ_1) of the ethylene π-HOMO and the s-orbital of the metal is concentrated on the olefin, as shown in Scheme 1, as long as the ionization potential of the metal is significantly lower than that of ethylene (10.51 eV, 242.4 kcal mol^{-1}) [9]. The corresponding antibonding combination, Ψ_2, is thus concentrated on the metal so that the excitation of an electron from Ψ_1 to Ψ_2 essentially corresponds to electron transfer from the olefin to the metal, which we have called the oxidative process. This redox process is intuitively attractive in the gas phase because it involves only a charge shift, not a separation of charges. The ethylene π*-LUMO can possibly interact weakly with a metal P- (or even d-) orbital, as shown in Ψ_3 of Scheme 1. Excitation of an electron from Ψ_2

Dedicated to Paul Schleyer on the occasion of his 75th Birthday

A. Alex
Medicinal Informatics, Structure and Design,
Pfizer Global Research and Development,
Sandwich Laboratories, Ramsgate Road,
Sandwich CT13 9NJ, UK

E. Hänsele · T. Clark (✉)
Computer-Chemie-Centrum der Universitaet
Erlangen-Nuernberg, Naegelsbachstrasse 25,
91052 Erlangen, Germany
E-mail: clark@chemie.uni-erlangen.de
Tel.: +49-9131-8522948
Fax: +49-9131-8526565

Table 3 Total energies (a.u), number of imaginary normal vibrations (NIMAG), lowest frequency (ν_1, cm^{-1}), zero-point vibrational energies (ZPE, kcal mol^{-1}) and complexation energies (metal + ethylene → complex, corrected using the unscaled CCSD-ZPE, kcal mol^{-1}) for the $^2A_1$ metal:ethylene complexes

Metal	CCSD					CCSD(T)	
	Total energy	NIMAG	ν_1	ZPE	$E_{comp.}$	Total energy	$E_{comp.}$
Li	−85.80643	0	19.8	32.07	−0.11	−85.81829	−0.16
Na	−240.22036	0	11.4	32.03	−0.13	−240.23220	−0.17
K	−677.70719	0	14.3	32.11	−0.12	−677.72244	−0.21
Rb	−102.20162	0	8.9	33.04	−0.62	−102.21401	−0.26
Cs	−98.24606	0	18.3	32.51	−1.24	−98.25845	−0.55
$Be^{•+}$	−92.71711	0	107.7	32.35	−40.56	−92.71940	−35.61
$Mg^{•+}$	−277.76548	0	81.7	32.90	−16.33	−277.77766	−16.59
$Ca^{•+}$	−114.89169	0	102.2	32.73	−11.80	−114.90847	−12.38
$Sr^{•+}$	−108.70019	0	90.3	32.62	−9.95	−108.71415	−10.46
$Ba^{•+}$	−103.45049	0	69.8	32.50	−9.62	−103.46653	−10.11
$Zn^{•+}$	−1856.14807	0	78.5	33.06	−22.70	−1856.16942	−23.70
$Cd^{•+}$	−245.07860	0	118.5	33.07	−22.00	−245.09489	−21.98
$Hg^{•+}$	−230.75905	0	112.7	33.19	−29.13	−230.77539	−29.41

Volt for the second ionization potentials of mercury and is independent of whether the metal is treated in an all-electron calculation or using pseudopotentials. It is known [35, 36] that correlation is important for reproducing the second ionization potentials of metals and that a basis set adequate to allow an accurate description of core–valence correlation is important [37]. Errors found can be expected, both for the 6-311+G(d,p) all-electron calculations with inflexible nonvalence basis functions and for the pseudopotential calculations.

$^2A_1$ structures and complexation energies

Table 2 shows some geometrical details of the $^2A_1$ C_{2v} complexes. The neutral complexes (those with alkali-metal atoms) all show the typical geometrical parameters of weak complexes, long C...M distances and an essentially unperturbed geometry for the ethylene moiety. The exception is the rubidium complex, which shows a larger deformation of the C_2H_4 unit. As expected from the theory of odd-electron bonds [38], the interaction between the metal and the olefin, and therefore also the perturbation of the olefin geometry, is far stronger for the singly charged (Group II and Group XII) complexes. The strengths of the interactions are shown in Table 3, which also give the total, zero-point and complexation energies.

As expected [38], the neutral complexes are all very weakly bound with complexation energies below 1 kcal mol^{-1}. The complexation energies of the cationic complexes should fall off exponentially with increasing difference between the first ionization potential of the metal and that of ethylene (ΔIP). Figure 1 shows a plot of ΔIP vs. $\log_{10}(E_{comp})$. There is a good linear correlation for all metals except beryllium, which can be expected to give a more negative complexation energy because it is the only first-row element [38]. Thus, the complexation energies between the metals and ethylene in the $^2A_1$ state behave exactly as expected for three-

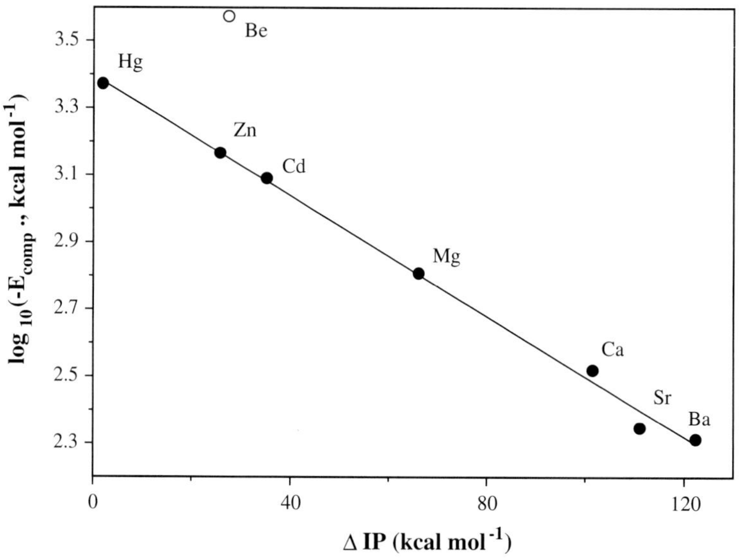

Fig. 1 Plot of the logarithm of the calculated complexation energies (CCSD(T), Table 3) against the differences in first ionization potential between the metal and ethylene (experimental values)

Table 4 CCSD/6-311+G(d,p) optimized carbon–carbon bond lengths (C=C, Å), carbon–metal distances (C...M, Å), C=C–H angles (∠CCH, °) and metal–C–C–H dihedral angles (∅M...CCH, °) for the $^2B_2$ metal:ethylene complexes

	Li	Na	K	Rb	Cs
C=C	1.435	1.444	1.437	1.351	1.346
C...M	2.070	2.441	2.788	3.466	3.761
∠CCH	119.6	119.2	119.2	121.7	121.7
∅M...CCH	104.1	105.6	105.8	91.9	91.5
	$Be^{\bullet+}$	$Mg^{\bullet+}$	$Ca^{\bullet+}$	$Sr^{\bullet+}$	$Ba^{\bullet+}$
C=C	1.464	1.480	1.438	1.433	1.406
C...M	1.756	2.163	2.423	2.591	2.768
∠CCH	119.3	118.6	119.1	119.2	120.1
∅M...CCH	102.4	106.0	105.5	105.4	105.4
	$Zn^{\bullet+}Zn^{+\cdot}$	$Hg^{\bullet+}$	$Cd^{\bullet+}$		
C=C	1.502	1.531	1.501		
C...M	2.057	2.271	2.263		
∠CCH	118.6	118.3	118.5		
∅M...CCH	104.5	104.7	105.1		

electron interactions. Note that the different behaviors of Groups II and XII with respect to the dependence of ΔE_{comp} on atomic number parallels the behavior of the relevant ionization potentials.

The good correlation shown in Fig. 1 suggests that the individual λ-values [35] of the original equation are very similar for the seven metals included in the correlation, which in turn suggests that they all have very similar symmetrical three-electron bond energies [39], which seems very reasonable.

$^2B_2$ structures and electron-transfer energies

Table 2 shows some geometrical details of the $^2A_1$ C_{2v} complexes.

With the exception of Rb and Cs, the complexes all have the expected structures for M^{2+}:ethylene$^{\bullet-}$ ion pairs. An ion pair of this type can be optimized at the UHF/6-311+G(d,p) level for M=Rb and Cs, but at the CCSD level, this collapses to the structure described in Table 4, which is a nonpolar $^2B_2$ complex between neutral ethylene and a ^2P metal atom.

All other Group I and II structures correspond to the ion-pair configuration with long (1.43–1.48 Å) C–C bonds, short C–metal distances and significantly nonplanar C_2H_4 moieties. The Group XII metals give significantly longer (1.50–1.53 Å) C–C bonds, suggesting that these complexes can be seen as metallacyclopropanes.

All C_{2v} structures except that for M = $Mg^{\bullet+}$ are found to be minima. We were unable to find a minimum of any symmetry corresponding to an M^{2+}:ethylene$^{\bullet-}$ ion pair and have therefore used the energy for the C_{2v} transition state in the remainder of the discussion. Note that this does not mean that we were unable to find a minimum but rather that distorting $C_2H_4Mg^+$ from the C_{2v} structure is accompanied by a change in the electronic character to that corresponding to a distorted $^2A_1$ state. The electron-transfer energies range from 2–3 (for Li and $Be^{\bullet-}$) to 53 $\left(Hg^{\bullet+}\right)$ kcal mol^{-1}. Perhaps most remarkably, $Be^{\bullet+}$ is found to be able to reduce ethylene as well as Li. In these two cases, we find the $^2A_1$ state to be slightly lower in energy than the $^2B_2$, but the assignment must be considered tentative because of the small calculated energy differences.

Table 5 Total energies (a.u), number of imaginary normal vibrations (NIMAG), lowest frequency (ν_1, cm^{-1}), zero-point vibrational energies (ZPE, kcal mol^{-1}) for the $^2B_2$ metal:ethylene complexes and electron-transfer ($^2A_1 \rightarrow {}^2B_2$complex, corrected using the unscaled CCSD–ZPE, kcal mol^{-1})

Metal	CCSD					CCSD(T)	
	Total energy	NIMAG	ν_1	ZPE	ΔE_{ET}	Total energy	ΔE_{ET}
Li	−85.80128	0	390.7	31.86	3.02	−85.81342	2.85
Na	−240.18512	0	156.5	30.68	20.76	−240.19748	20.44
K	−667.66442	0	254.3	30.64	24.92	−677.68492	22.06
Rb	−102.15853	0	93.7	32.17	26.17	−102.17117	26.01
Cs	−98.20663	0	77.1	32.32	24.56	−98.21916	24.46
$Be^{\bullet+}$	−92.70336	0	489.5	32.30	8.58	−92.71584	2.18
$Mg^{\bullet+}$	−277.71385	1	251.6i	31.67	32.40	−277.72617	32.31
$Cd^{\bullet+}$	−114.86120	0	322.1	31.99	18.39	−114.87907	17.71
$Sr^{\bullet+}$	−108.66642	0	288.3	31.73	20.31	−108.68125	19.87
$Ba^{\bullet+}$	−103.44042	0	228.9	31.76	5.58	−1856.08811	4.91
$Zn^{\bullet+}$	−1856.08811	0	440.2	33.07	37.63	−1856.11062	36.91
$Cd^{\bullet+}$	−245.00481	0	380.6	32.53	45.76	−245.02186	45.29
$Hg^{\bullet+}$	−230.67299	0	346.9	32.72	53.53	−230.69040	52.86

Table 6 Mulliken and natural population analysis (NPA [44]) charges calculated for the $^2B_2$ complexes

M	Atomic charges		Spin	M	Atomic charges		Spin	M	Atomic charges		Spin
	Mulliken	NPA			Mulliken	NPA			Mulliken	NPA	
Li	0.431	0.817	0.11	$Be^{\bullet +}$	0.664	1.528	0.36	$Zn^{+\bullet}$	0.793	1.397	−0.22
Na	0.629	0.841	−0.05	$Mg^{\bullet +}$	0.942	1.510	−0.20	$Cd^{\bullet +}$	0.955	1.300	−0.44
K	0.894	0.921	−0.05	$Ca^{\bullet +}$	1.460	1.792	0.04	$Hg^{\bullet +}$	0.911	1.109	−0.48
Rb	0.036	0.058	0.96	$Sr^{\bullet +}$	1.203	1.790	0.27				
Cs	−0.037	0.006	1.04	$Ba^{\bullet +}$	1.318	1.720	0.40				

Discussion

The nature of the $^2B_2$ states

The calculated charges and spin densities are shown in Table 6. With the exceptions of M = Rb and Cs, for which the $^2B_2$-state simply represents an $s \rightarrow p$ excitation on the metal atom, the NPA-charges confirm the electron-transfer interpretation of the nature of the $^2B_2$ states, although this is not as clear from the Mulliken charges. The Group XII metals also give lower total charges on the metal ions than Group II, supporting the idea that the $^2B_2$ states for these metals have considerable metallacyclopropane character. However, the calculated spin densities demonstrate clearly that electron (and spin) transfer to the ethylene moiety has occurred. Once again, the high negative spin densities found for the Group XII metals suggest different bonding to the Group II complexes. Figure 2 shows the calculated (CCSD/6-311+G(d,p)) spin density for $^2B_2$ Li:ethylene. The unpaired electron clearly occupies the ethylene π^* orbital, polarized toward the lithium ion.

The situation is similar for $^2B_2$ Be:ethylene$^{\bullet +}$, as shown in Fig. 3, except that the beryllium is more heavily involved in the C–Be bonding, making the electron transfer less complete than in the lithium case.

An analogous plot for $^2B_2$ Zn:ethylene$^{\bullet +}$, is shown in Fig. 4. It is generally similar to that for M = Li but shows more significant contributions from the metal ion.

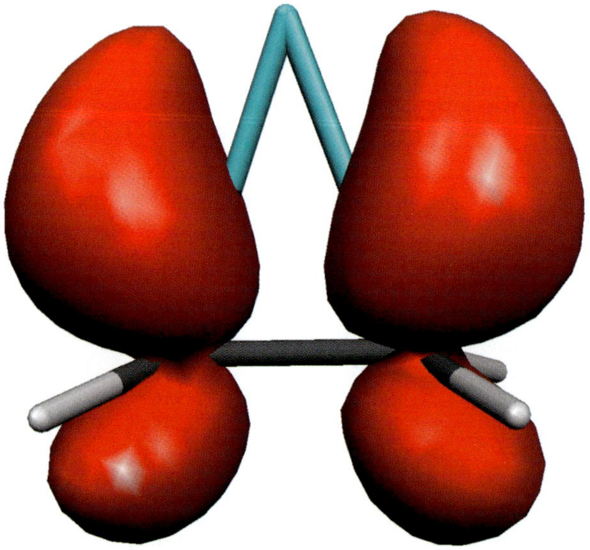

Fig. 2 Calculated (CCSD/6-311+G(d,p)) spin density for $^2B_2$ Li:ethylene. The isodensity level is 0.005 a.u

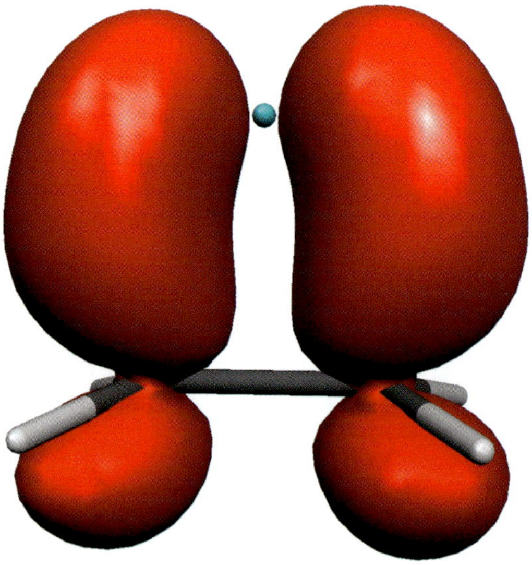

Fig. 3 Calculated (CCSD/6-311+G(d,p)) spin density for $^2B_2$ Be:ethylene$^{\bullet +}$. The isodensity level is 0.005 a.u

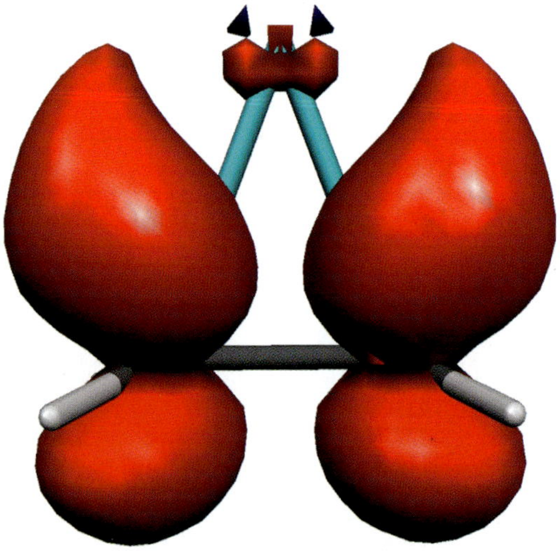

Fig. 4 Calculated (CCSD/6-311+G(d,p)) spin density for $^2B_2$ Zn:ethylene$^{\bullet +}$. The isodensity level is 0.005 a.u

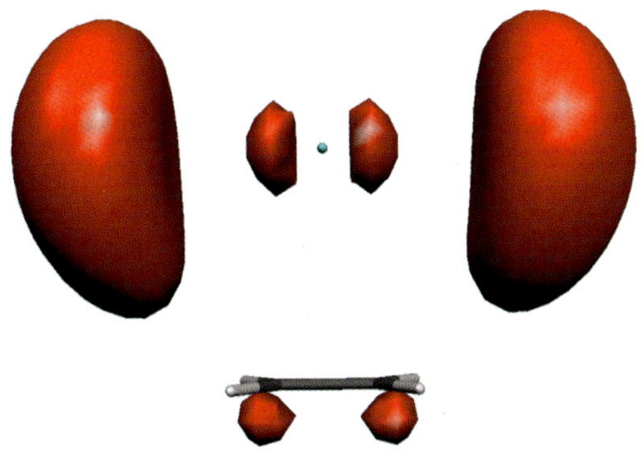

Fig. 5 Calculated (CCSD/6-311+G(d,p)) spin density for $^2B_2$ Cs:ethylene. The isodensity level is 0.0005 a.u

Finally, Fig. 5 shows an analogous plot for the $^2B_2$ state of Cs:ethylene. As suggested above, this state does not involve electron transfer but is rather a complex of the ^2P state of the metal atom.

Electron-transfer energies

The electron-transfer process from the metal atom or ion to the ethylene moiety can be visualized as a combination of the processes shown schematically in Scheme 2. Conceptually, the electron-transfer process consists of the following individual steps

- Separation of the $^2A_1$ complex to give the separated ethylene and metal components ($-\Delta E_{\mathrm{Comp.}}$)
- Ionization of the metal atom or ion (IP_M)
- Reduction of the C_2H_4 moiety ($-EA_{C_2H_4} = +41.4$ kcal mol^{-1} [40, 41])
- Formation of the ion pair (E_{Coulomb})

The electron-transfer energy, ΔE_{ET} is thus given by

$$\Delta E_{\mathrm{ET}} \simeq -\Delta E_{\mathrm{Comp}} + \mathrm{IP(M)} - \mathrm{EA}_{C_2H_4} + E_{\mathrm{Coulomb}} \quad (1)$$

Thus, because $EA_{C_2H_4}$ is constant and by introducing the approximation that E_{Coulomb} is inversely proportional to the C–M distance in the $^2B_2$ complex, we can propose that $\Delta E_{\mathrm{ET}} - \mathrm{IP(M)} + \Delta E_{\mathrm{Comp.}}$ should be proportional to the reciprocal of the C–M distance and that the intersect should correspond to $EA_{C_2H_4}$. Figure 6 shows the corresponding plot for all metals that undergo electron transfer in the $^2B_2$ state.

Although Groups I and XII only contain three valid points, the correlations are at least convincing for Groups I and II. The intersects are clearly not equal, even at the 99% confidence level, so the electron affinity of ethylene is clearly affected by the proximity of a metal ion. However, the intersects are of the correct order. Similarly, a naïve view would expect the slope for Groups II and XII to be twice that for Group I (because the metal ion is doubly charged for groups II and XII). The factor between Groups I and II is found to be 2.4, whereas the slope for Group XII is too poorly defined to be able to make comparisons.

Thus, Fig. 6 supports qualitatively the simple relationship expressed in Eq. 1 and thus the interpretation of the electron-transfer process depicted in Scheme 2. The simple model is, however, not adequate for quantitative predictions unless the slope and the intersect are treated as empirical parameters for each group of the periodic table. Nevertheless, Fig. 6 does support the proposal that the Coulomb ion-paring energy plays a dominant role in determining the reducing ability of metal atoms and ions. This is of course not a proof of the dominance of the Coulomb term. It is, however, difficult to imagine another attractive force with an r^{-1} dependence in these systems. The poor correlation found for Group XII may either be a consequence of only having three points very close to each other, but may also reflect the difference in the basis sets compared with the calculations for the other metals.

The ground states of Li:C_2H_4 and Be:$C_2H_4^{\bullet+}$

Our calculations suggest that the ground states of Li:C_2H_4 and Be:$C_2H_4^{\bullet+}$ are both $^2A_1$. These results

Scheme 2 Schematic energy diagram showing the components of the electron-transfer energy

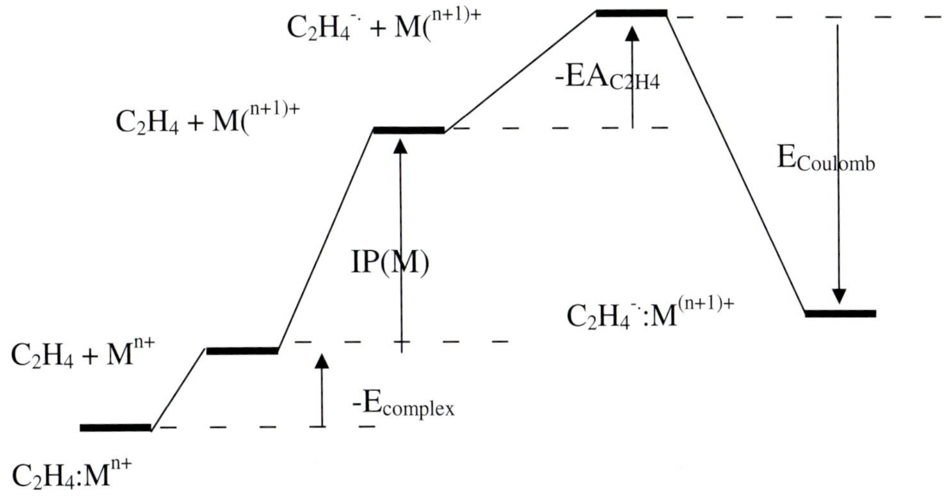

Group	Slope	Intersect	R^2
I	-342 ± 15	45.5 ± 6.3	0.998
II	-818 ± 27	89.7 ± 12.1	0.997
XII	-1053 ± 535	80.0 ± 201.2	0.801

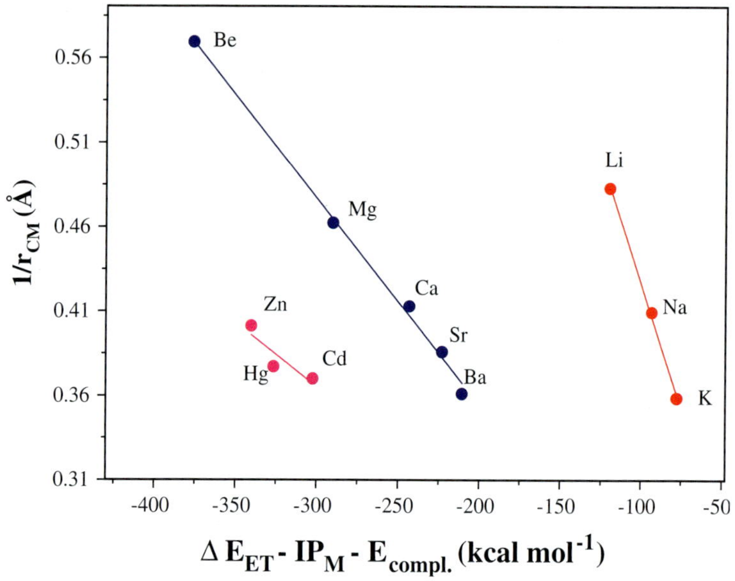

Fig. 6 Plot of $\Delta E_{ET} - \mathrm{IP(M)} + \Delta E_{Comp}$ vs. $1/r_{CM}$ for the metals that give electron transfer in the $^2B_2$ state. CCSD(t)/6-311+G(d,p)//CCSD/6-311+G(d,p) calculated energies with an unscaled CCSD/6-311+G(d,p) ZPE correction were used throughout. The regression equations are listed above

contrast with earlier calculations [10, 12] and experiments in an argon matrix, [13–14] which found $^2B_2$ to be the ground state. Although the energy difference found here is very small (<3 kcal mol^{-1}) for both metals, so that we cannot assign the ground states unequivocally, our calculations are the highest level yet reported for these systems and at least indicate that the ground state in vacuo is not as clearly $^2B_2$ as the earlier work suggests. In our earlier paper [11], we found a significant solvent effect favoring the $^2B_2$ state over the $^2A_1$, but were unable to perform geometry optimizations with solvent simulation. We have therefore now investigated the geometries and relative energies of the two metal complexes in argon solvent using the PCM model for the CCSD/6-311+G(d,p) geometry optimizations [42]. We originally suggested [11] that the cavity effect would be important in making the $^2A_1$ state more compact because of its very flat energy hypersurface. The results of the PCM optimizations, shown in Table 7, support this hypothesis. The C–Li distance in the $^2A_1$ complex decreases from 4.651 Å in vacuo to 2.837 Å in the argon solvent. This is primarily an effect of the cavity compressing the very flexible C...M distance. The total solvation energy of the $^2A_1$ complex is -2.9 kcal mol^{-1}, compared with -5.8 kcal mol^{-1} found for the $^2B_2$ ion pair. Thus, the argon matrix, as expected, favors the ionic structure slightly, making the two configurations essentially energetically degenerate in the argon matrix. Adding the triples correction to the CCSD energy would make the calculated energy difference even smaller. Thus, our calculations suggest that the $^2A_1$ and $^2B_2$ states are very close in energy in vacuo and that solvation in an argon matrix favors the $^2B_2$ ion pair by roughly 2.9 kcal mol^{-1}. Our results are thus compatible with the experimental detection of a $^2B_2$ state but not with the results of earlier calculations [10, 12], which suggest that $^2B_2$ is clearly the ground state in vacuo. We suggest, however, that the differences are primarily due to the smaller basis sets used in the earlier calculations.

The geometries of the Be : ethylene$^{•+}$ complexes are less strongly affected by solvation than the very weakly bound $^2A_1$Li:ethylene complex. The C...Be distances become slightly longer in the argon solution. As for Li,

Table 7 Calculated (CCSD/6-311+G(d,p)) geometries and energies for Li:C_2H_4 and Be:$C_2H_4^{•+}$ within C_{2v} symmetry using a PCM [42] simulation for argon as solvent. The geometrical variables are designated as for Tables 2 and 4

	$^2A_1$	$^2B_2$
M = Li		
C=C (Å)	1.343	1.436
C...M (Å)	2.837	2.103
∠CCH (°)	121.5	119.6
∅M...CCH (°)	90.5	104.1
Total energy (a.u.)	−85.81106	−85.81048
Solvation energy (kcal mol^{-1})	−2.91	−5.77
Relative energy (kcal mol^{-1})	0.00	0.36
M = Be$^{•+}$		
C=C (Å)	1.361	1.459
C...M (Å)	2.138	1.769
∠CCH (°)	121.2	119.3
∅M...CCH (°)	92.0	102.3
Total energy (a.u.)	−92.75367	−92.74736
Solvation energy (kcal mol^{-1})	−22.94	−27.59
Relative energy (kcal mol^{-1})	0.00	3.97

the $^2B_2$ state is stabilized preferentially relative to the $^2A_1$, in this case by roughly 5 kcal mol^{-1}. This preferential stabilization would suffice to make the $^2B_2$ ion pair the more stable configuration at the CCSD(T) level.

Summary and conclusions

Our results suggest that, despite more sophisticated binding possibilities such as contributions from metallacyclopropane structures, the simple redox/electrostatic model given in Scheme 2 and Eq. 1 does a remarkably good job of rationalizing electron-transfer energies between ethylene and the Group I metal atoms or Group II and XII radical cations considered here. The dominant effect of the Coulomb ion-pairing energy means that, for instance, $Be^{\bullet +}$ and $Li^{\bullet}$ are similar in their abilities to reduce ethylene. Our results suggest that diffuse-augmented basis sets favor the $^2A_1$ states relative to the $^2B_2$ ion pairs because BSSE with smaller basis sets favors the ion pairs by stabilizing the $C_2H_4^{\bullet -}$ moiety with excess basis functions from the metal. PCM calculations reveal a small preferential stabilization for the $^2B_2$ ion pairs in argon matrices. Finally, NPA charges and Mulliken spin densities give more reliable descriptions of the electronic structures of these complexes than Mulliken charges, which underestimate the charge separation.

Acknowledgements This work was supported by the Deutsche Forschungsgemeinschaft as part of Sonderforschungsbereich 583, Redox-Active Metal Complexes: Control of Reactivity via Molecular Architecture. We thank Dr. Matthias Hennemann for the spin-density plots.

References

1. Alex A, Clark T (1992) J Am Chem Soc 114:506–510
2. Alex A, Clark T (1992) J Am Chem Soc 114:10897–10902
3. Clark T (1996) Ab initio calculations on electron-transfer catalysis by metal ions. In: Mattay J (ed) Topics in current chemistry, vol 77, electron transfer II. Springer, Berlin Heidelberg New York, pp 1–24
4. Kochi JK (1991) In: Dötz KH, Hoffmann RW (eds) Organic synthesis via organometallics. Vieweg, Braunschweig, pp 95–117
5. Chanon M, Julliard M, Poite JC (1989) Paramagnetic organometallic species in actiovation/selectivity catalysis, NATO ASI Series. Kluwer, Dordrecht
6. Trogler WC (1990) J Organomet Chem Lib 22:306–337
7. Astruc D (1991) Acc Chem Res 24:36–42
8. Chanon M, Rajzmann M, Chanon F (1990) Tetrahedron 46:6193–6299
9. Williams BA, Cool TA (1991) J Chem Phys 94:6358–6366
10. Trenary M, Schaefer III HF, Kollmann PA (1978) J Chem Phys 68:4047–4050
11. Hänsele E, Clark T (1991) Z Phys Chem 171:21–31
12. Alikhani ME, Hannachi Y, Manceron L, Boutelier Y (1995) J Chem Phys 1103:1028–1036
13. Manceron L, Andrews L (1985) J Am Chem Soc 107:563–568
14. Manceron L, Andrews L (1988) J Am Chem Soc 110:3840–3846
15. Rodriguez-Monge L, Larsson S (1996) J Chem Phys 105:7857–7863
16. Bernardi F, Bottoni A, Fossey J, Sorba J (1989) J Mol Struct THEOCHEM 52:301–309
17. White JC, Cave RJ, Davidson ER (1988) J Am Chem Soc 110:6308–6314
18. Balaji V, Jordan KD (1988) J Phys Chem 92:3101–3105
19. Frisch MJ, Trucks GW, Schlegel HB, Scuseria GE, Robb MA, Cheeseman JR, Montgomery JA Jr, Vreven T, Kudin KN, Burant JC, Millam JM, Iyengar SS, Tomasi J, Barone V, Mennucci B, Cossi M, Scalmani G, Rega N, Petersson GA, Nakatsuji H, Hada M, Ehara M, Toyota K, Fukuda R, Hasegawa J, Ishida M, Nakajima T, Honda Y, Kitao O, Nakai H, Klene M, Li X, Knox JE, Hratchian HP, Cross JB, Bakken V, Adamo C, Jaramillo J, Gomperts R, Stratmann RE, Yazyev O, Austin AJ, Cammi R, Pomelli C, Ochterski JW, Ayala PY, Morokuma K, Voth GA, Salvador P, Dannenberg JJ, Zakrzewski VG, Dapprich S, Daniels AD, Strain MC, Farkas O, Malick DK, Rabuck AD, Raghavachari K, Foresman JB, Ortiz JV, Cui Q, Baboul AG, Clifford S, Cioslowski J, Stefanov BB, Liu G, Liashenko A, Piskorz P, Komaromi I, Martin RL, Fox DJ, Keith T, Al-Laham MA, Peng CY, Nanayakkara A, Challacombe M, Gill PMW, Johnson B, Chen W, Wong MW, Gonzalez C, Pople JA (2004) Gaussian 03. Gaussian Inc, Wallingford
20. Cizek J (1969) Adv Chem Phys 14:35–89
21. Purvis GD, Bartlett RJ (1982) J Chem Phys 76:1910–1918
22. Scuseria GE, Janssen CL, Schaefer HF III (1988) J Chem Phys 89:7382–7387
23. Scuseria GE, Schaefer HF III (1989) J Chem Phys 90:3700–3703
24. McLean AD, Chandler GS (1980) J Chem Phys 72:5639–5648
25. Krishnan R, Binkley RS, Seeger R, Pople JA (1980) J Chem Phys 72:650–654
26. Blaudeau J-P, McGrath MP, Curtiss LA, Radom L (1997) J Chem Phys 107:5016–5021
27. Wachters AJH (1970) J Chem Phys 52:1033–1036
28. Hay PJ (1977) J Chem Phys 66:4377–4384
29. Raghavachari K, Trucks GW (1989) J Chem Phys 91:1062–1065
30. Clark T, Chandrasekhar J, Spitznagel GW, Schleyer PvR (1983) J Comp Chem 4:294–301
31. Frisch MJ, Pople JA, Binkley JS (1984) J Chem Phys 80:3265–3269
32. Dunning TH Jr, Hay PJ (1976) In: Schaefer HF III (ed) Modern theoretical chemistry, vol 3. Plenum, New York, pp 1–28
33. Wedig U, Dolg M, Stoll H, Preuss H (1986) In: Veillard A (ed) Quantum chemistry: the challenge of transition metals and coordination chemistry. Reidel, Dordrecht and references therein
34. Pople JA, Head-Gordon M, Raghavachari K (1987) J Chem Phys 87:5968–5675
35. Kaldor U (1987) J Chem Phys 87:4693–4695
36. Raghavachari K, Trucks GW (1989) J Chem Phys 91:2457–2460
37. Iron MA, Oren M, Martin JML (2003) Mol Phys 101:1345–1361
38. Clark T (1988) J Am Chem Soc 110:1672–1678
39. Hiberty PC, Humbel S, Archirel P (1994) J Phys Chem 98:11697–11704
40. Jordan KD, Burrow PD (1978) Chem Phys Lett 36:594–598
41. Jordan KD, Burrow PD (1978) Acc Chem Res 11:341–348
42. Cossi M, Scalmani G, Rega N, Barone V (2002) J Chem Phys 117:43–54
43. Emsley J (1989) The elements. Clarendon Press, Oxford
44. Reed AE, Curtiss LA, Weinhold F (1988) Chem Rev 88:899–926

J Mol Model (2006) 12: 631–639
DOI 10.1007/s00894-005-0038-2

ORIGINAL PAPER

Cinzia Chiappe · Christian Silvio Pomelli
Dieter Lenoir · Carsten Wattenbach

The first intermediates in the bromination of bicyclo[3.3.1]nonylidenebicyclo[3.3.1]nonane, combination of experiments and theoretical results

Received: 26 April 2005 / Accepted: 13 July 2005 / Published online: 10 November 2005

Abstract Bicyclo[3.3.1]nonylidenebicyclo[3.3.1]nonane (**1**) and adamantylideneadamantane (Ad = Ad) are two caged olefins with closely related structures at the double bond. Both compounds react instantaneously with Br_2 in chlorinated hydrocarbon solvents to give mixtures of olefin–Br_2 aggregates identified as the 1:1 π-complex and bromonium tribromide, bromonium pentabromide ion pairs. The stoichiometry, formation constants and the electronic spectra of all the species present at equilibrium (π-complex and bromonium ions), obtained by addition of bromine to alkene **1**, have been determined in 1,2-dichloroethane at 25° C and compared with the values that characterize the corresponding aggregates arising from Ad = Ad. The absence of the two bridging CH_2 groups in **1** significantly affects all the formation constants. Moreover, at variance with Ad = Ad, olefin **1** reacts with bromine to give, depending on reagent concentration, a substitution product. DFT (B3LYP) and ONIOM computations of 1:1 Br_2–olefin complexes for **1** and Ad = Ad confirm that the association energy is larger for the complex **1**–Br_2. The higher stability of this species seems to be correlated to the greater IP of **1** with respect to Ad = Ad which is able to compensate the reduced polarizability. The experimental value of the formation constant found for the complex **1**–Br_2, 643 vs 289 M^{-1} further supports the primary role exerted by dispersion interactions in alkene-Br_2 π-complexes.

Keywords Electrophilic bromination · π-Complex · Ionic intermediates · Bromonium ion

Dedicated to Professor Dr. Paul von Ragué Schleyer on the occasion of his 75th birthday.

C. Chiappe (✉)
Dipartimento di Chimica Bioorganica e Biofarmacia, Università di Pisa, 56126 Pisa, Italy
E-mail: cinziac@farm.unipi.it

C. S. Pomelli
Dipartimento di Chimica e Chimica Industriale, Università di Pisa, 56126 Pisa, Italy

D. Lenoir · C. Wattenbach
Institut für Ökologische Chemie, GSF Forschungszentrum, Postfach 1129, 85758 Neuherberg, Germany

Introduction

Electrophilic bromine addition to olefins is a classical widely- investigated reaction in organic chemistry [1–6]. Much information about the mechanistic features of this seemingly simple reaction has been obtained using crowded olefins, in which the rapid formation of the ionic intermediate and its subsequent trapping are retarded by the substituents on the double bond [5, 7–14]. A significant contribution to these studies has been made by adamantylideneadamantane (Ad = Ad), an allylically bridgehead tetrasubstituted olefin that shows unique behavior among all the alkenes investigated [15]. Its highly congested structure absolutely prevents the progress of Br_2 addition beyond the stage of bromonium ion formation and, in halogenated solvents, an equilibrium between free Ad = Ad, free Br_2, Ad = Ad–Br_2 π-complex, and bromonium tribromide, pentabromide, and heptabromide ion pairs is established instantaneously, see Scheme 1 [16, 17]. The unusual stability of the bromonium ion of Ad = Ad is kinetically controlled; as a consequence of the rehybridation of double bond carbon atoms, the two adamantylidene moieties in the bromonium ion are oriented in a such way as to prevent the rear attack of the counteranion (Br^- or Br_3^-).

Surprisingly, despite the large number of studies performed on this topic, no other olefin has been found to give a bromonium ion of comparable stability. For example, *d,l*-d_3-trishomocubylidene-d_3-trishomocubane, is an allylically bridged tetrasubstituted olefin with a structure very similar to Ad = Ad. The reduced dimension of the two cages with respect to Ad = Ad, however, drastically changes the reactivity of the system. *d,l*-d_3-trishomocubylidene-d_3-trishomocubane reacts with Br_2 to give the normal addition product; the ionic intermediate can neither be isolated nor detected [18]. On the other hand, even the simple introduction of two methyl

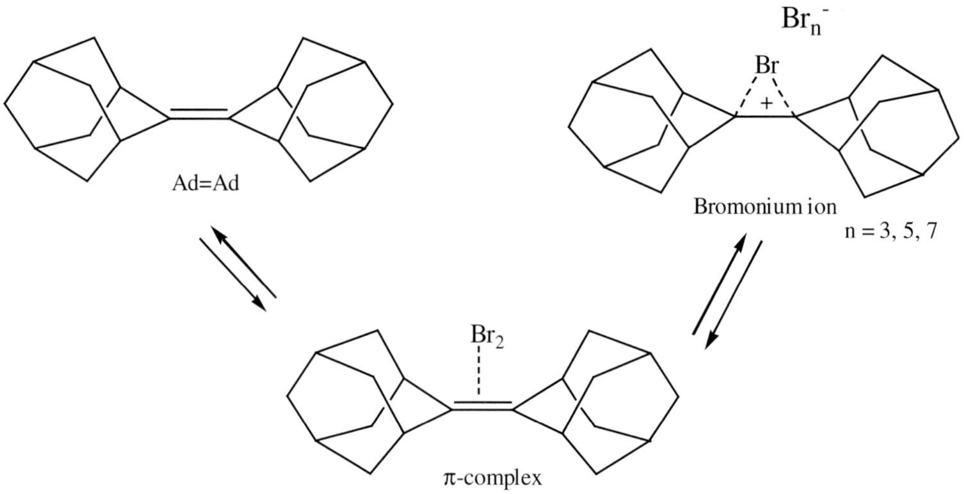

Sch. 1 Reaction of adamantylideneadamantane with Br_2 in chlorinated solvents

groups on the allylic carbons of Ad = Ad gives an olefin that reacts with bromine; in this case a substitution product has been isolated [19]. Analogously, the introduction of substituents on the homoallylic carbon of Ad = Ad generally reduces the stability of the corresponding bromonium ions; it has been been shown that several Ad = Ad derivatives react with Br_2 to give substitution products [20].

Herein, we report the results of a theoretical and experimental investigation of the reaction of bicyclo[3.3.1]nonylidenebicyclo[3.3.1]nonane (**1**) with bromine (Scheme 2). This olefin, which has a closely related structure to Ad = Ad, gives, with Br_2, a stable π-complex (**2**) and a quite stable bromonium ion (**3**, **4**).

The stability constants of all the species present at equilibrium (π-complex, and bromonium ions) have been determined in this work and compared with those characterizing Ad = Ad. At variance with Ad = Ad, olefin **1** reacts slowly with bromine to give a substitution product. We report the structure of the substitution product. Finally, we give details of a theoretical description of the bonding in the π-complex, in the halonium ion and an interpretation for the formation of the substitution product.

Results and discussion

Spectral data

Preliminarily, to verify the ability of **1** to give a sufficiently stable bromonium ion, equal volumes of 1,2-dichloroethane (DCE) solutions of olefin **1** (ca 1.5×10^{-3} M) and Br_2 (3×10^{-3} M) were mixed at 25° C.[1] In analogy with Ad = Ad, many of the colored mixtures obtained from **1** were stable for at least 20 min and therefore they could be investigated by conventional spectrophotometric techniques. Spectral data were collected in the 250–500 nm wavelength range working at 2.0×10^{-4} to 1.0×10^{-2} M Br_2 and 4×10^{-4} to 7×10^{-3} M olefin. All spectra exhibited a maximum around 270 nm, near the λ_{max} values found previously, both for the Br_3^- ion and for many 1:1 olefin–Br_2 complexes (including that of Ad = Ad), and the observed dependence of the 270-nm absorption on reagent concentrations, when these solutions were progressively diluted, pointed to the presence of two species of different stoichiometry, probably the π-complex **2** and the bromonium tribromide ion pair **3**. Furthermore, an examination of the entire UV spectrum revealed that an increasing excess of Br_2 produced a shoulder around 320 nm, attributable to the formation of a bromonium pentabromide salt (**4**). In analogy with Ad = Ad, all these data showed that several species of different stoichiometry are present at equilibrium in DCE solutions of **1** and Br_2.

It is worth mentioning that the stability of mixtures obtained by mixing **1** and bromine, depended on reagent concentration; diluted solutions containing a large excess of Br_2 ($[\mathbf{1}]=5\times10^{-4}$, 1.5×10^{-3} $[Br_2]<2\times10^{-2}$) or a moderate excess of olefin ($[Br_2]=5\times10^{-4}$, $2\times10^{-3}<[\mathbf{1}]<6\times10^{-3}$) were sufficiently stable to be investigated using a conventional UV–vis spectrophotometer. More concentrated solutions containing an excess of **1** ($[Br_2]=6\times10^{-4}$, $1\times10^{-2}<[\mathbf{1}]<0.6$) showed a disappearance of the coloration. All the spectrophotometric measurements related to solutions of **1** and Br_2 that did not show any detectable disappearance of the absorbances were fitted using the NLLSQ computational procedure [16] previously used for Ad = Ad. Fitting parameters were the formation constants $K_{m,n}$ of the A_mB_n complexes, defined from olefin and bromine (A stands for olefin, B for Br_2, the (m,n) pair having the value 1:1, 1:2, and 1:3), and the extinction coefficients of the 1:1 π-complex, the spectra of salts **3** and **4** being fixed at the values reported for $Bu_4N^+Br_3^-$ and $Bu_4N^+Br_5^-$, respectively.

The presence of more species (**2**, **3**, and **4**) in equilibrium must be considered to obtain good fits (rms

[1] It has been reported that olefin **1** gives a iodonium triflate sufficiently stable to be investigated by NMR, whereas the corresponding bromonium triflate had to be maintained at −60° C to prevent decomposition (see Ref. [5])

Sch. 2 Structure of adamantylidenadamantane (Ad = Ad) and bicyclo[3.3.1]non-ylidenbicyclo[3.3.1]nonane (**1**)

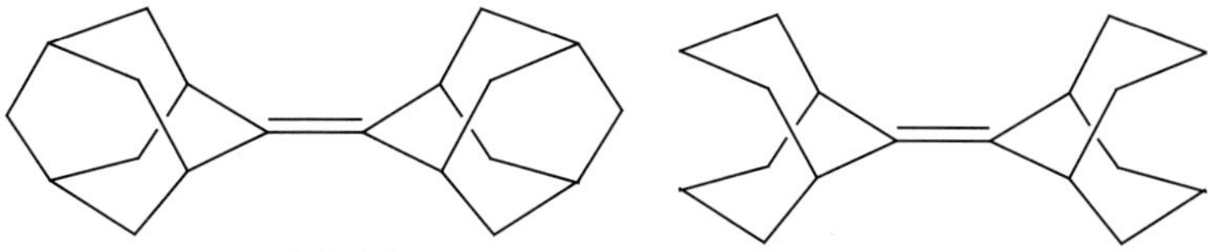

Ad=Ad

1

deviations < 0.03 absorbance units) of the spectrophotometric data (Scheme 3). In Table 1 the equilibrium constants of species **2**, **3**, and **4** and the extinction coefficient the complex **2** at the absorption maximum are reported. Furthermore, the values of the corresponding species for Ad = Ad–Br_2 system have been reported [16] for comparison. In analogy with Ad = Ad, the formation of **3** from **1** was too fast to be followed even with a stopped-flow apparatus. Therefore, it was impossible to check experimentally if **3** was formed directly from **1** and Br_2, or through the complex **2**. For this reason, the formation constant of **3** was defined in the computational procedure, as in the case of Ad = Ad, from **1** and Br_2, with $m = 1$ and $n = 2$. Although the formation constant of bromonium pentabromide salt **4** was again defined in the computational procedure from **1** and Br_2, with $m = 1$ and $n = 3$, there is, however, no doubt that **4** arises from **3** and Br_2. The pertinent formation constant is $K'_{13} = K_{13}/K_{12}$ and the value of this constant (18.9 M^{-1}) is, as expected, not too different from that found for the bromonium pentabromide of Ad = Ad, formed from the corresponding bromonium tribromide [16], and with the value characterizing the formation of $Bu_4N^+Br_5^-$ from $Bu_4N^+Br_3^-$ and Br_2, 14.3 M^{-1} [21].

The comparison of the values of the formation constants computed in the present work for complexes olefin **1**–Br_2, with those previously evaluated for Ad = Ad–Br_2, shows that the stability of all the intermediates is considerably affected by the lack of the two bridging CH_2 groups. The formation constant for the π-complex is significantly higher in the case of **1** than for Ad = Ad, whereas that of the bromonium ion tribromide is lower of one order of magnitude.

Reaction products

In analogy with Ad = Ad, olefin **1** reacts with an excess of Br_2 to give an ionic intermediate sufficiently stable to allow the UV investigation. However, when an equimolar amount of bromine at 0° C, in DCE or dichloromethane-d_2, was added to a solution of **1** of concentration lower than 10^{-3} M, after the disappearance of the color,

Sch. 3 Reaction of olefin **1** with Br_2 in 1,2-dichloroethane under controlled reagent concentrations

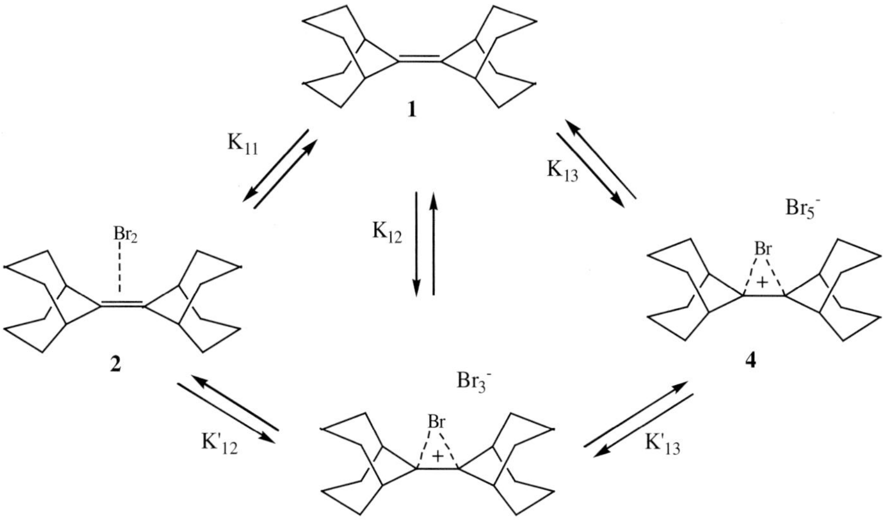

Table 1 Formation constants and molar extinction coefficients at 270 nm and 25° C for the Br_2–**1** complexes obtained from the NLLSQ fitting of UV–vis spectral data

Alkene	K_{11} (M^{-1})	ε_{270} (M^{-1} cm^{-1})	K_{12} (M^{-2})	K_{13} (M^{-3})	K'_{13} (M^{-1})
1	643 (43)	2,934 (64)	2.96 (0.1)×10^4	5.51 (0.5)×10^5	18.9
Ad = Ad	289	17,780	3.23×10^5	7.2×10^6	22.2

a non-ionic reaction product was recovered. On the basis of its mass spectrum ($C_{18}H_{27}Br$) and NMR spectra, we propose structure 5. The ^{13}C NMR–DEPT analysis shows the presence of a tetrasubstituted double bond and of a CH carbon at 50.16 ppm, attributable to a CH carbon bearing a bromine atom. The spectrum is, moreover, characterized by other two signals due to CH carbons, at 33.05 and 33.3 ppm, and by six signals due to CH_2 carbons at 17.08, 22.69, 34.42, 34.72, 35.83, and 40.27 ppm. The total number of signals in the ^{13}C NMR spectrum suggests the formation of a bromo adduct characterized by a structure having a symmetry plane. Therefore, between the two adducts bearing a bromine atom on a secondary carbon (CHBr), reported in Scheme 4, we suggest structure 5.

Although Ad = Ad and derivatives give, with suitable electrophiles which combine an electrophilic and a sufficient basic moiety, homoallylic substitution products of the type of **6** [22–25], compound **1** seems to yield, at least under the reaction conditions, the substitution product at the more distant CH_2 carbon.[2]

Theoretical investigation of complexes and bromonium ions

To obtain more detailed information about the factors that determine the different stability and reactivity of the intermediates formed by bromine addition to olefin **1** and Ad = Ad, we have performed a theoretical investigation on olefin **1**, the corresponding olefin–Br_2 π-complex, **2**, and the bromonium ion, **3**. The results have been compared with those previously obtained for Ad = Ad and its derivatives [20]. To find a level of calculation that gave an appropriate description of these systems with a reasonable computational cost, especially for π-complexes, the following strategy was chosen: the geometries of olefin, π-complex, and bromonium ions were refined at a B3LYP/CEP-121G level with supplementary basis functions on all atoms as previously reported [20] for Ad = Ad derivatives. On the final geometries a frequency calculation was performed. For the π-complex alone, an ONIOM calculation, on the model system constituted by the bromine atoms and the quaternary (sp_2) carbons described at the MP2 level, was performed in order to compare the data with those previously reported for the π-complex Ad = Ad–Br_2. All the calculations were performed using the Gaussian Package [26]. Geometries of complex **2**, of the corresponding bromonium ions and of the bromonium tribromide ion pair **3**, are reported in Figs. 1, 3, 4, 5, and 6, whereas Tables 2 and 3 summarize the energies of the investigated species. The calculated intermolecular distances are given in Table 4.

Figure 1 shows that the complex **2** and the analogous complex of Ad = Ad have the angular geometry (T-shape) generally reported [27] for these systems. The two structures are characterized, however, by different distances between the axial protons at homoallylic positions. The distance between the homoallylic protons directed toward Br_2 is 4.647 Å in the complex **2** and 5.082 Å in the Ad = Ad complex; at the same time the distances between the homoallylic protons opposite to Br_2 in the same complexes are 2.240 and 2.343 Å, respectively. These distances are moreover, in both cases, significantly different from those characterizing the same H-atoms in the starting olefins (3.342 Å in **1** and 3.699 Å in Ad = Ad). At variance with ethylene, for which practically no rehybridization has been calculated [20] on going from olefin to the complex, in **1** as well as in Ad = Ad, a significant rehybridization at the carbons already occurs in the π-complex, suggesting that an "inner" character being more important than that in complexes of linear non-congested alkenes.

In Table 2 are reported the energetic results for the **1**–Br_2 π-complex (**2**) and the more significant parameters have been compared with those previously calculated [20] for the Ad = Ad–Br_2 system. The meaningful energetic values are those at the ONIOM level. At this level of calculation, the π-complex **2** is slightly more stable than the corresponding Ad = Ad–Br_2 complex, in agreement with the experimental results obtained in solution, see Table 1.

To shed further light on the prominent factors that affect the stability of these complexes, we have calculated[3] the ionization potential (IP) and polarizability (α) for olefin **1** (IP 7.44 eV; α 202.9 a.u.) and for Ad = Ad (IP 7.37 eV; α 214.0 a.u). Recently, we have shown that for alkyl-tetrasubstituted olefins a good linear relation may be found [28] by plotting the constant of formations of the π-complex olefin–Br_2 (log K_f) against the calculated induced dipole-induced dipole dispersion interactions (d_6, atomic units) defined as reported in Eq. 1:

$$d_6 = \frac{3}{2}\alpha_1\alpha_2\frac{IP_1 IP_2}{IP_1 + IP_2} \qquad (1)$$

where indexes 1 and 2 are for bromine (α_1 30.35 a.u., IP_1 0.387 eV) and olefin, respectively.

On the basis of the calculated values of IP and α for olefins **1** and Ad = Ad values of d_6 of 1628.5 and 1554.8 a.u. can be evaluated, respectively. The correlation between the formation constants of several olefin–Br_2 π-complexes and the respective d_6 values [28], including also the complex of olefin **1**, is reported in Fig. 2. The increased formation constant (K_f) of **2**, with respect to the analogous complex of Ad = Ad, may therefore be attributed to the higher IP value of **1**, which is able to overcompensate the reduced polarizability of this olefin with respect Ad = Ad.

[2] It has been reported (see Ref. [24]) that **1** does not react with NBS whereas with NCS gives the 4(e)-chlorobicyclo[3.3.l]nonylidenebicyclo[3.3.l]nonane, characterized by 16 signals in the ^{13}C NMR spectrum

[3] Both values are calculated at the B3LYP/CEP-121G(*d,p*) level of calculation

Sch. 4 Possible substitution products obtained from alkene **1**

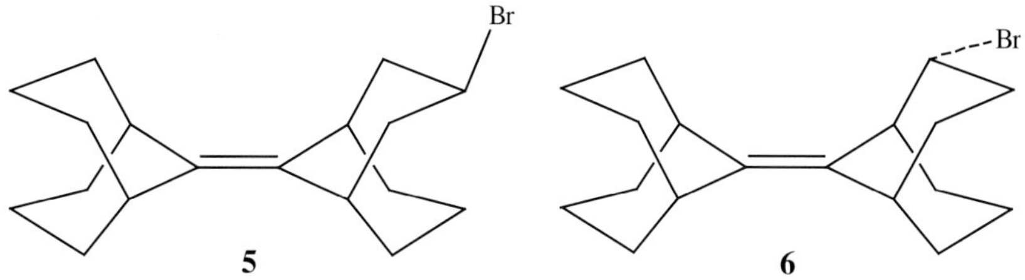

5 **6**

Table 2 Numerical results on the π-complex **2**

System	E(B3LYP)		E(ONIOM)		ZPE(B3LYP)
1	−118.620541		−117.759099		0.423799
Br_2	−26.764366		−26.607046		0.000722
π-complex, **2**	−145.387113		−144.386469		0.426185
ΔG_{form}	−1.38	(2.95)[a]	−12.76	(−9.66)[a]	1.04
Deformed[b] **1**	−118.613626		−117.7533588		
ΔE_{def} C=C	4.34	(3.87)[a]	3.60	(3.48)[a]	

Absolute energies (a.u.), relative energies (kcal mol^{-1})
[a]Values in parenthesis are refereed to the corresponding complex of Ad=Ad
[b]Energy of olefin **1** in the "deformed" structure of the complex **2**, for definition see Ref. [20]

Table 3 Numerical results of the bromonium ion systems

System	E(B3LYP)	ZPE(B3LYP)	Δ(E+ZPE)
Chair bromonium ion	−131.755154	0.426402	–
Boat bromonium ion	−131.748709	0.426258	–
Br_3^-	−40.285980	0.001167	–
Chair bromonium ion + Br_3^-	−172.041134	0.427569	0.00
Boat bromonium ion + Br_3^-	−172.034689	0.426425	3.70
Chair T-shaped complex	−172.151946	0.427090	
Boat bromonium ion– Br_3^- complex	−172.122648	0.427408	−18.19

Finally, to obtain information about the possible pathway through which olefin **1** reacts with Br_2 to give a substitution product, we have performed geometrical optimization of the corresponding bromonium ion and of the bromonium ion of Ad=Ad and we have compared the relative stabilities and geometrical features of these species.

Two different conformations have to be considered for the bromonium ion of olefin **1**. Since in vacuo calculations have been performed by us, we cannot directly compare charged and non-charged systems [20]. π-Complexes and bromonium ions are very different from the physical point of view: the absence of significant dispersion interactions in bromonium ions allow us to perform B3LYP calculations and avoid the more expensive ONIOM calculations.

Reported in Table 3 are the absolute energies of the bromonium ion of **1** in different conformations and the energies of the complexes between the bromonium ion and Br_3^-; all these species are represented in Figs. 3, 4, and 5. Note that the comparison between neutral complexes (ion pairs) instead of bromonium ions reduces the error introduced by disregarding the solvent effect into the calculations.

Figure 3 represents the "classical" bromonium ion with the olefin skeleton in its more stable conformation. In the second structure (Fig. 4) one of the side-chain rings is passed from the chair to the boat conformation. In the third one (Fig. 5) we consider the bromonium tribromide ion pair, i.e., the complex between the structure of Fig. 4 and a Br_3^- ion. Finally, represented in Fig. 6 is a chair bromonium tribromide complex with an aspecific interaction of Br_3^- with Br^+ that represents the reactive pathway leading back to the π-complex.

The most stable form of the bromonium ion arising from **1** is that reported in Fig. 3 (chair conformation of bromonium ion), but its preference over the other conformation (boat conformation reported in Fig. 4) is small.

It is, however, worth mentioning that the formation of the complex between the boat conformation of the bromonium ion and the Br_3^- anion, tends to a structural situation (Fig. 5) characterized by a quasi-collinear position of Br1, H1, H2, and Br2. The approach of the Br atom of tribromide, reported as 2 in Fig. 5, toward the proton 2, shift the bromine atom of bromonium nearer to C1, giving an ionic intermediate with higher bromocarbenium character. We have previously shown [20] that the ionic intermediates arising from bromine addition to the double bond of tetrasubstituted olefins (including Ad=Ad derivatives), which are characterized by a more β-bromocarbenium character, are generally

Fig. 1 Geometry of bicyclo[3.3.1]nonylidenebicyclo[3.3.1]nonane-Br_2 (**2**) and Ad = Ad–Br_2 complexes

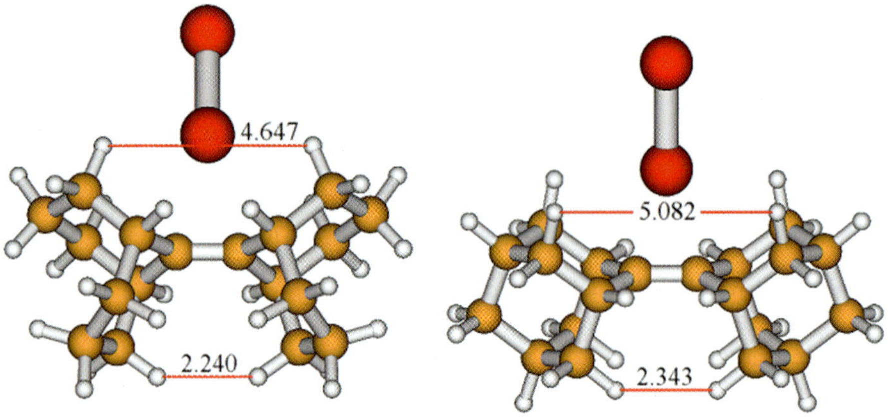

more prone to evolve toward substitution products. In agreement with this statement, the boat bromonium ion arising from **1** has significantly different Br1–C distances (Table 4 and Figs. 4, 5) and therefore a significantly different charge development on the two carbons. This structure can be compared with the isomeric T-shaped 2:1 bromine–**1** complex reported in Fig. 6. The C1, C2–Br1 distances for this latter T-shaped complex shows that this species has a nature very similar to the π-complex and this arrangement of bromine atoms and **1** may be considered a regression from the bromonium ion toward the π-complex. The structure in Fig. 5 is 18.19 kcal mol^{-1} is more stable than the T-shaped complex of Fig. 6 as reported in Table 3, and this value is in agreement with the tendency of this olefin to react with bromine to give a substitution product.

Considering the structure reported in Fig. 5, it is indeed possible to hypothesize that the abstraction of the proton labeled as H2 by the base (Br_3^- or Br) and the shift of Br1 from the C1 carbon to the γ position may be responsible of the formation the substitution product **5**. The two processes, which are represented in Scheme 5 through a synchronous mechanism, can really occur at different times. Proton abstraction might precede Br^+ shift or, alternatively, the initial expulsion of Br^+ as HBr could give a μ-hydridobridged carbocation, stabilized by the interaction with the double bond through the σ-bonds, able to give the product **5** by subsequent reaction with Br_3^- (Scheme 6).

Calculations are in due course to verify these latter hypotheses. However, the reduced tendency of **1** to give the substitution product on increasing bromine concentration (i.e., under conditions which shift all the equilibria of Scheme 3 toward the bromonium pentabromide ion pair) suggests that the basicity of the counteranion plays a role in the mechanism of product formation. This feature is more in agreement with the mechanistic pathway of Scheme 5.

Conclusion

The comparison of the results reported here for the bromination of **1** with the previous ones for Ad = Ad

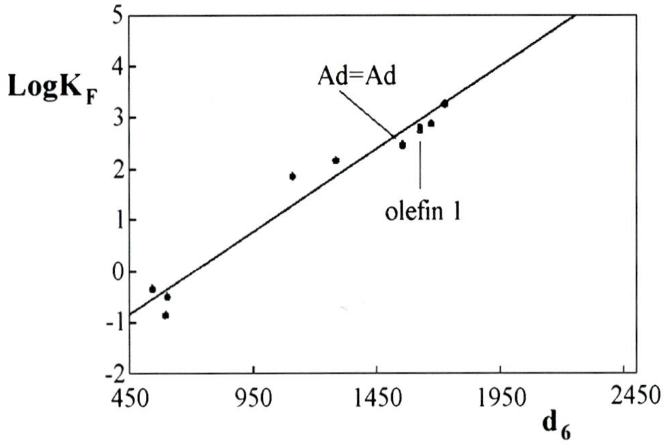

Fig. 2 Correlation between π-complex formation constant (log K_f) and induced dipole-induced dipole dispersion interactions (d_6)

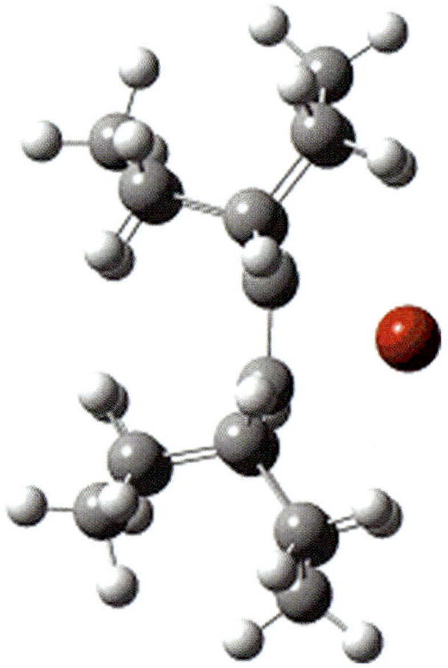

Fig. 3 Bromonium ion of **1** with both the side chains in their standard chair conformations

Table 4 Comparative bond distances (Å)

Distances	π-Complex **2**[a]	Chair conform. brom. ion	Boat conform. brom. ion	Boat complex
Br1–C1	2.87	2.19	2.15	2.09
Br1–C2	2.87	2.19	2.23	2.41
Br1–H1	4.31	4.36	2.78	2.80
Br2–H2	–	–	–	2.82

The atom labels are defined in Fig. 5
[a]The distance Br1–π (middle point of the C1–C2 bond) is 2.78 Å (for Ad = Ad–Br_2 is 2.41 Å)

shows a marked effect due to the absence of the two bridging CH_2 groups on the reactivity of the double bond. It modifies the stability of all intermediates involved on the reaction coordinate.

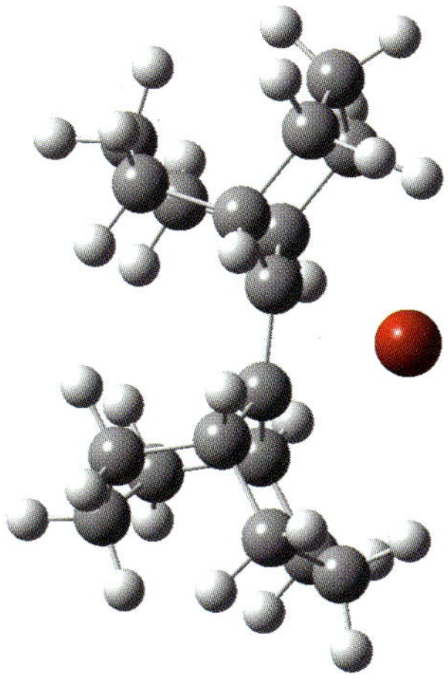

Fig. 4 Bromonium ion of **1** with one of the cyclic side chains in a boat conformation

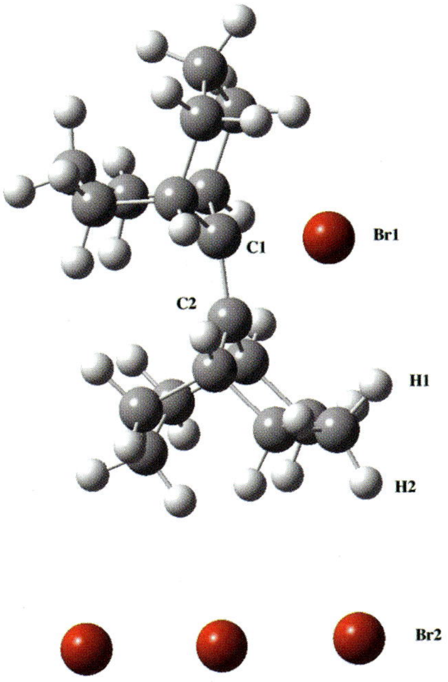

Fig. 5 Complex of the boat conformation bromonium ion with Br_3^-

Analogously to Ad = Ad, olefin **1** reacts with bromine instantaneously to form a 1:1 π-complex (**2**) which further rapidly evolves to the ionic intermediate. The formation constant of the π-complex **2** (643 M^{-1} at 25° C) is two times higher than that reported for the analogous complex of Ad = Ad. This increase is consistent with the different polarizability and the donor ionization potential of **1**. The previously proposed equation [28], which takes into account both effects, is also able in this case to describe this complex better than the Mulliken analysis.

The bromonium tribromide ion pair arising from olefin **1** is characterized by a lower stability constant with respect to the same species formed from Ad = Ad and Br_2, in agreement with the higher tendency of **1** to give a substitution product with bromine. This peculiar behavior of **1**, which distinguish this olefin from Ad = Ad, may be considered a consequence of the lack of the bridging CH_2 groups which, allowing the fused cyclohexanes to perform other conformations, permits proton abstraction and Br^+ shift. Generally, ionic intermediates having a greater β-bromocarbenium character are more prone to evolve to products. DFT calculations show that the boat conformation of **3** is characterized by a significantly high difference of charge on the two carbons of the bromonium ring; a situation which does not occur in the bromonium ion arising from Ad = Ad.

Experimental section

Bicyclo[3.3.1]nonylidenebicyclo[3.3.1]nonane (**1**) was prepared as previously reported [29]. 1,2-Dichloroethane (DCE, Fluka, puriss. pa ACS) was used as supplied without further purification. Bromine was drawn from 1-ml vials (C. Erba, RPE grade >99.5%) opened immediately before use.

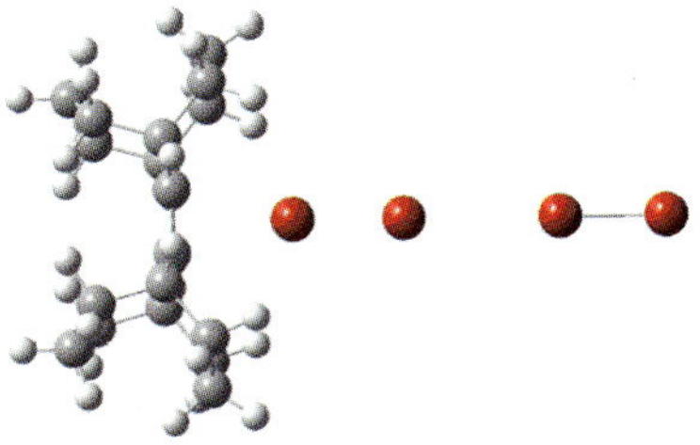

Fig. 6 T-shaped complex of chair bromonium ion

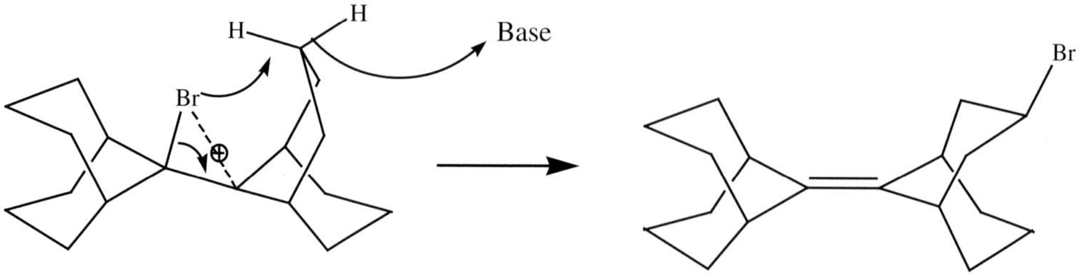

5

Sch. 5 Suggested mechanistic transformation of the bromonium ion intermediate into the substitution product **5**

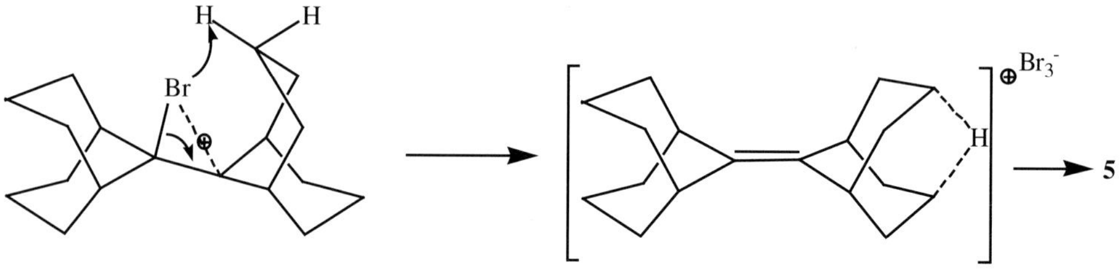

Sch. 6 Formation of **5** through a μ-hydridobridged carbocation

Spectrophotometric measurements

1,2-Dichloroethane solutions of **1** and Br_2 were prepared by weighing the reagents in accurately calibrated volumetric flasks. The Br_2 solutions were stored in the dark and checked spectrophotometrically for their concentrations. They were discarded when the absorptions around the Br_2 UV minimum (280–350 nm) were found to be higher than expected.

Working-strength solutions were prepared by mixing solvent and aliquots of the reagent stock solutions withdrawn with precision microburets. The analytical concentrations of Br_2 and olefin are reported in the text. The UV spectra were registered at 25° C in 1 or 0.1 cm cells in the 250–420 nm range. An already described, a nonlinear least-squares procedure was applied to fit the spectrophotometric data. The root-squares deviation was 0.032 absorbance units.

Product study

To precooled DCE (or dichloromethane-d_2) solutions (0.5–2 ml) of **1** (ca 30 mg), an equimolar amount of Br_2 was added at 0° C in the dark. After the complete disappearance of color, the samples were directly analyzed by NMR, when the reactions were carried out in the deuterated solvent, or after removal of the volatile components by means of a rotary evaporator when the reactions were carried out in DCE. ^{1}H NMR (CD_2Cl_2, δ) 3.95 (tt, 1H, C*H*Br), 3.10 (br, 2H); 2.85 (br s, 2H), 2.5–1.2 (br, 22H); ^{13}C NMR- DEPT (CD_2Cl_2, δ) 138.0 (>C<), 126.1 (>C<), 50.16 (*C*HBr), 40.27 (CH_2), 35.83 (CH_2), 34.72 (CH_2), 34.42 (CH_2), 22.69 (CH_2), 17.08 (CH_2), 33.35 (CH) and 33.3 (CH). 21.7. Mass spectrum, *m/e* 322:324 (1:1).

Computational methodology

All calculations were performed with the Gaussian 03 suite of programs [26]. All the geometries were optimized at the B3LYP/CEP-121G level with supplementary basis functions on all atoms as previously reported [20]. A frequency calculation was performed on the final geometries. Only for the π-complexes was an ONIOM calculation with the bromines and the sp_2 carbons described at the MP2 level performed. All the calculations were performed on an APPRO 1124 workstation with two Athlon MP1800+ processor, 1 Gb RAM and 18 Gb SCSI Hard Disk with Suse 9.1 O.S. The Gaussian Package was compiled with MP support with a Portland pgf 77 5.1 fortran compiler.

References

1. de la Mare PBD, Bolton R (1982) Electrophilic additions to unsaturated systems, 2nd edn. Elsevier, New York, NY, pp 136–197
2. V'yunok KA, Ginak AI (1981) Russ Chem Rev (English translation) 50:151–163
3. Schmid GH (1989) In: Patai S (ed) The chemistry of double-bonded functional groups, Supplement A, vol 3, Part 1. Wiley, New York, NY, p 699
4. Ruasse MF (1993) Adv Phys Org Chem 28:207–291
5. Lenoir D, Chiappe C (2003) Chem Eur J 9:1037–1044
6. Brown RS (1997) Acc Chem Res 30:131–137
7. Brown RS, Slebocka-Tilk H, Bennet AJ, Bellucci G, Bianchini R, Ambrosetti R (1990) J Am Chem Soc 112:6310–6316
8. Bennet AJ, Brown RS, McClung RED, Klobukowski M, Aarts GHM, Santarsiero BD, Bellucci G, Bianchini R (1991) J Am Chem Soc 113:8532–8534
9. Bellucci G, Bianchini R, Chiappe C, Lenoir D, Attar A (1995) J Am Chem Soc 117:6243–6248
10. Bellucci G, Chiappe C, Bianchini R, Lenoir D, Herges R (1995) J Am Chem Soc 117:12001–12002

11. Brown RS, Nagorski RW, Bennet AJ, McClung RED, Aarts GHM, Klobukowski M, McDonald, Santarsiero BD (1994) J Am Chem Soc 116:2448–2456
12. Nagorski RW, Slebocka-Tilk H, Brown RS (1994) J Am Chem Soc 106:419–420
13. Slebocka-Tilk H, Motallebi S, Nagorski RW, Turner P, Brown RS, McDonald R (1995) J Am Chem Soc 107:8769–8776
14. Slebocka-Tilk H, Gallagher D, Brown RS (1996) J Org Chem 61:3458–3466
15. Strating J, Wieringa JH, Wynberg H (1969) J Chem Soc Chem Commun 907–908
16. Bellucci G, Bianchini R, Chiappe C, Marioni F, Ambrosetti R, Brown RS, Slebocka-Tilk H (1989) J Am Chem Soc 111:2640–2647
17. Bellucci G, Bianchini R, Chiappe C, Ambrosetti R, Catalano D, Bennet AJ, Slebocka-Tilk H, Aarts G, Brown RS (1993) J Org Chem 58:3401–3406
18. Bellucci G, Bianchini R, Chiappe C, Gadgil VR, Marchand AP (1993) J Org Chem 58:3575–3577
19. Chiappe C, De Rubertis A, Lemmen P, Lenoir D (2000) J Org Chem 65:1273–1279
20. Chiappe C, De Rubertis A, Jaber A, Lenoir D, Wattenbach C, Pomelli CS (2002) J Org Chem 67:7066–7074
21. Bellucci G, Bianchini R, Chiappe C, Ambrosetti R (1989) J Am Chem Soc 111:199–202
22. Bolster J, Kellogg RM, Meijer EW, Wynberg H (1979) Tetrahedron Lett 20:285–286
23. Lopez L, Troisi L, Mele G (1991) Tetrahedron Lett 32:117–120
24. Meijer EW, Kellogg RM, Wynberg H (1982) J Org Chem 47:2005–2009
25. Huang X, Batchelor RJ, Einstein FWB, Bennet AJ (1994) J Org Chem 59:7108–7116
26. Frisch MJ, Trucks GW, Schlegel HB, Scuseria GE, Robb MA, Cheeseman JR, Montgomery JA Jr, Vreven T, Kudin KN, Burant JC, Millam JM, Iyengar SS, Tomasi J, Barone V, Mennucci B, Cossi M, Scalmani G, Rega N, Petersson GA, Nakatsuji H, Hada M, Ehara M, Toyota K, Fukuda R, Hasegawa J, Ishida M, Nakajima T, Honda Y, Kitao O, Nakai H, Klene M, Li X, Knox JE, Hratchian HP, Cross JB, Bakken V, Adamo C, Jaramillo J, Gomperts R, Stratmann RE, Yazyev O, Austin AJ, Cammi R, Pomelli C, Ochterski JW, Ayala PY, Morokuma K, Voth GA, Salvador P, Dannenberg JJ, Zakrzewski VG, Dapprich S, Daniels AD, Strain MC, Farkas O, Malick DK, Rabuck AD, Raghavachari K, Foresman JB, Ortiz JV, Cui Q, Baboul AG, Clifford S, Cioslowski J, Stefanov BB, Liu G, Liashenko A, Piskorz P, Komaromi I, Martin RL, Fox DJ, Keith T, Al-Laham MA, Peng CY, Nanayakkara A, Challacombe M, Gill PMW, Johnson B, Chen W, Wong MW, Gonzalez C, Pople JA (2004) Gaussian 03, Revision C.02. Gaussian Inc. Wallingford, CT
27. Chiappe C, Lenoir D, Pomelli CS, Bianchini R (2004) Phys Chem Chem Phys 6:3235–3241 (and references cited therein)
28. Chiappe C, Detert H, Lenoir D, Pomelli CS, Ruasse MF (2003) J Am Chem Soc 125:2864–2865
29. McMurry JE, Fleming MP (1974) J Am Chem Soc 96:4708–4708

J Mol Model (2006) 12: 641–648
DOI 10.1007/s00894-005-0095-6

ORIGINAL PAPER

Pradip K. Biswas · Jiansen Niu
Tobias Frederico · Valentin Gogonea

Numerical simulation and graphical analysis of in vitro benign tumor growth: application of single-particle state bosonic matter equation with length scaling

Received: 2 May 2005 / Accepted: 29 November 2005 / Published online: 23 March 2006

Abstract We describe the application of a non-linear single-particle state bosonic condensate equation to simulate multicellular tumor growth by treating it as a coupling of two classical wave equations with real components. With one component representing the amplitude of the cells in their volume growth phase and the other representing the amplitude of the cells in their proliferation or mitosis phase, the two components of the coupled equation feed each other during the time evolution and are coupled together through diffusion and other linear and non-linear terms. The features of quiescent and necrotic cells, which result from poor nutrient diffusion into a tumor, have been found to correspond quite well to experimental data when they are modeled as depending on higher cell density. Classical hallmarks of benign tumor growth, such as the initial rapid growth, followed by a dramatic collapse in the proliferating cell count and a strong re-growth thereafter appear quite encouragingly in the theoretical results. A tool for graphical analysis of the tumor simulation results has been developed to provide morphological information about tumors at various growth stages. The model and the graphical analysis can be extended further to create an effective tool to predict/monitor tumor growth.

Keywords Multicellular tumor · Computer simulation · Heisenberg equation · Length scaling

Dedicated to Professor Dr. Paul von Ragué Schleyer on the occasion of his 75th birthday

P. K. Biswas · J. Niu · V. Gogonea
Department of Chemistry, Cleveland State University,
Cleveland, OH 44115, USA

T. Frederico
Departamento de Física,
Instituto Tecnológica de Aeronáutica,
CTA, Sao José dos Campos, SP, 12228-900, Brazil

V. Gogonea (✉)
Department of Immunology, Lerner Research Institute,
Cleveland Clinic Foundation,
9500 Euclid Avenue,
Cleveland, OH 44195, USA
e-mail: v.gogonea@csuohio.edu
Tel.: +1-216-8759717
Fax: +1-216-6879298

Introduction

Tumors arise as a consequence of unrestricted cell division. This unrestricted cell division is supposed to be possible due to a mutation of the *p53* transcription factor, which otherwise functions as a tumor suppressor gene in normal tissues. In response to DNA damage, *p53* activation results either in the induction of the cyclin-dependent kinase inhibitor *p21* and cell cycle arrest or in the induction of genes that induce programmed cell death through the transcriptional factor *p53*. One molecular mechanism through which unrestricted cell growth can be achieved is through mutation of *p53*. Mutated non-functional *p53* can no longer induce either cell cycle arrest or programmed cell death in response to DNA damage [1].

Thus, we can associate the functional role of the *p53* protein in cells (the biological building blocks) with that of the Pauli exclusion principle, which restricts the fermions (the elementary building blocks of matter) in occupying the same quantum state. On the other hand, bosons do not obey the Pauli exclusion principle, and a single-particle bosonic state can be occupied by any number of bosons. Thus, the statistical behavior of fermions and bosons resembles that of normal and cancer cells, where the number of normal cells is regulated by *p53* protein while the population of cancer cells is unregulated. In the same way as fermions and bosons are distinguished by their spins (half-integral for fermions and integral for bosons), normal and cancer cells may be distinguished by the presence of a wild type or mutated *p53* gene.

Guided by this analogy, a mathematical equation relating to a single-particle bosonic state was applied to the growing cancer cells in a tumor state with relative length-scaling [2, 3].

The primary impetus in applying such a model to tumor growth was that the growth dynamics of a single-particle state bosonic condensate matter and that of a multicellular tumor of cancer cells may follow the same mathematical equation but with different length scales. Such modeling may provide us crucial information about tumor morphology and be applied in simulations of tumor-drug interactions. However, the basic problem in applying a single-particle state atomic (bosons) equation to the aggregate of tumor cells is that while the atoms are quantum particles and their wave functions may be represented by a complex function, the biological cells are classical objects and must be represented by real functions. Here we describe how the Gross-Pitaevskii (GP) equation (a mean-field approximation version of the complex single-particle state Heisenberg equation) could be regarded as a coupling of two real classical wave equations and employed to model the morphology and growth dynamics of benign tumors successfully.

In vitro simulations of benign tumors in three dimensions have revealed that they are finite-size and inhomogeneous aggregates of cancer cells, which appear in a bound spheroid form [4–6]. In vitro tumor culture data [4] reveals a strong oscillatory pattern for the proliferating cell count with intricate features of partial collapse and regrowth. Although the collapse may be attributed to the start of a necrotic process [4], there is no current explanation, either from mathematical models or biological data, to explain the subsequent spurt in growth, since in an in vitro growth angiogenesis or neo-vascularization [7] are not possible and the necrotic waste, which is a toxic product, is unlikely to trigger a spurt in growth. During avascular growth (benign tumor) or its in vitro simulation, the diffusion of nutrients attenuates in the central region as the tumor grows in size. Consequently, the cells in the central region grow slowly and some enter a quiescent phase, where they no longer proliferate. With further increase in tumor size, where nutrient diffusion into the central region is further reduced, some of the quiescent cells enter a necrotic phase and the tumors develop a central necrotic core [4]. As the necrotic contribution increases, the avascular tumors enter a dormant phase, striking a balance between the rates of mitosis and necrosis. Experimentally, it has been found that before entering the dormant phase, the cell count shows a great deal of fluctuations [4] mentioned above. This strong fluctuation, which has been described in [2, 3], may result from squeezing and relaxation of the tumor during its oscillations, and has never been addressed by any other theoretical model.

In previous theoretical modeling of tumors, the emphasis was given to the simulation of tumor morphology, immune response and estimation of the effect of administered drugs [8–16]. The simplest tumor model uses ordinary first-order differential equations in an empirical form and simulates cell number or volume growth by using various mitosis and necrosis rates and immunological reactions [8, 9]. Such models do not consider any of the physical interactions occurring in the cell aggregate. The second type of study employs spatio-temporal (diffusion-type) equations [10–12], which include cell mobility or kinetic factors, or use a cellular automaton model [13–15] to simulate the tumor structure by minimizing the energy of the system with each additional cell added to the aggregate. These calculations explicitly consider cell-cell interactions, and show that these interactions play a crucial role in determining the dormant size (for benign tumors), the morphology and the growth dynamics of a tumor [13–15]. However, these results do not reproduce the oscillatory nature of the in vitro proliferation dynamics. This points towards the presence of non-linear potentials in the cell aggregate and the necessity of treating the tumor as a *bound state* of cancer cells. As far as the non-linear effects are concerned, their effects have been studied elaborately in connection with immunological reactions in the growth profile of a tumor [16]. Apart from the non-linear cell-cell interaction, the behavior of the tumor's growth might be severely affected by the influence of the surrounding medium. In the non-linear matter wave model [2, 3] described below, both the non-linear cell-cell interactions and the effect of the external field is taken into account considering that the aggregate of cells in the tumor is in an energy minimized bound state arising out of their kinetic energy, interaction potential energy, and the potential energy from the external field of the surrounding medium and water.

Methodology

Single-particle state Gross-Pitaevskii (GP) equation and its dimensionless form

The time-evolution dynamics of a single-particle state bosonic condensate with feeding of atoms to the condensate from the thermal cloud and loss of atoms from the condensate through recombinations is described by the extended GP equation [17, 18]:

$$i\hbar\frac{\partial}{\partial t}\Phi = \left[-\frac{\hbar^2}{2m}\nabla^2 + \frac{4\pi\hbar^2 a}{m}|\Phi|^2 + V_{ext} + i\frac{\gamma}{2}\hbar\omega - 2i\xi\left(\frac{4\pi\hbar a}{m\omega}\right)^2\hbar\omega|\Phi|^4\right]\Phi \tag{1}$$

where the two-body atom-atom interaction is described by the s-wave scattering length a and is represented as: $v(r, r') = 4\pi\hbar^2 a/m \times \delta(r - r')$; where m is the mass of the atoms; δ is the Dirac delta function. For two-body attractive interactions, a is negative in sign, while for repulsive interactions the sign of a is positive. γ and ξ are parameters concerning feeding to and dissipation from the condensate. The term proportional to γ provides an exponential growth for the order parameter Φ and hence for the number density $|\Phi|^2$. The $|\Phi|^4$ term leads to a clustering of two atoms into a dimer formation from three-body interactions. This results in a loss of atoms from the condensate wave function Φ in the form of dimer.

$$N(t) = \int |\Phi(r,t)|^2 d^3 r \tag{2}$$

$$\langle R^2(t) \rangle = \frac{\int r^2 |\Phi(r,t)|^2 d^3 r}{\int |\Phi(r,t)|^2 d^3 r} \tag{3}$$

Dividing Eq. (1) by $\hbar\omega$ and introducing a dimensionless length: $x = (\sqrt{2}/l_0) r$ (where $l_0 = \sqrt{\hbar/m\omega}$), and time: $\tau = \omega t$, we recast Eq. (1) as

$$i\hbar \frac{\partial}{\partial\tau}\Phi = \left[-\nabla_x^2 + \eta|\Phi|^2 + \mathcal{V}(x) + i\frac{\gamma}{2} - 2i\xi\eta^2|\Phi|^4\right]\Phi \tag{4}$$

Here $V(x) = V_{ext}/\hbar\omega$ and $\eta = 4\pi a l_0^2$. Using a dimensionless order parameter: $\phi(x,\tau) = \sqrt{4\pi|a|} l_0 \Phi$ we arrive at the following complex wave equation [2, 3]:

$$i\frac{\partial}{\partial\tau}\phi(x,\tau) = \left[-\nabla_x^2 - |\phi|^2 + V(x) + i\gamma' - i\xi'|\phi|^4\right]\phi(x,\tau) \tag{5}$$

where ϕ is a complex matter field representing the atoms in the condensate; $\gamma' = \gamma/2$, $\xi' = 2\xi$, and the negative sign of the scattering length a (attractive interaction) is explicitly taken into account in Eq. (5).

Equivalence of GP equation with two coupled equations with real components

For application of the GP equation to a classical system we interpret ϕ as a two component field with real components ϕ_1 and ϕ_2 where $\phi = \phi_1 + i\phi_2$. In terms of these real components, one can rewrite Eq. (5) in the form of two coupled equations as follows:

$$\begin{aligned}\frac{d}{d\tau}\phi_1(x,\tau) = &-\nabla_x^2\phi_2(x,\tau) - |\phi|^2\phi_2(x,\tau)\\ &+ V(x)\phi_2(x,\tau) + \gamma'\phi_1(x,\tau)\\ &- \xi'|\phi|^4\phi_1(x,\tau)\end{aligned} \tag{6}$$

$$\begin{aligned}\frac{d}{d\tau}\phi_2(x,\tau) = &\nabla_x^2\phi_1(x,\tau) + |\phi|^2\phi_1(x,\tau) - V(x)\phi_1(x,\tau)\\ &+ \gamma'\phi_2(x,\tau) - \xi'|\phi|^4\phi_2(x,\tau)\end{aligned} \tag{7}$$

Rearranging, we rewrite them as:

$$\begin{aligned}\frac{d}{d\tau}\phi_1(x,\tau) = &\gamma'\phi_1(x,\tau) - \xi'|\phi|^4\phi_1(x,\tau)\\ &- \left[\nabla_x^2 - V(x) + |\phi|^2\right]\phi_2(x,\tau)\end{aligned} \tag{8}$$

$$\begin{aligned}\frac{d}{d\tau}\phi_2(x,\tau) = &\gamma'\phi_2(x,\tau) - \xi'|\phi|^4\phi_2(x,\tau)\\ &+ \left[\nabla_x^2 - V(x) + |\phi|^2\right]\phi_1(x,\tau)\end{aligned} \tag{9}$$

From Eqs. (8) and (9), we see that the first term of the right hand side (r.h.s) can give rise to the so-called first-generation tumor model, providing exponential growth with the growth rate determined by the parameter γ'. The second term on the r.h.s accounts for depletion in the cell number that resembles necrosis and in this model this necrotic contribution is taken as proportional to the square of number density ρ (where $\rho = |\phi|^2$ in dimensionless units mentioned above); the proportionality constant being the parameter ξ'. The rest are coupling terms concerning diffusion, external potential, and cell-cell interaction, respectively. The model, as it stands now, introduces a simplification in terms of having the same growth and depletion rates for both ϕ_1 and ϕ_2 components. This simplification allows us to combine the two equations into a complex field equation (Eq. 5), which is solved numerically. For simplicity of the numerical task, we restrict the application to in vitro growth of multicellular tumors in three dimensions, which reveal a spherical symmetry [4]. Assuming the external potential $V(x)$ to be spherically symmetric, as is the case for an in vitro tumor growth, and defining $\varphi = \varphi \times x$, we arrive at the following equation in one dimension:

$$i\frac{\partial}{\partial\tau}\varphi(x,\tau) = \left[-\frac{d}{dx^2} - \frac{|\varphi|^2}{x} + V(x) + i\gamma' - i\xi'\frac{|\varphi|^4}{x^4}\right]\varphi(x,\tau) \tag{10}$$

In terms of φ, the number of particles in the bound state and the mean-square radius of the bound state are given by:

$$N(t) = \int |\Phi(r,t)|^2 d^3r = \frac{4\pi}{8\pi|a|}\frac{l_0}{\sqrt{2}} \times \int |\varphi(x,\tau)|^2 dx = \frac{1}{2\sqrt{2}|a|} l_0 \times n(\tau) \quad (11)$$

$$\langle R^2(t)\rangle = \frac{\int r^2|\Phi(r,t)|^2 d^3r}{\int |\Phi(r,t)|^2 d^3r} = \frac{l_0^2}{2} \times \frac{\int x^2|\varphi(x,\tau)|^2 dx}{\int |\varphi(x,\tau)|^2 dx} = \frac{l_0^2}{2}\langle x^2(\tau)\rangle \quad (12)$$

where $n(\tau) = \int |\varphi(x,\tau)|^2 dx$.

Length scaling of the coupled matter wave equation

To simulate the tumor, we study a trial value of l_0 of the form: [2, 3, 19]

$$L_0 = l_0 \times \frac{\text{effective size of cell in tumor}}{\text{effective size of atom in condensate}} \quad (13)$$

Similar correlations between classical statistical mechanics and multicellular systems have been studied previously by Arlotti et al. [20] The length scaling redefines the cell number and the tumor radius as:

$$N(\tau) = \frac{1}{2\sqrt{2}|a|} L_0 \times n(\tau) = \left(\frac{l_0}{\sqrt{2}d}\right) \times \left(\frac{D}{2|a|}\right) \times n(\tau) : \left(\frac{l_0}{\sqrt{2}d}\right) \times n(\tau) \quad (14)$$

$$R(\tau) = \frac{L_0}{\sqrt{2}}\sqrt{\langle x^2(\tau)\rangle} = \left(\frac{l_0}{\sqrt{2}d}\right) \times D \times \sqrt{\langle x^2(\tau)\rangle} \quad (15)$$

The value of D, the effective size of a cell, is obtained from the measured data of [4] while the value of l_0/d is fixed by selecting an appropriate solution of the GP Eq. (5) for the short-range, attractively interacting, and externally trapped atomic condensate [17, 18].

Time-independent form of the coupled matter wave equation

Considering $V(x) = V_{ext}/h\omega$ we write down the stationary form of Eq. (10) as:

$$\left[-\frac{d}{dx^2} - \frac{|\psi|}{x} + V(x)\right]\psi = \beta\psi \quad (16)$$

where $\beta = \mu/\omega$; μ representing the chemical potential (average single-particle energy). This Eq. (16) has various stable, metastable, or unstable solutions [21] correspond-

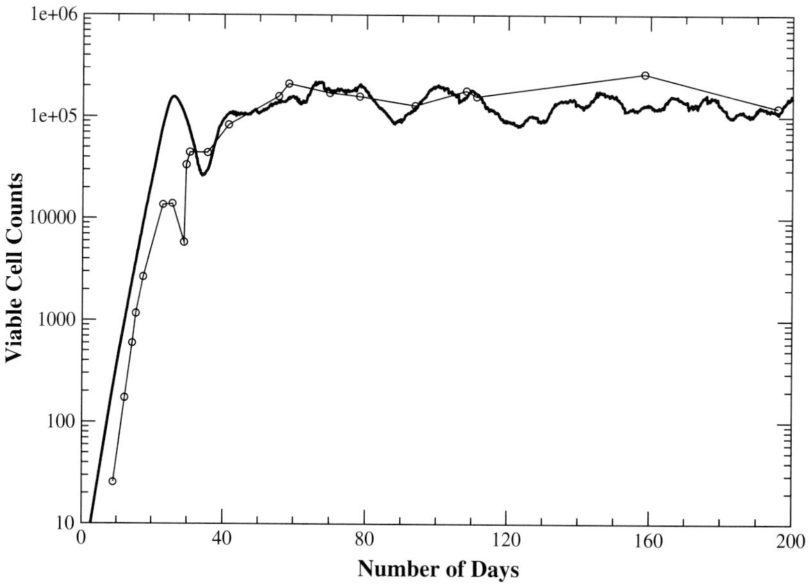

Fig. 1 In vitro growth profile of a benign tumor. Culture data of Folkman and Hochberg [4] for the multicellular spheroids developed in agar with V79 Chinese hamster lung cells (*circles*); simulation results as obtained from the coupled non-linear matter wave equations (*thick solid line*)

ing to various values of β. Each solution corresponds to a particular value of $N|a|/l_0$ (see Eq. 11).

Numerical methods

The time-independent Eq. (16) is solved numerically using the Runge-Kutta method [22] as described in the work of Edward and Burnett [23], and Gammal et al. [24] and employed in previous calculations [2, 3]. This equation has various solutions depending on the number of particles [21, 24]. We choose a minimum-energy solution with only a few particles and then propagate the solution in time by feeding it to Eq. (5). The time-dependent Eq. (5) is discretized using the Crank-Nicholson difference [22] and the resulting set of tri-diagonal equations are solved using the Gaussian elimination method [22] as implemented in [2, 3].

Results and discussions

In this work, we focus on the simulation of the experimental data for the growth of multicellular spheroids that developed when V79 Chinese hamster lung cells are

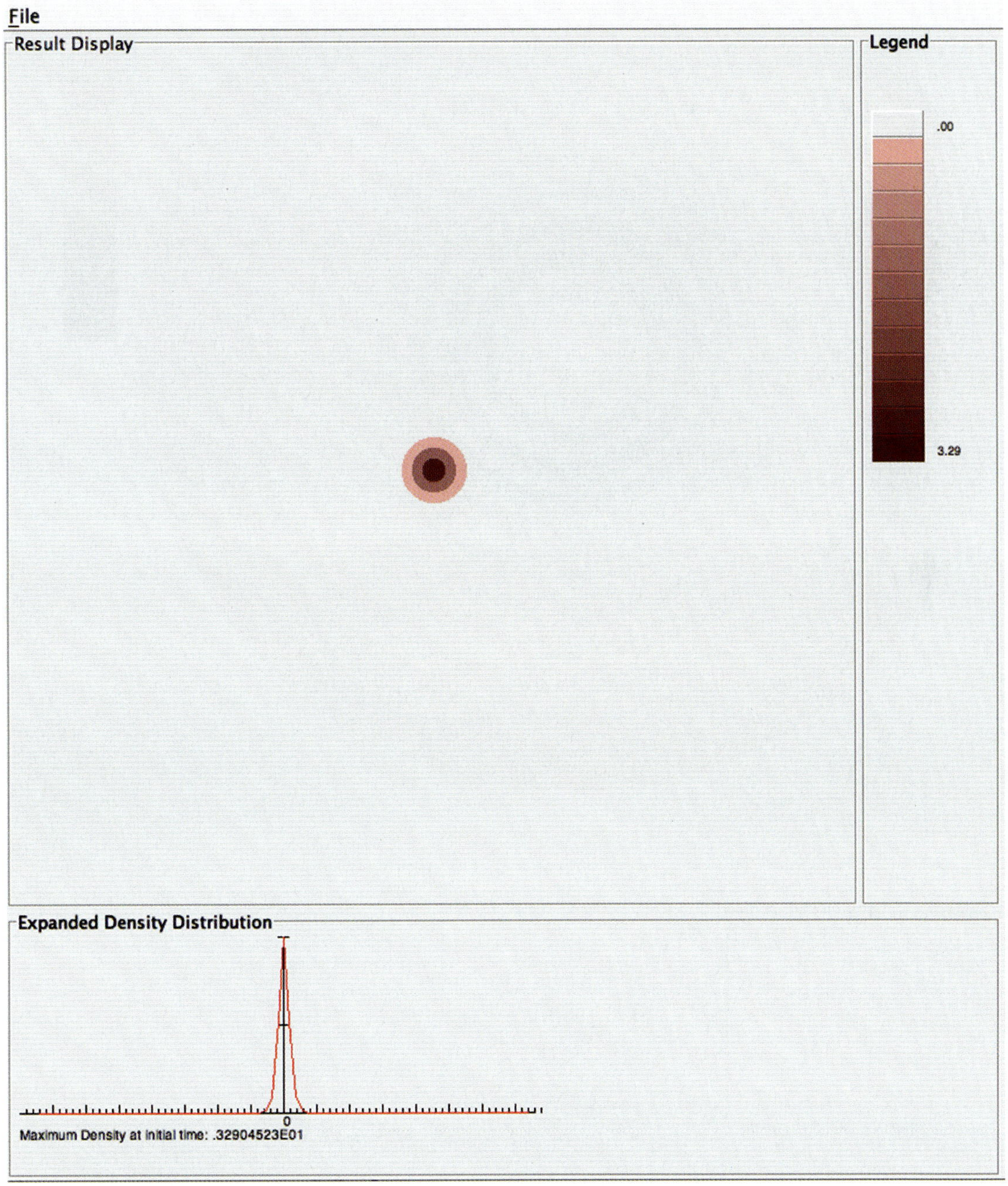

Fig. 2 Screen shot from the graphical analysis tool showing clustering of cells in the initial tumor (*up*); cell density profile (*down*)

grown in soft-agar. We approximate the hydration effect of agar on the cell aggregate as a constant gauge potential experienced by the cells. We model it by $V(x) = V_0$. Firstly, we solve the time-independent form of the single-particle state Gross-Pitaevskii Eq. (16) for different bound-state solutions as a function of various sets of values for V_0 and β (the constant gauge potential can be absorbed in the redefinition of β). Despite the constant external gauge potential, the stationary solution is stable due to the attractive nature of the cell-cell interactions. Then, we propagate each solution in time, employing Eq. (5) and plot the number of cells (as obtained from Eq. 14) against time. In Fig. 1, the experimental result for the V79 cell colony and the theoretical results for the proliferating cell count obtained with V_0=1.5, γ=0.5, and ξ=0.001 are shown. It can be seen that there is significant agreement between the two

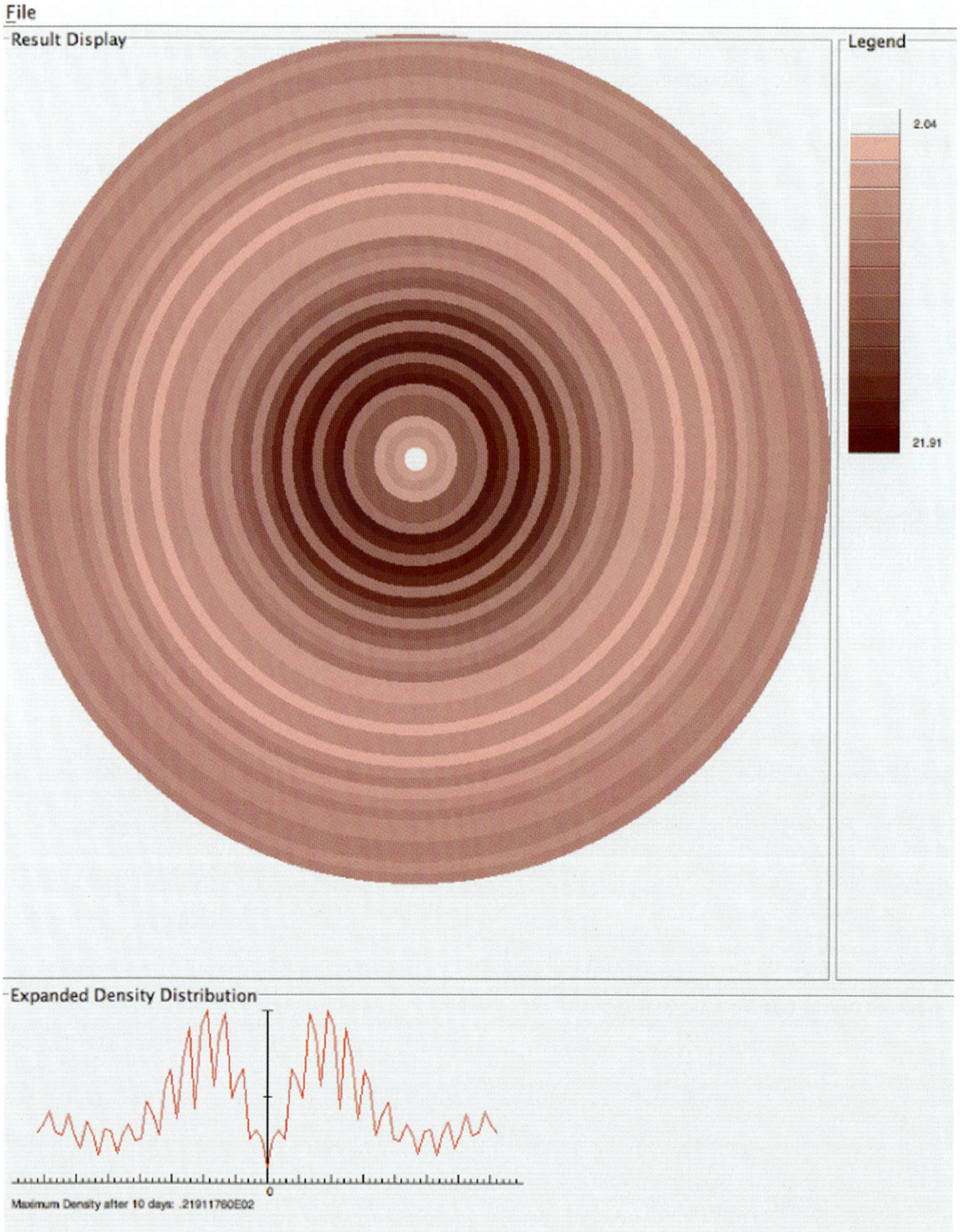

Fig. 3 Screen shot from the graphical analysis tool showing simulation results after 10 days: clustering of cells of the tumor (*up*); cell density profile (*down*)

sets of results once the time scale ($t = \tau/\omega$) is chosen such that ω corresponds to a value $\omega \sim 1$ per day, that is $\omega = 1/(24 \times 60 \times 60)\,\text{sec}^{-1}$.

From the measured proliferation count (solid curve with circles; Fig. 1), we find that 1) initially the number of cells increases in a steep exponential fashion; 2) it attains a local peak at around day-23; 3) next it suffers a partial collapse; 4) after completing the collapse, it again starts a strong exponential growth at around day-29. The second experimental growth phase had a short duration and gradually stabilizes with a viable cell count around $N \sim 10^5$. From the thick solid curve of Fig. 1, we see that all the characteristic features of tumor growth described above are present in the theoretical results, including the approximate positions of the peaks, collapse, and re-growth.

To understand the collapse and re-growth phenomena described above, we study the density profile of the cell aggregate. We find that as the cell number increases, the central density (in units of $|\varphi(x,t)/x|^2$) also increases rapidly, leading to a greater contribution to the necrotic process. When the central density reaches a maximum of 10^4 near day-24, the increased attractive potential triggers a visible partial collapse of the cell aggregate; with the central region being affected the most. In the process of collapse, there appears an increase in the clustering of cells and consequent necrotic contribution leading to the formation of a necrotic core [4, 13]. The contribution to the necrotic core diminishes the attractive potential in tumor (in its bound state), which results in a volume expansion of the tumor. As the volume expands, the cell-density decreases, which in turn paves the way for a greater nutrient diffusion. A greater nutrient diffusion might be responsible for the second exponential growth, which according to the model is possible because of an oscillation of tumor volume. While the model explicitly reveals the oscillation in tumor size, the impetus in the growth arises in part from a sharp decrease in the necrotic contribution and also in part from the diffusion term (although, the diffusion term here represents cell diffusion).

To achieve a more critical understanding of the cell proliferation kinetics, we have developed a graphical analysis tool to study the tumor growth. Figure 2 represents the initial tumor as given by the solution of the time-independent Eq. (16). From the plot of the density distribution of the cells we find that initially the cell growth and proliferation occur mostly at the center, leading the cell density to be the highest at the center. This eventually leads to necrosis in this central region, as the higher cell density retards the nutrient diffusion. In this model, the above scenario has been captured by introducing a term in the equation that provides attenuation of proliferating cells as a function of $(density)^2$. From Fig. 3, we see that this term leads to the modeling of a central necrotic core. Also from the diagram, we find that the central necrotic core is surrounded by densely proliferating cells and the cell density resembles an onion-type structure. The graphical analysis enables us to visualize and understand the density profile of the proliferating cells, which is crucial information in order to optimize the effect of drugs. Our current work is addressing this latter issue.

Summary and conclusions

We present here an application of the non-linear Schrödinger equation to the cell aggregate of a multicellular tumor considering it as a coupling of two wave equations with real components. The simulation results reproduce the experimentally observed oscillatory pattern in the growth profile of a tumor and explain the spurt in cell growth in the aftermath of a partial collapse in terms of a volume expansion and resulting increment in the nutrient diffusion due to a lesser attractive potential of the cell aggregate. The study indicates that the necrotic contribution during the growth of a tumor could be preceded by a small localized clustering of cells, thus reducing the probability of nutrient diffusion inside the cluster. The graphical analysis described here provided us with invaluable information about the time evolution of the density profile of the proliferating cells, which is essential in developing effective strategies for administering drugs.

Acknowledgements Authors thank Drs. Crystal Weyman and Joseph Fontes for reading the manuscript and providing their valuable suggestions. We acknowledge the financial support from the Department of Energy (grant number: DE-FG02-03ER15462). TF acknowledges the partial financial support from FAPESP and CNPq of Brazil.

References

1. Lodish H, Berk A, Matsudaira P, Kaiser CA, Krieger M, Scott MP, Lawrence Zipursky S, Darnell J (2004) Molecular Cell Biology, 5th edn. Freeman WH and Company, New York, p 941
2. Biswas PK, Pacheco MTT (2005) Math comp modeling 41:1299–1306
3. Biswas PK, Germano JSE (2004) Int J Nonl Sc Numerical Siml 5:149–156
4. Folkman J, Hochberg M (1973) J Expt Medicine 138:745–753
5. Sutherland RM, McCredie JA, Inch WR (1971) J Natl Cancer Inst 46:113–120
6. Dalen H, Burki HJ (1971) Exp Cell Res 65:433–438
7. Folkman J (1971) New England J of Medicine 285:1182–1186
8. Miklavcic D, Jarm T, Karba R, Sersa G (1995) Math Comp Simulation 39:597–602 and references therein
9. Kuznetsov VA, Knott GD (2001) Math Comp Modeling 33:1275–1287 and references therein
10. Tracqui P, Cruywagen GC, Woodward DE, Bartoo GT, Murray JD, Alvord EC Jr (1995) Cell Prolif 28:17–31
11. Burgess PK, Kulesa PM, Murray JD, Alvord EC Jr (1997) J Neuropath Expt Neur 56:704–713
12. Maggelakis SA, Adam JA (1990) Math Comp Model 13:23–38
13. Stott EL, Britton NF, Glazier JA, Zajac M (1999) Math Comp Modeling 30:183–198
14. Dormann S, Deutsch A (2002) In Silico Biology 2:393–406
15. Mombach JCM (1993) Phys Rev E 48:598–602
16. Bellomo N, Firmani B, Guerri L (1999) Appl Math Letters 12:39–44

17. Kagan Yu, Muryshev AE, Shlyapnikov GV (1998) Phys Rev Lett 81:933–937
18. Filho VS, Gammal A, Frederico T, Tomio L (2000) Phys Rev A 62:033605/1–033605/4
19. Biswas PK, Pacheco MTT (2003) Los Alamos National Laboratory, Preprint Archive, Condensed Matter. 1–13, arXiv: cond-mat/0306539
20. Arlotti L, Bellomo N, de Angelis E (2002) Math Models Meth Appl Sci 12:567–591
21. Gammal A, Frederico T, Tomio L, Chomaz Ph (2000) J Phys B 33:4053–4067
22. Press WH, Flannery BP, Teukolsky SA, Vetterling WT (1996) Numerical recipes in fortran 77—the art of scientific computing, 2nd edn. Cambridge University Press, p 42, pp 840–844
23. Edwards M, Burnett K (1995) Phys Rev A 51:1382–1386
24. Gammal A, Frederico T, Tomio L (1999) Phys Rev E 60:2421–2424

J Mol Model (2006) 12: 649–652
DOI 10.1007/s00894-005-0044-4

ORIGINAL PAPER

Roger W. Alder · Anthony P. Davis

The design of organic catalysis for epoxidation by hydrogen peroxide

Received: 29 April 2005 / Accepted: 30 September 2005 / Published online: 23 December 2005

Abstract The potential of various organic species to catalyze epoxidation of ethene by hydrogen peroxide is explored with B3LYP/6-31G* DFT calculations.

Keywords Catalysis · Epoxidation · Oxidation · Peroxides

Introduction

Hydrogen peroxide, 1, is a cheap and environmentally-friendly oxidant, since the only by-product is water, but it is not reactive enough for many applications. Catalysts to promote the epoxidation of alkenes, oxidation of sulfides, etc., by hydrogen peroxide are potentially of great value. While there are many examples of promotion of H_2O_2 reactions by transition metal ions, in this paper we discuss a different approach, based on the following simple idea. Water oxide [1], 2, is unknown, but would be expected to be an extremely powerful oxidant. Although water oxide is never likely to be involved as a reaction intermediate, a compound that could remove a proton from one oxygen of hydrogen peroxide as it attacked a substrate, while more or less simultaneously donating a proton to the other oxygen, thus creating a water molecule, could be a powerful catalyst for reactions of H_2O_2, as shown for a hypothetical RC(X)YH catalyst in 3 below.

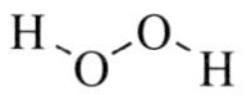

1

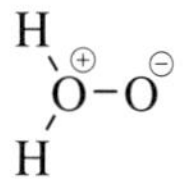

2

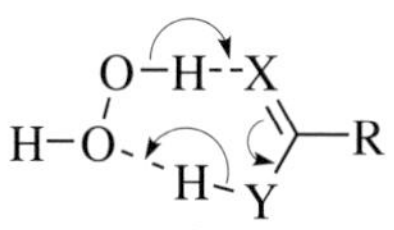

3

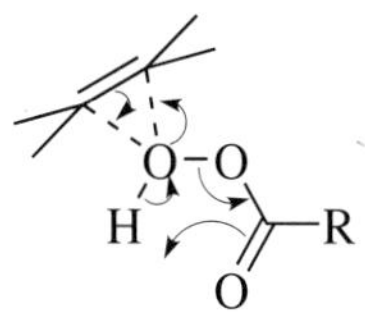

4

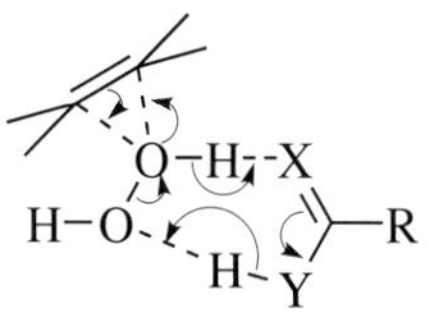

5

Traditionally, alkene epoxidation has been carried out with peroxycarboxylic acids, although many other methods have been developed in recent years, especially for enantioselective epoxidation. Epoxidation with peroxycarboxylic acids is not very sensitive to solvent polarity and the accepted "butterfly" mechanism [2, 3] is concerted and involves a complex series of electron shifts [4, 5], as shown in 4 above. In this paper the process shown in 5 above for epoxidation of ethene is modeled, and the energetics are compared with those calculated for 4. The main objective is to define the best choice of proton transfer catalyst, but we also explore the potential for further hydrogen-bonding assistance, with a view to the eventual design of hosts that might catalyze hydrogen peroxide reactions with control of stereochemistry. Since the dielectric properties inside the host cannot be predicted, we have restricted our study to the gas phase, and have not attempted to model environmental effects.

Computational methods

All DFT calculations were performed with the Jaguar program package [6], using Becke's three-parameter exchange functional [7] with the correlation functional of Lee, Yang and Parr (B3LYP) [8]. All species were characterized by full

Dedicated to Professor Dr. Paul von Ragué Schleyer on the occasion of his 75th birthday

Electronic Supplementary Materials Supplementary material is available for this article at http://dx.doi.org/10.1007/s00894-005-0044-4.

R. W. Alder (✉) · A. P. Davis
School of Chemistry, University of Bristol,
Cantock's Close,
BS8 1TS, Bristol, UK
e-mail: rog.alder@bris.ac.uk
Tel.: +44-117-9287657
Fax: +44-117-9298611

geometry optimization with the standard 6-31G* basis set, and minima and transition states were characterized by analytical frequency calculations. The single imaginary frequency of the transition states was animated in the Spartan '04 program [9], and it was confirmed that this vibration involved the stretching of the O–O bond with one oxygen moving more or less symmetrically towards the alkene-carbon atoms, and with one hydrogen moving towards the putative water molecule. IRC calculations were carried out in Spartan '04 for the reactions catalyzed by one and two formic acid molecules, and these confirmed that the transition states reported below do evolve to the starting complexes and epoxide products.

Protein Database (PDB) files for the species illustrated in the Figures are available as Electronic Supplementary Material.

Table 1 Activation parameters for epoxidation of ethene from B3LYP/6-31G* calculations

Peroxy acid or catalyst+H_2O_2	ΔH‡ (kJ mol^{-1})	ΔG‡ (kJ mol^{-1})	PA of conjugate anion (kJ mol^{-1})
CF_3CO_3H	44.6	84.4	
HCO_3H	59.4	101.8	
CF_3CO_2H+H_2O_2	60.7	109.5	1,481
HCO_2H+H_2O_2	73.9	120.7	1,393
$HCONH_2$+H_2O_2	103.3	151.9	
Formamidine+H_2O_2	109.9	157.6	
CH_3SO_3H+H_2O_2	56.9	111.8	
7 R^1=H R^2=CN+H_2O_2	98.7		1,424
7 R^1=CN R^2=H+H_2O_2	83.8		1,343
8 R^1=R^2=H+H_2O_2	83.3		1,472
8 R^1=H R^2=CN+H_2O_2	48.4		1,245

Results and discussion

What form should the proton acceptor/proton donor catalyst take? Optimization of a complex of formic acid with hydrogen peroxide leads to the structure shown in Fig. 1a. The two hydrogen bonds are close to linear and of comparable length. Thus, the RC(X)YH species shown in 3 and 5 above appears to be a good starting point for catalysis design. Some alternative geometries for the catalyst are explored later.

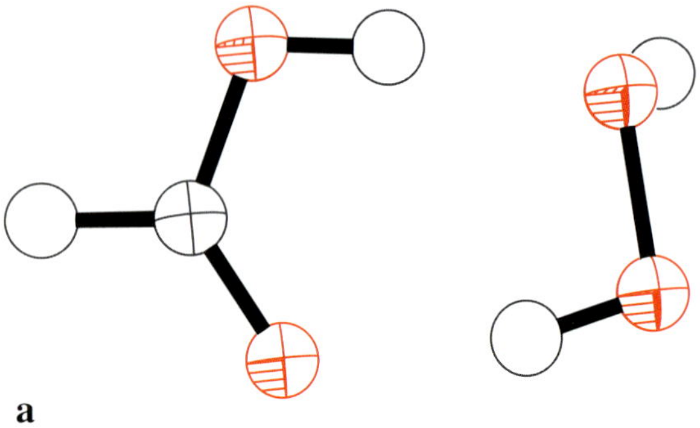

a

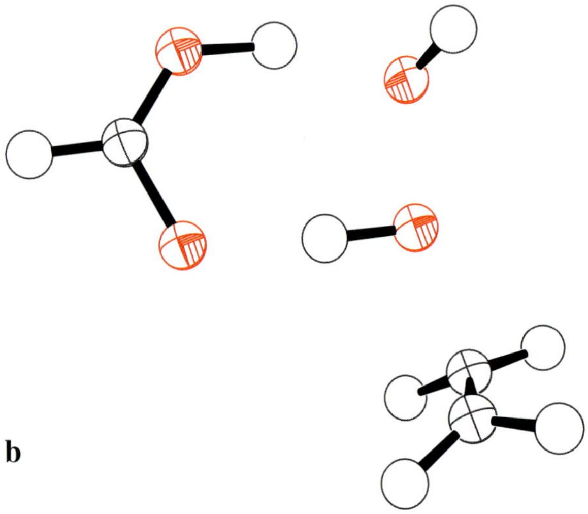

b

Fig. 1 **a** Hydrogen peroxide complex with formic acid; **b** Transition state for epoxidation of ethene with H_2O_2 catalyzed by HCO_2H

Transition states have been located at the B3LYP/6-31G* level for the epoxidation of ethene by hydrogen peroxide catalyzed by various HC(X)YH species. Activation parameters calculated relative to ethene and an optimized complex of hydrogen peroxide with the catalyst are given in Table 1 and the transition state for the case of formic acid is shown in Fig. 1b. As can be seen, the latent epoxide ring is more or less perpendicular to the general plane containing the hydrogen peroxide and formic acid molecules; this type of transition-state geometry has been described as "spiro", as opposed to "planar".

A priori, it is not obvious whether it is desirable to have a strongly basic X-group to promote deprotonation of the epoxidizing oxygen atom or to go for a strongly acidic Y–H group to protonate the developing hydroxide ion. However, it can be seen from Table 1 that more acidic species are better catalysts, although methanesulfonic acid is not a significant improvement on trifluoroacetic acid. Proton donation to hydrogen peroxide is clearly more important than proton removal in the transition state. This is reasonable on the grounds that normal alkenes like ethene are attacked by electrophilic reagents. Full proton transfer would create $H_3O_2^+$ [10–12], a powerfully electrophilic species formed in solutions of hydrogen peroxide in very strong acids that is known to be an extremely powerful (but very indiscriminate) oxidant [10–14].

Further evidence supporting the importance of proton transfer to the non-epoxidizing oxygen atom may be gleaned from examination of the geometry of the transition states. Thus, in the transition state for epoxidation catalyzed by formic acid, the OH bond in formic acid has lengthened to 1.13 Å, compared with 1.00 Å in the initial complex, and the distance from this proton to the oxygen atom in H_2O_2 has shortened from 1.77 to 1.32 Å. Furthermore, animation of the single imaginary frequency shows that motion of this proton is strongly coupled to the O–O stretching process. On the other hand, proton loss from the epoxidizing oxygen atom of the hydrogen peroxide has hardly begun at the transition state, since the OH bond has only stretched by 0.02 Å, although the distance from the proton to the carbonyl group of formic acid has shortened from 1.81 to 1.63 Å.

In order to provide some calibration of the activation parameters in Table 1, the transition state for epoxidation of ethene by peroxyformic acid was optimized using Jaguar. The accepted [15] $\Delta H^{\ddagger}$ value was reproduced and $\Delta G;^{\ddagger}$ was found to be 101.8 kJ mol^{-1} (see Table 1). This transition state has been the focus of much work [16, 17], with lively debate concerning spiro *vs* planar transition state geometries. As a result of high level ab initio and CASSCF calculations [15], it now appears that the spiro geometry is preferred, and that the economical B3LYP DFT method is a useful and reasonably accurate method that can be applied to studies of larger systems. The corresponding epoxidation of ethene with peroxytrifluoroacetic acid (see Table 1) resulted in lowered activation parameters, in qualitative agreement with the known high reactivity of CF_3CO_3H.

Comparison of the activation parameters for the mechanism shown in 5 (Table 1) with those for epoxidation *via* 4 with the corresponding peroxyacids, shows that the catalyzed process to be poorer by about 15 kJ mol^{-1} in $\Delta H^{\ddagger}$. The catalyzed process 5 with carboxylic acids as catalysts is clearly a reasonable possibility, worthy of further optimization. When carboxylic acids are mixed with hydrogen peroxide, an equilibrium is set up involving formation of the peroxycarboxylic acid. Equilibration is quite slow, however, taking about 1 h at ambient temperature when 30% H_2O_2 is added to a large excess of 88% HCO_2H [18, 19]. Since this reaction almost certainly involves nucleophilic attack at the carbonyl group of the carboxylic acid, equilibration is much slower for acetic and other simple carboxylic acids [20], although it could be faster for strong acids like CF_3CO_2H. The slow rates for this equilibration rule out its incorporation into a viable catalytic cycle. Most discussions of the promotion of hydrogen peroxide reactions by acids concentrate on the formation of new peroxidic species [21], rather than the proton transfer process proposed here, although there have been several recent reports of electrophilic activation of hydrogen peroxide by perfluorinated alcohols [22–24] and phenol [25]. In the absence of strong acids, it is likely that this activation involves clusters of strongly hydrogen-bonded solvent molecules, as suggested by Berkessel for catalysis by HFIP (1,1,1,3,3,3-hexafluoro-2-propanol) [24].

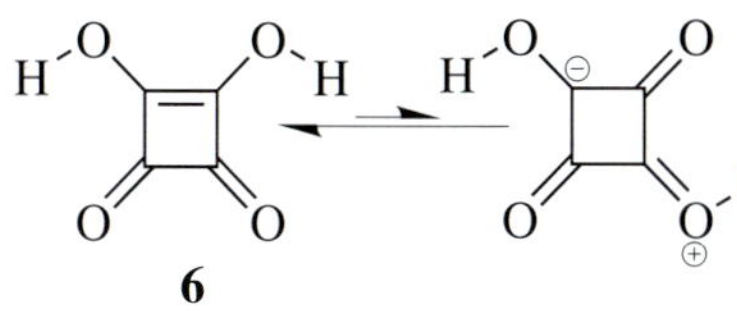

6

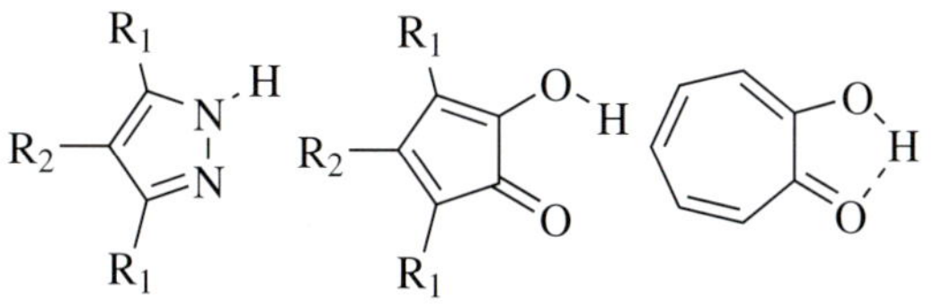

7 **8** **9**

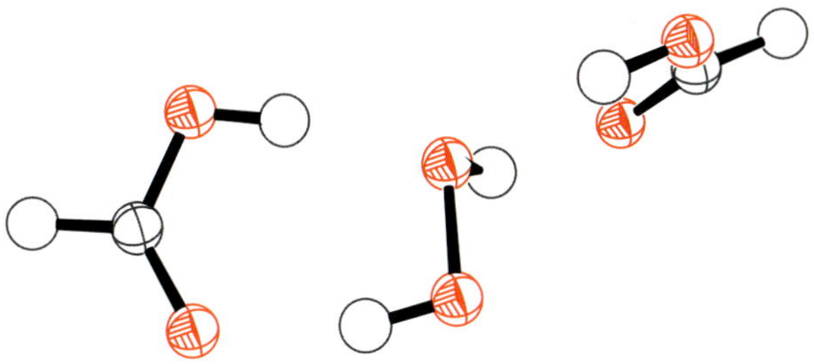

a

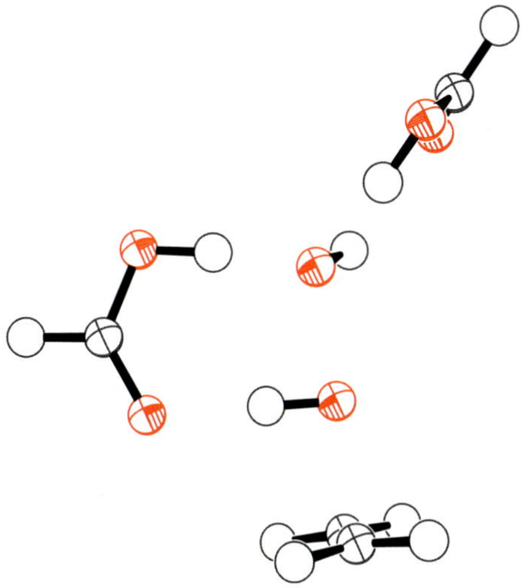

b

Fig. 2 **a** Hydrogen peroxide complexed to two formic acid molecules; **b** Transition state for epoxidation of ethane by hydrogen peroxide catalyzed by two formic acid molecules. The "primary" acid which takes part in proton transfer appears at the *left* while the "secondary" acid which contributes through static H-bonding appears at the *upper right*

While the RC(X)YH species shown in 3 and 5 above appears to be a good starting point for catalysis design, it seemed desirable to see whether separating the proton-donating and proton-accepting groups by more (or fewer) atoms would offer any significant advantages. For a reliable comparison with carboxylic acids, the catalyst must be symmetrical with respect to the proton exchange. Thus, squaric acid derivatives, 6, are unsatisfactory, although they might be attractive in other respects. Derivatives of pyrazole 7 and of 2-hydroxycyclopentadienone 8 were selected for comparison with carboxylic acids. Tropolone 9 is of course much more stable and readily available than 9 but it forms a

Table 2 Epoxidation of ethene by H_2O_2 catalyzed by two acids; see Fig. 2 for the definition of primary and secondary

Primary acid	Secondary acid	$\Delta H\ddagger$ (kJ mol^{-1})	$\Delta G\ddagger$ (kJ mol^{-1})
HCO_2H	HCO_2H	67.67	109.79
HCO_2H	CF_3CO_2H	56.96	111.13
CF_3CO_2H	HCO_2H	59.57	113.41
CF_3CO_2H	CF_3CO_2H	56.05	101.79

strong intramolecular H-bond, as do enolized β-diketones, and so these are unlikely to be good catalysts. In view of the clear importance of acidity for catalysis, we sought derivatives whose anions were of comparable proton affinity to formate and trifluoroacetate. The results are summarized in Table 1.

While 8, R^1=H, R^2=CN yields the lowest value for $\Delta H^\ddagger$ so far, this is due to its extreme acidity, rather than any superiority derived from having two carbons separating the proton donating and proton accepting groups. This conclusion is backed up by examination of the transition state with 8, R^1=H, R^2=CN as catalyst; the proton transfer from the catalyst is more or less complete, so this is almost a reaction of $H_3O_2^+$. Comparison based on relative PA-values shows that the carboxylic acids are somewhat superior to the 2-hydroxycyclopentadienone derivatives with the pyrazole derivatives a very poor third. Thus, the original catalyst design appears to be the best choice.

Extra hydrogen bonding: towards catalytic hosts for H_2O_2

The water molecule that is created in the catalyzed process should be stabilized by the provision of further hydrogen bonding, especially by proton donation to the oxygen. This is a likely component of the catalysis by perfluorinated alcohols and phenol cited above. Since our eventual aim is the design of hosts for H_2O_2 which might catalyze its reactions and ultimately provide steric control of the stereochemistry of epoxidation, we sought extra stabilization of the transition state associated with a well-defined geometry.

A second formic acid was therefore positioned to take up this role (the formic acid that carries out the proton transfer role will be called the primary acid). First, we optimized hydrogen peroxide complexed to two formic acid molecules This structure is shown in Fig. 2a. Ethene was then introduced, and a transition state located *via* QST-guided calculation in Jaguar, Fig. 2b. The activation parameters, $\Delta H^\ddagger$ 67.7 and $\Delta G^\ddagger$ 109.8 kJ mol^{-1} relative to ethene and the complex of hydrogen peroxide with two formic acid molecules, represent a significant improvement on the process catalyzed by a single formic acid, suggesting that hosts based on this design could form the basis of a successful process. In order to explore the optimum acidity properties of the two carboxylic acids, trifluoromethyl substituents were introduced in both positions. The results are summarized in Table 2.

As expected, two trifluoroacetic acids provide the best catalysis, but it is surprising that $\Delta H^\ddagger$ is lower when a single trifluoroacetic acid is used in the secondary role, rather than the primary role. A possible explanation of this is that the primary acid molecule must act as both proton donor and acceptor, while that in the secondary role is a pure proton donor, since the hydrogen bond from the incipient water molecule to the carbonyl group of this acid is very long and weak.

Conclusions and implications for host design

In the $H_2O_2 \cdot 2HCO_2H$ complex, the carbonyl carbons are separated by 6.30 Å, and the two H–C bonds subtend an angle of 160°. These parameters remain fairly constant as CF_3 groups are introduced. In the transition state for epoxidation of ethene in the presence of two formic acids, the separation of the carbonyl carbons is reduced to 5.97 Å, and two H–C bonds subtend a smaller angle of 115°. As CF_3 groups are introduced, this angle remains fairly constant, but the C…C distance decreases significantly, to 5.69 Å in the transition state involving two trifluoroacetic acids.

These geometrical parameters provide a useful guide to the design of potential hosts; the separation of around 6 Å requires quite a large host, but not so big as to be outside the scope of reliable calculations and further exploration is underway. Clearly it would be desirable to design a host which was somewhat strained as the H_2O_2 complex, but where this strain would be relieved as the transition state was attained.

References

1. Oiestad EL, Harvey JN, Uggerud E (2000) J Phys Chem A 104: 8382–8388
2. Bartlett PD (1950) Record of Chemical Progress 11:47–51
3. Bartlett PD (1957) Record of Chemical Progress 18:111
4. Herges R (1994) J Chem Inf Comput Sci 34:91–102
5. Herges R (1994) Angew Chem-Int Edit Engl 33:255–276
6. Jaguar (2002) Jaguar 4.2 edn. Schrödinger Inc, Portland OR, USA
7. Becke ADJ (1993) Chem Phys 98:5648–5652
8. Lee CT, Yang WT, Parr RG (1988) Phys Rev B 37:785–789
9. Spartan '04 ed (2004) Wavefunction Inc, Irvine CA, USA
10. Alder RW, Whiting MC (1964) J Chem Soc 4707–4712
11. Christe KO, Wilson WW, Curtis EC (1979) Inorg Chem 18: 2578–2586
12. Olah GA, Berrier AL, Prakash GKS (1982) J Am Chem Soc 104:2373–2376
13. Olah GA, Yoneda N, Parker DG (1977) J Am Chem Soc 99:483–488
14. Olah GA, Parker DG, Yoneda N (1978) Angew Chem-Int Edit Engl 17:909–931
15. Bach RD, Glukhovtsev MN, Gonzalez C (1998) J Am Chem Soc 120:9902–9910
16. Houk KN, Liu J, DeMello NC, Condroski KR (1997) J Am Chem Soc 119:10147–10152
17. Freccero M, Gandolfi R, Sarzi-Amade M, Rastelli A (2004) J Org Chem 69:7479–7485
18. Toennies G, Homiller RP (1942) J Am Chem Soc 64:3054–3056
19. Monger JM, Redlich O (1956) J Phys Chem 60:797–799
20. Swern D (1949) Chem Rev 45:1–68
21. Vayssie S, Elias H (1997) Liebigs Ann-Recl 2567–2572
22. Neimann K, Neumann R (2000) Org Lett 2:2861–2863
23. Berkessel A, Andreae MRM (2001) Tetrahedron Lett 42:2293–2295
24. Berkessel A, Adrio JA (2004) Adv Synth Catal 346:275–280
25. Wahlen J, de Vos DE, Jacobs PA (2003) Org Lett 5:1777–1780

J Mol Model (2006) 12: 653–664
DOI 10.1007/s00894-006-0111-5

ORIGINAL PAPER

Christopher A. Brown · Michael L. McKee

Rearrangements in icosahedral boranes and carboranes revisited

Received: 29 April 2005 / Accepted: 27 July 2005 / Published online: 30 June 2006

Abstract The structure, stability, and intermolecular rearrangements between *ortho*-, *meta*-, and *para*-$C_2B_{10}H_{12}$ and $B_{12}H_{12}{}^{2-}$ were investigated using the hybrid density functional B3LYP/6-31G(d) for vibrational frequencies, as well as B3LYP/6-311+G(2d,p) for single-point electronic energies. The general trends in free energies of rearrangement between *ortho*-$C_2B_{10}H_{12}$ to *meta*-$C_2B_{10}H_{12}$ and *meta*-$C_2B_{10}H_{12}$ to *para*-$C_2B_{10}H_{12}$ presented here are consistent with experimental reaction temperatures. In addition, the majority of the rearrangements can be viewed in terms of concerted diamond–square–diamond steps and triangular face rotations.

Keywords Deltahedron · Carboranes · Icosahedral · Triangular face rotation · Diamond–square–diamond

Introduction

The roots of modern boron cluster chemistry originated in a few seminal works by Wade [1, 2], Hart and Lipscomb [3–6], and Williams [7–9]. These compounds, which are characterized by their electron deficiencies, form non-classical three-center two-electron bonds and have been the topic of numerous texts [10–14]. This leads to the very familiar deltahedron geometries for *closo*-$C_xB_{n-x}H_{n-x+2}$ with *n*+1 skeletal electron pairs, *nido*-$C_xB_{n-x}H_{n-x+4}$ with *n*+2 skeletal electron pairs, and *arachno*-$C_xB_{n-x}H_{n-x+6}$ with *n*+3 skeletal electron pairs, determined empirically by Wade's Rules [1, 2, 15, 16]. Although a great deal of time has been spent in comparing and contrasting carborane systems of varying size in terms of stability, rearrangement, and reactivity [7, 8, 17–23], a special effort has been expended in understanding *closo*-$B_{12}H_{12}{}^{2-}$, *closo*-$C_2B_{10}H_{12}$, and their respective derivatives [21, 24–29], which is due primarily to their unusually high stability. An eclectic array of mechanisms have been quite naturally suggested for the intermolecular rearrangement of *closo*-$C_2B_{10}H_{12}$, from those involving highly symmetric cubeoctahedral and anticubeoctahedral transition states to those with relatively simple triangular face rotations (TFRs) and diamond–square–diamond (DSD) steps [4, 5, 20, 25, 30–36]. Rearrangements in substituted icosahedral $C_2B_{10}H_{12}$ have been studied experimentally by way of ^{10}B-labeling, halogen substitution, methyl substitution, and even tethered linkages [3, 24, 28, 36, 37] in the hopes that the electronic structure of the cage will be preserved.

Despite previous theoretical work on the stability of large carboranes and their derivatives and the aforementioned experimental work on intermolecular rearrangements in *closo*-$C_2B_{10}H_{12}$, there has been relatively little rigorous computational work in verifying whether any of the proposed mechanisms play a role in these intermolecular rearrangement reaction pathways. The present work confirms the presence of a TFR pathway in both the carborane and the borane case.[1] Another pathway, which consists of concerted DSD steps and the opening of the cage into a pseudo-*nido* intermediate, is also described for both cases.

Methodology

The following calculations were done using the Gaussian03 program [38]. Species were optimized and vibrational frequencies were calculated at the B3LYP/6-31G(d) level, leading to their characterization as stationary points. In

[1] From this point on, carborane will be used synonymously with icosahedral $C_2B_{10}H_{12}$ and borane with icosahedral $B_{12}H_{12}{}^{2-}$.

Dedicated to Professor Dr. Paul von Ragué Schleyer on the occasion of his 75th birthday.

This article has already been published online first (DOI: http://dx.doi.org/10.1007/s00894-005-0039-1). Due to an oversight, this version contained several mistakes. The article is herewith republished in its entirety as an "erratum".

Electronic Supplementary Material Electronic Supplementary Material found in http://dx.doi.org/10.1007/s00894-006-0111-5

C. A. Brown · M. L. McKee (✉)
Department of Chemistry and Biochemistry,
Auburn University,
179 Chemistry Building,
Auburn, AL 36849, USA
e-mail: mckeeml@auburn.edu
e-mail: cabrown@fas.harvard.edu

addition, zero-point energies, integrated specific heat corrections at constant pressure, and entropies were calculated at this level. Once a continuous reaction pathway was determined, single-point electronic energies were calculated for the pertinent species at the B3LYP/6-311+G(2d,p) level. The enthalpies for the species at 500 and 800 K were approximated to be the same as the enthalpy at 298 K. These values were then used to determine the absolute free energy at 298, 500, and 800 K. The change in free energies at 298, 500, and 800 K for the species in the carborane rearrangement are relative to *ortho*-$C_2B_{10}H_{12}$ and, similarly, the species in the borane rearrangement are relative to $B_{12}H_{12}{}^{2-}$.

Results and discussion

The reaction pathways between *ortho*-, *meta*-, and *para*-$C_2B_{10}H_{12}$, and $B_{12}H_{12}{}^{2-}$ are shown in Figs. 1 and 2, with representative geometries in Figs. 3 and 4, respectively. At first glance, some common features between Figs. 1 and 2 should be apparent. The arrangement of the vertices and edges for both $C_2B_{10}H_{12}$ and $B_{12}H_{12}{}^{2-}$ are very similar (Figs. 1 and 2). Both have two distinct pathways operating the TFR pathway and the non-TFR pathway (the non-TFR pathway is actually quite complicated and cannot be succinctly described). In addition to the graphical similarities in the schemes, the representative geometries in Figs. 3 and 4 are analogs. The *ortho*- to *meta*-$C_2B_{10}H_{12}$ case mimics the *meta*- to *para*-$C_2B_{10}H_{12}$ case, which mimics the $B_{12}H_{12}{}^{2-}$ case. These similarities will prove to be useful when discussing the mechanisms of the pathways in detail.

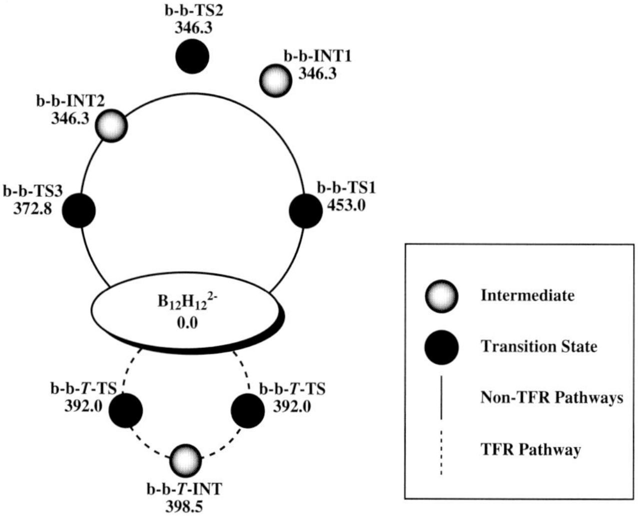

Fig. 2 Rearrangement pathways for $B_{12}H_{12}{}^{2-}$. Values indicate free energies in kilojoules per mole (kJ/mol) at 800 K. The stationary points *b–b*-TS2 and *b–b*-INT1 are above the curve because the free energy barrier is essentially zero, which suggests that $B_{12}H_{12}{}^{2-}$ is converted directly to *b–b*-INT2 (through *b–b*-TS1)

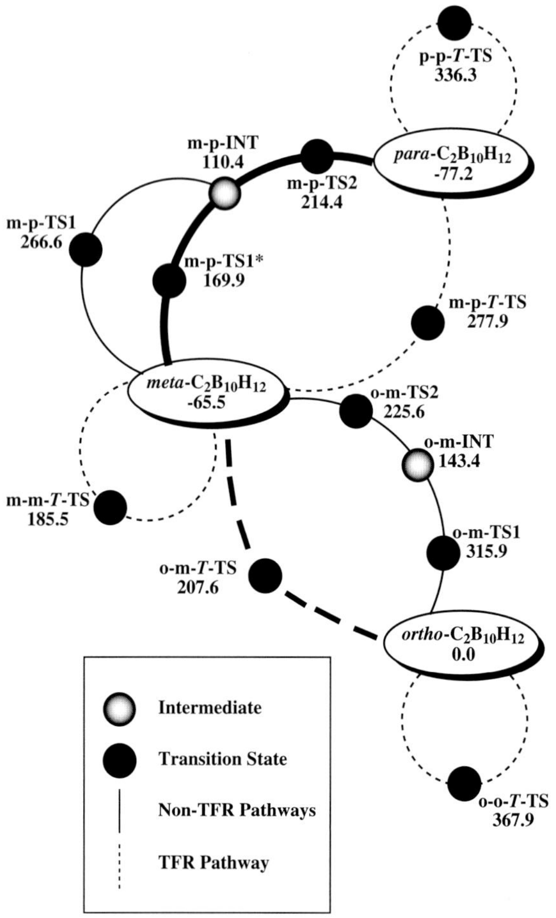

Fig. 1 Rearrangement pathways for $C_2B_{10}H_{12}$. Values indicate free energies in kilojoules per mole (kJ/mol) at 800 K. *Bold lines* (*dashed and solid*) indicate the lowest free energy pathway from *ortho*- to *meta*-$C_2B_{10}H_{12}$ (*bold and dashed*) and from *meta*- to *para*-$C_2B_{10}H_{12}$ (*bold and solid*)

General trends

Before discussing rearrangement mechanisms, it is important to establish some confidence by way of general trends between the stabilities and geometries of the three carborane isomers. Pertinent electronic and thermodynamic data can be found in Tables 1 and 2. Notice that the ΔG (298 K) for the three isomers decreases from *ortho*- to *meta*- to *para*-$C_2B_{10}H_{12}$, which is consistent with experimental observation [39]. Furthermore, the approximate transition temperatures of 450 °C for the *ortho*- to *meta*-$C_2B_{10}H_{12}$ rearrangement and 620 °C for the *meta*- to *para*-$C_2B_{10}H_{12}$ rearrangement are qualitatively consistent with the barrier heights in Fig. 1 [39–41]. Besides trends in stability, two recent gas-phase electron diffraction (GED) studies have made it possible to compare the calculated unique bond distances in the three isomers to the

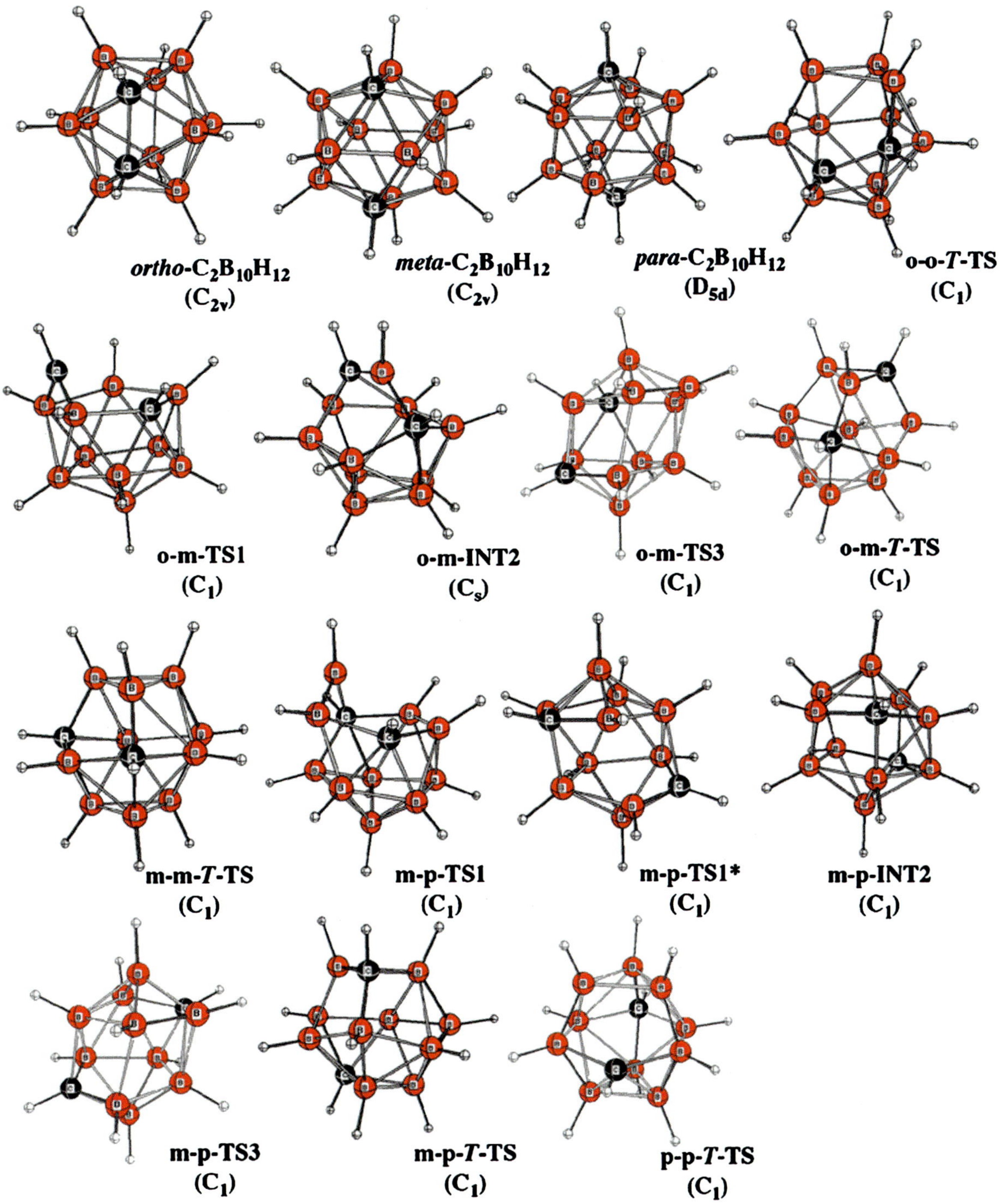

Fig. 3 Geometries for *ortho*-, *meta*-, and *para*-$C_2B_{10}H_{12}$. The nomenclature is as follows: the *first two letters* separated by a *dash* indicate which carboranes are being rearranged (e.g., m–p corresponds to a *meta* to *para* rearrangement species). An *italicized T* indicates a TFR pathway species. The *last letters* (and possible *number*) indicates the type of stationary point of the species

experimentally determined unique bond distances (see Table 3 and Fig. 5) [42, 43]. In almost all of the bonds, the difference between the calculated distance and the most recent experimentally determined distance (GED at 463 K) is around 0.01 Å.

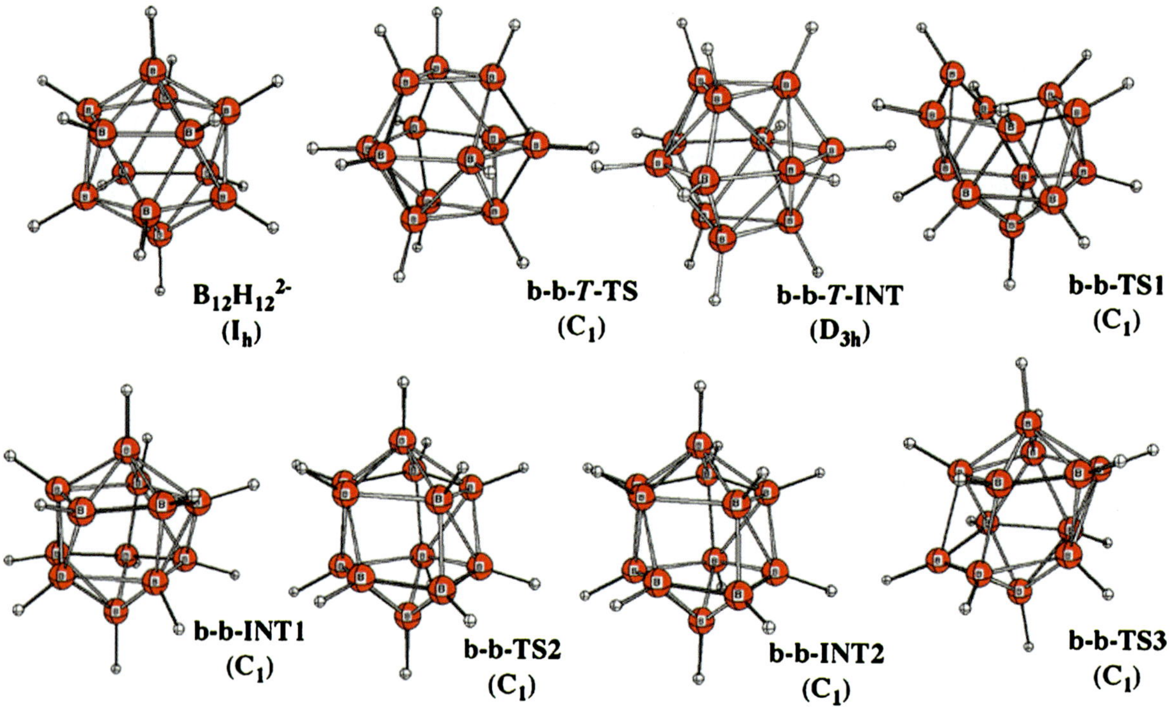

Fig. 4 Geometries of $B_{12}H_{12}{}^{2-}$species. The nomenclature system is analogous to the carborane case. Consult Fig. 3 for a description

Table 1 Total electronic energies (in hartrees), zero-point energies (kJ/mol), heat capacity corrections to 298 K, and entropies (J/mol K) for species on the $C_2B_{10}H_{12}$ and $B_{12}H_{12}{}^{2-}$ potential energies surfaces

Structures	Total energy	Zero-point energy	∫ Cp	Entropy
ortho-$C_2B_{10}H_{12}$	−332.17455	466.1	22.7	347.1
meta-$C_2B_{10}H_{12}$	−332.20059	468.0	22.4	345.5
para-$C_2B_{10}H_{12}$	−332.20500	468.0	22.4	345.7
o–*m*-TS1	−332.04031	445.7	26.6	372.2
o–*m*-INT	−332.10921	455.1	26.8	373.7
o–*m*-TS2	−332.07899	451.5	25.4	363.8
o–*m*-*T*-TS	−332.08224	448.5	26.7	373.5
o–*o*-*T*-TS	−332.01780	444.1	27.6	380.3
m–*m*-*T*-TS	−332.09144	449.5	26.7	372.2
m–*p*-TS1	−332.06051	448.2	26.3	370.2
m–*p*-TS1*	−332.10091	452.5	25.1	362.4
m–*p*-INT	−332.12258	455.5	26.6	371.5
m–*p*-TS2	−332.08274	451.5	25.6	365.8
m–*p*-***T***-TS	−332.05341	446.7	27.3	378.8
p–*p*-*T*-TS	−332.03032	444.5	27.7	379.4
$B_{12}H_{12}^{2-}$	−305.77394	439.1	23.1	348.7
b–*b*-*T*-TS	−305.60815	418.6	28.3	383.6
b–*b*-*T*-INT	−305.60974	419.7	29.6	373.2
b–*b*-TS1	−305.58691	419.4	27.4	377.1
b–*b*-INT1	−305.63168	423.9	26.0	367.4
b–*b*-TS2	−305.63167	423.9	26.0	367.4
b–*b*-INT2	−305.63167	423.9	26.0	367.5
b–*b*-TS3	−305.61911	422.4	26.6	374.5

Table 2 Relative enthalpies (kJ/mol), entropies (J/K·mol), and free energies (kJ/mol) for species on the $C_2B_{10}H_{12}$ and $B_{12}H_{12}{}^{2-}$ potential energy surfaces

Structures	ΔH (0 K)	ΔH (298 K)	ΔS	ΔG (298 K)	ΔG (500 K)	ΔG (800 K)
ortho-$C_2B_{10}H_{12}$	0.0	0.0	0.0	0.0	0.0	0.0
meta-$C_2B_{10}H_{12}$	−66.5	−66.8	−1.5	−66.3	−66.0	−65.5
para-$C_2B_{10}H_{12}$	−78.0	−78.3	−1.4	−77.9	−77.6	−77.2
o–m-TS1	332.1	336.0	25.1	328.5	323.4	315.9
o–m-INT	160.6	164.7	26.6	156.8	151.4	143.4
o–m-TS2	236.3	239.0	16.7	234.0	230.6	225.6
o–m-T-TS	224.7	228.8	26.4	220.9	215.5	207.6
o–o-T-TS	389.6	394.5	33.3	384.6	377.9	367.9
m–m-T-TS	201.6	205.6	25.1	198.1	193.1	185.5
m–p-TS1	281.6	285.1	23.1	278.2	273.6	266.6
m–p-TS1*	179.8	182.2	15.4	177.6	174.5	169.9
m–p-INT	125.9	129.9	24.4	122.6	117.7	110.4
m–p-TS2	226.5	229.4	18.7	223.8	220.0	214.4
m–p-T-TS	298.7	303.3	31.7	293.8	287.4	277.9
p–p-T-TS	357.1	362.1	32.3	352.5	346.0	336.3
$B_{12}H_{12}^{2-}$	0.0	0.0	0.0	0.0	0.0	0.0
b–b-T-TS	414.7	419.9	34.9	409.5	402.5	392.0
b–b-T-INT	411.6	418.2	24.5	410.8	405.9	398.5
b–b-TS1	471.4	475.7	28.4	467.2	461.5	453.0
b–b-INT1	358.3	361.3	18.7	355.7	351.9	346.3
b–b-TS2	358.3	361.3	18.7	355.7	351.9	346.3
b–b-INT2	358.3	361.3	18.8	355.7	351.9	346.3
b–b-TS3	389.8	393.4	25.8	385.7	380.5	372.8

Ortho- to *meta*-carborane rearrangement[2]

TFR pathway

Figure 6 illustrates the *ortho*- to *meta*-$C_2B_{10}H_{12}$ TFR mechanism where the structures of the stationary points are similar to those in the *ortho*-$C_2B_{10}H_{12}$ to *ortho*-$C_2B_{10}H_{12}$ degenerate TFR, the *meta*-$C_2B_{10}H_{12}$ to *para*-$C_2B_{10}H_{12}$ TFR, and the borane TFR rearrangement.[3] The left column in Fig. 6 displays the geometric progression through the reaction and the right column shows a two-dimensional projection of the pertinent atoms and bonds of the corresponding geometry on the left. As can be seen in the left column, a TFR acts by taking an equilateral triangle of three atoms, lifting them up slightly, rotating them by 120 ° (counter-clockwise in this case), and lowering them back down. The projective graphs display unequivocally that the mechanism is, in fact, a sequence of concerted diamond–square–diamonds. The transition state graph corresponding to *o–m-T*-TS shows the three simultaneous square faces.

Non-TFR pathway

Figure 7 illustrates the non-TFR mechanism. The first step in this mechanism is for one of the *ortho*-$C_2B_{10}H_{12}$ carbons, situated at the apex of a pentagonal cap, to break three of its bonds, while performing a DSD transition on the lower diamond. The square face in the DSD transition can be seen in the lower part of the graph corresponding to *o-m*-TS1. As the carbon migrates and forms the bond in the DSD, the species approaches a pseudo-*nido* geometry, which is displayed graphically in brackets. This geometry, in which the carbon lies approximately in a hexagonal plane, is never realized because other stabilizing forces exert themselves and propel the species towards the geometry in *o-m*-INT. The geometry displays two pentagonal faces in lieu of the hexagonal face of the pseudo-*nido* geometry. The last transition state *o-m*-TS2 is due to two adjacent DSD transitions on the opposite side of the cluster. In one direction, the species minimizes to *o-m*-INT. In the other direction, the molecule minimizes to *meta*-$C_2B_{10}H_{12}$.

Meta- to *para*-carborane rearrangement

TFR pathway

See "TFR pathway" subsection under "Ortho- to meta-carborane rearrangement".

[2] David Wales has two examples of carborane rearrangements that illustrate the necessary movement: (1) rearrangement from the icosahedral 1,2-carborane to a higher energy C_S isomer (http://www.ch.cam.ac.uk/SGTL/teach/djw1.html) and (2) rearrangement from the 1,7-icosahedral carborane to a higher energy isomer based on an anticuboctahedron (http://www.ch.cam.ac.uk/SGTL/teach/djw2.html).

[3] All of these pathways operate in essentially the same manner, except for slight deviations in geometry due to juxtaposition of boron with carbon and vice versa.

Table 3 Comparison of calculated distances [B3LYP/6-31G(d)] with distances from two gas-phase electron diffraction studies for *ortho*-, *meta*-, and *para*-$C_2B_{10}H_{12}$

Bond	B3LYP/6-31G(d)	GED [42]	GED [43]
ortho-$C_2B_{10}H_{12}$			
C(1)–C(2)	1.625	1.624(8)	1.653(49)
C(1)–B(3)	1.720	1.734(7)	1.711(14)
C(1)–B(4)	1.700	1.700(6)	1.711(14)
B(3)–B(4)	1.780	1.788(6)	1.802(13)
B(4)–B(5)	1.786	1.794(8)	1.802(13)
B(3)–B(8)	1.765	1.774(9)	1.789(9)
B(5)–B(9)	1.778	1.787(6)	1.789(9)
B(7)–B(8)	1.781	1.796(6)	1.789(9)
B(8)–B(9)	1.793	1.808(8)	1.789(9)
B(9)–B(12)	1.784	1.787(9)	1.789(9)
meta-$C_2B_{10}H_{12}$			
C(1)–B(2)	1.694	1.678(5)	1.720(9)
C(1)–B(4)	1.715	1.730(4)	1.720(9)
C(1)–B(5)	1.712	1.704(8)	1.720(9)
B(2)–B(3)	1.789	1.786(9)	1.831(52)
B(2)–B(6)	1.768	1.771(6)	1.791(15)
B(4)–B(5)	1.783	1.801(5)	1.791(15)
B(4)–B(9)	1.781	1.780(6)	1.817(13)
B(5)–B(9)	1.778	1.783(6)	1.817(13)
B(4)–B(8)	1.771	1.778(9)	1.817(13)
B(9)–B(10)	1.793	1.795(9)	1.817(13)
C(1)···C(7)	2.608	2.575(9)	
para-$C_2B_{10}H_{12}$			
C–B	1.708	1.698(3)	1.710(11)
B–B tropical	1.790	1.784(1)	1.792(7)
B–B meridianal	1.767	1.774(4)	1.772(13)
C(1)···C(12)	3.057	3.029(5)	3.05(5)

Non-TFR pathway

The non-TFR mechanism for the *meta*- to *para*-rearrangement (Fig. 8) mimics the *ortho*- to *meta*-$C_2B_{10}H_{12}$ rearrangement except that there is a second possible transition state *m–p*-TS1*, which is lower in energy than the analogous *ortho*- to *meta*-$C_2B_{10}H_{12}$ transition state (*m–p*-TS1, Table 1). The structural change from *meta*-$C_2B_{10}H_{12}$ to *m–p*-TS1 is exactly the same as in the *ortho*- to *meta*-$C_2B_{10}H_{12}$ case, except that only a boron vertex moves from the apex of a pentagonal cap opening a face in the cluster. Just as with the *ortho*- to *meta*-$C_2B_{10}H_{12}$ rearrangement, a "*nido*" geometry is never reached, leading instead to *m–p*-INT. The more thermodynamically favorable pathway is through *m–p*-TS1* which can be described as two adjacent concerted DSD steps. The second half of the rearrangement is analogous to the *ortho*-$C_2B_{10}H_{12}$ to *meta*-$C_2B_{10}H_{12}$ rearrangement: two concerted adjacent DSD steps occur on the opposite side of the cluster (see *m–p*-TS2) and *para*-$C_2B_{10}H_{12}$ results.

Borane rearrangement

TFR pathway

See "TFR pathway" subsection under "Ortho- to meta-carborane rearrangement". There is one exception in the $B_{12}H_{12}{}^{2-}$ case. There is a D_{3h} intermediate *b–b-T*-INT that is lower than *b–b-T*-TS in enthalpy at 298 K by 1.7 kJ/mol. When entropic effects are taken into account, the intermediate *b–b-T*-INT is actually higher in free energy at 298, 500, and 800 K.

Non-TFR pathway

Figure 9 illustrates the non-TFR mechanism in the $B_{12}H_{12}{}^{2-}$ rearrangement. The first transition state *b–b*-TS1 is essentially the same as *o–m*-TS1 and *m–p*-TS1 except that there are no carbon atoms present. A boron atom located at

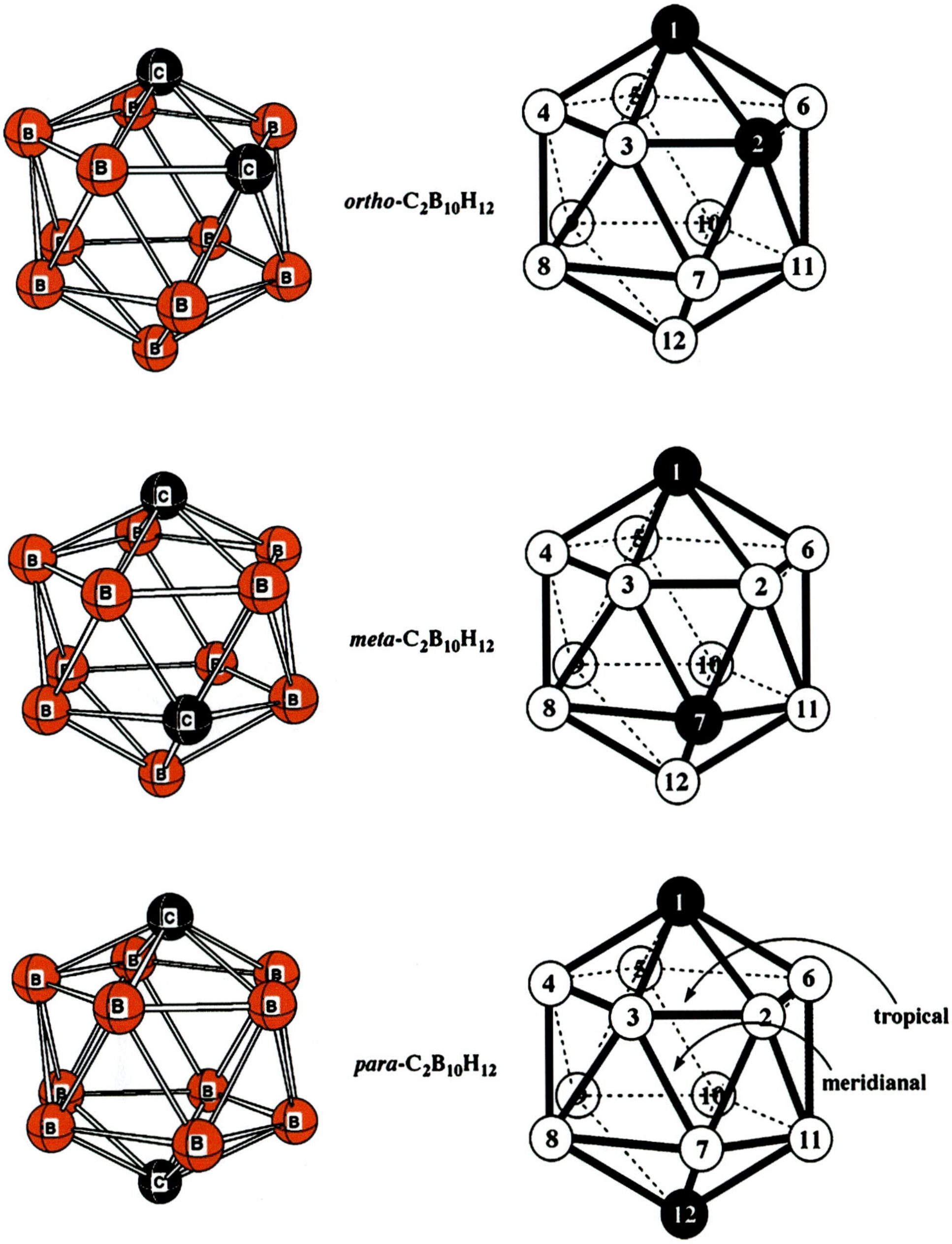

Fig. 5 Numbering system with respective geometries for Table 3

the apex of a pentagonal cap migrates, breaking three bonds, and performs a DSD step on the lower diamond. The geometry again progresses towards a pseudo-*nido* geometry but does not quite achieve it. Another bond begins to form, which acts to prevent the cluster from opening, resulting in *b*–*b*-INT1. An interesting feature of the pathway is that there is an additional minimum (*b*–*b*-TS2) and transition state (*b*–*b*-INT1) along the reaction pathway that disappear when zero-point corrections are included. From *b*–*b*-TS2, two adjacent concerted DSD steps are needed to reach *b*–*b*-TS3 which then leads directly to icosahedral $B_{12}H_{12}{}^{2-}$.

The free energy of activation in the gas phase for rearrangement of the boron vertices at 800 K is 398.5 kJ/mol via *b*–*b*-*T*-INT (Fig. 2). The free energy barrier is much higher than calculated for rearrangement in the isoelectronic carboranes (Fig. 1). In fact, the barrier is higher than for fragmentation of a B–H bond, which implies that $B_{12}H_{12}{}^{2-}$ should undergo decomposition rather than rearrangement. In a recent study of the boron hydride monoanions, $B_{12}H_{12}{}^{2-}$ was determined to be the smallest boron hydride dianion with a positive adabatic ionization potential [22]. It is still unlikely that both electrons would remain attached at elevated temperatures.

Experimental verification

There is one very interesting anomaly that emerges with respect to carborane rearrangements. The lowest free

Fig. 6 A localized graphical picture that describes the progression of the TFR pathway. The *ortho*- to *meta*-$C_2B_{10}H_{12}$ TFR pathway is shown

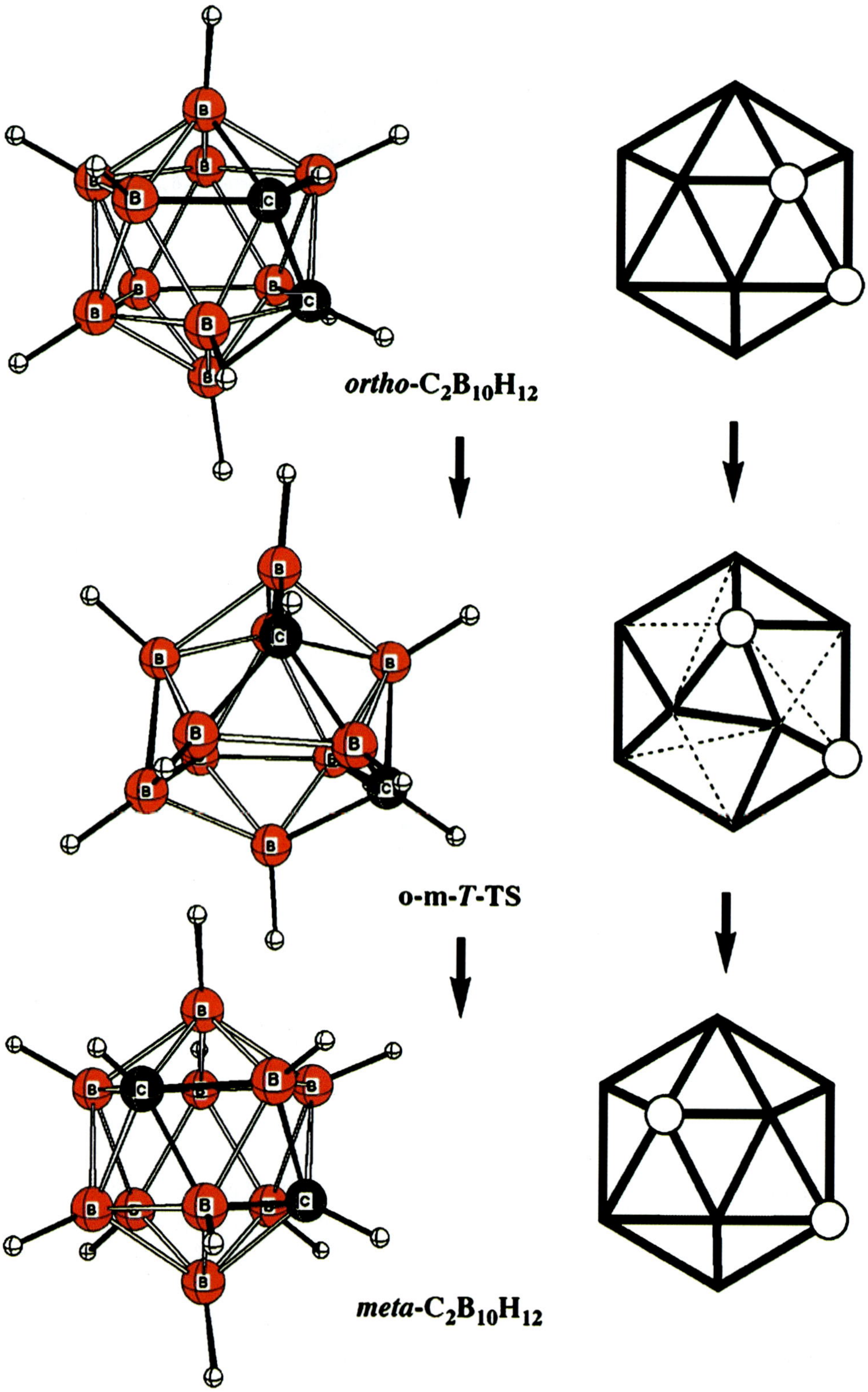

energy carborane barrier is a degenerate rearrangement *meta*- to *meta*-$C_2B_{10}H_{12}$, which is 185.5 kJ/mol at 800 K. This process may prove useful in rationalizing products obtained from rearrangement of the substituted carborane. In addition, because the barrier is lower than the barrier for rearrangement to *ortho*- and *para*-$C_2B_{10}H_{12}$, if one were able to synthesize a substituted *meta*-$C_2B_{10}H_{12}$ with a methyl group on a boron atom adjacent to carbon, then, at a moderate temperature, one would expect to see migration of the methyl group to a boron position non-adjacent to carbon. We offer this prediction as a means of testing the rearrangement mechanism in the gas phase.

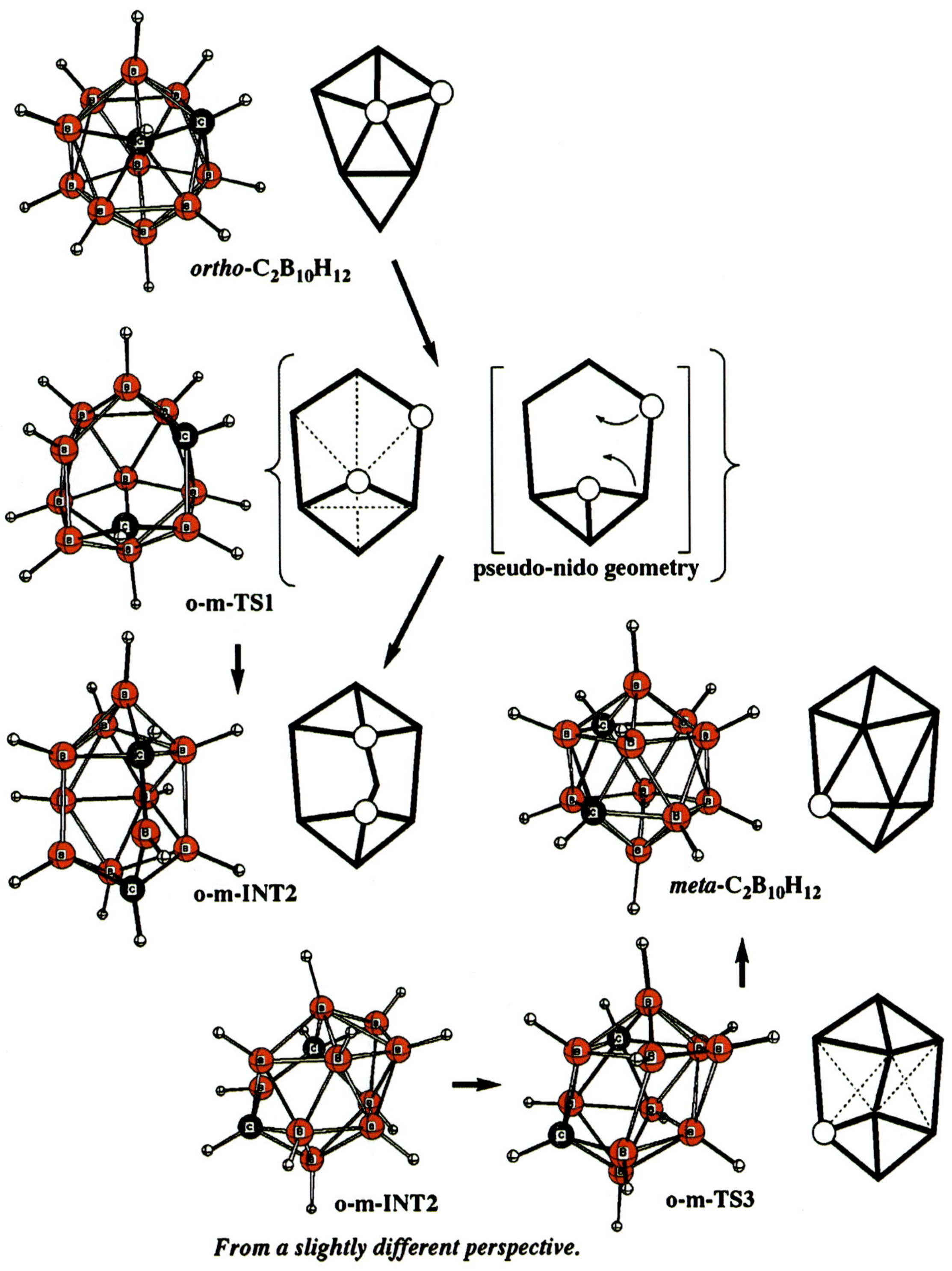

Fig. 7 The *ortho*- to *meta*-$C_2B_{10}H_{12}$ non-TFR rearrangement pathway

Conclusion

The rearrangement mechanism in $C_2B_{10}H_{12}$ and $B_{12}H_{12}{}^{2-}$ has been explored using density functional theory methods. From *ortho*-$C_2B_{10}H_{12}$, the lowest free energy pathway to *meta*-$C_2B_{10}H_{12}$ is via a triangular-face-rotation mechanism with a barrier of 207.6 kJ/mol. From *meta*- to *para*-$C_2B_{10}H_{12}$, the lowest-energy rearrangement mechanism is non-TFR which might loosely be described as a *closo* to *nido* to *closo* mechanism as the transition state is much more open than the reactant or product. The free energy barrier is 279.9 kJ/mol, 72.3 kJ/mol lower in free energy

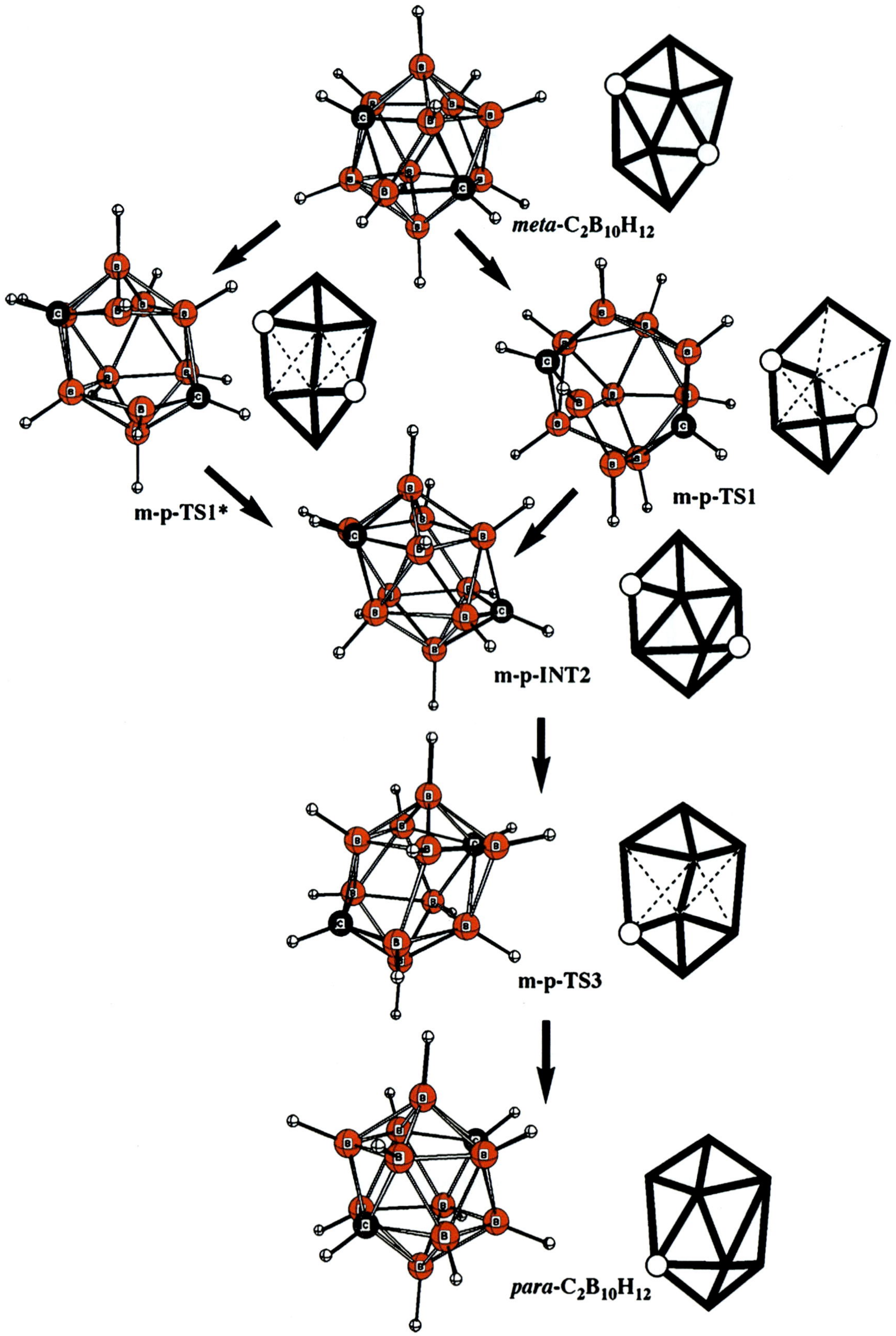

Fig. 8 The *meta*- to *para*-$C_2B_{10}H_{12}$ non-TFR rearrangement pathway

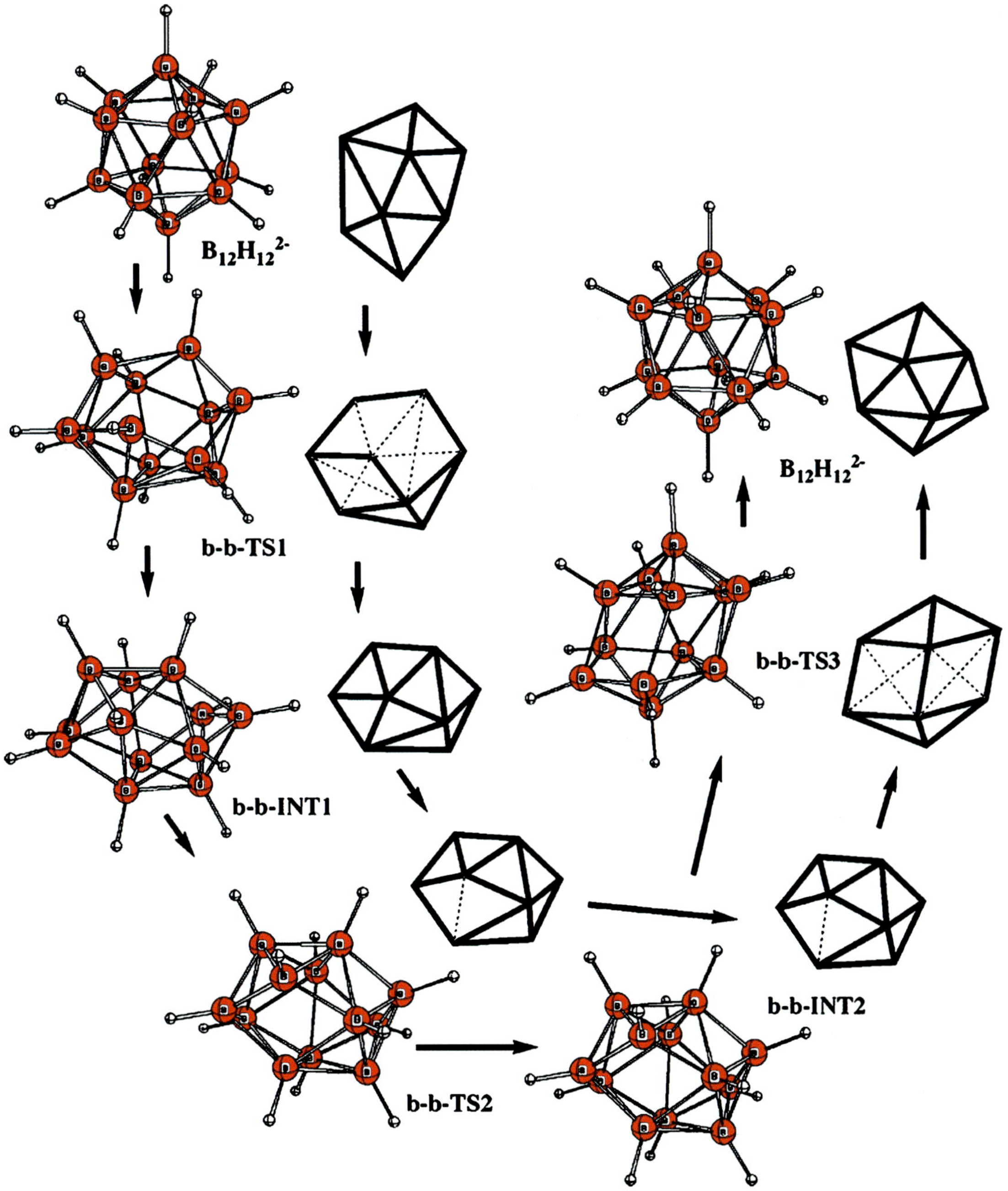

Fig. 9 The $B_{12}H_{12}{}^{2-}$ to $B_{12}H_{12}{}^{2-}$ non-TFR rearrangement pathway

than the *ortho* to *meta* process. The rearrangement of $B_{12}H_{12}{}^{2-}$, which would scramble labeled vertices, has a free energy barrier of 398.5 kJ/mol. For $B_{12}H_{12}{}^{2-}$, the preferred process in the gas phase would be to emit an electron or cleave a B–H bond.

Acknowledgements We acknowledge the Alabama Supercomputer Center for providing time on their SGI Altix 350. C. A. B. would like to thank the College of Science and Mathematics at Auburn University for an undergraduate research fellowship. Above all, we appreciate the numerous contributions that Prof. Schleyer has made to chemistry.

References

1. Wade K (1976) Adv Inorg Chem Radiochem 18:1–66
2. Wade K (1971) Chem Commun 1:792–793
3. Hart HV, Lipscomb WN (1973) Inorg Chem 12:2644–2649
4. Hart HV, Lipscomb WN (1969) J Am Chem Soc 91:771–772
5. Lipscomb WN (1966) Science 153:373–378
6. Lipscomb WN (1959) Adv Inorg Chem Radiochem 1:117–156
7. Williams RE (1971) Inorg Chem 10:210–214
8. Williams RE, Gerhart FJ (1965) J Am Chem Soc 87: 3513–3515
9. Williams RE (1976) Adv Inorg Chem Radiochem 18:67–142
10. Driess M, Nöth H (eds) (2004) Molecular clusters of the main group elements. Wiley-VCH Verlag GmbH KGaA, Weinheim
11. Grimes RN (1970) Carboranes. Academic, New York

12. Lipscomb WN (1963) Boron hydrides. Benjamin, New York
13. Mingos DMP, Wales DJ (1990) Introduction to cluster chemistry. Prentice Hall, New York
14. Mutterties EL, Knoth WH (1968) Polyhedral boranes. Marcel Dekker, New York
15. Fox MA, Wade K (2003) Pure Appl Chem 75:1315–1323
16. Kiani FA, Hofmann M (2004) Inorg Chem 43:8561–8571
17. Casanova J (ed) (1998) The borane, carborane, carbocation continuum. Wiley, New York
18. Ceulemans A, Goijens G, Nguyen MT (1994) J Am Chem Soc 116:9395–9396
19. Deng L, Chan H, Xie Z (2005) Angew Chem Int Ed 44:2–5
20. Kaczmarcyzk A, Dobrott RD, Lipscomb WN (1962) Proc Natl Acad Sci U S A 48:729–733
21. Laromaine A, Teixidor F, Viñas C (2005) Angew Chem Int Ed 44:2–4
22. McKee ML, Wang Z, Schleyer PvR (2000) J Am Chem Soc 122:4781–4793
23. McKee ML (1992) J Am Chem Soc 114:879–881
24. Barnett-Thamattoor L, Wu JJ, Ho DM, Jones M Jr (1996) Tetrahedron Lett 37:7221–7224
25. Gimarc BM, Warren DS, Ott JJ, Brown C (1991) Inorg Chem 30:1598–1605
26. McKee ML (1994) J Phys Chem 98:13243–13248
27. McKee ML (1992) J Am Chem Soc 114:5856–5858
28. Wong HS, Lipscomb WN (1975) Inorg Chem 6:1350–1357
29. Ye Q, Li J, Jones M Jr, Wu H, Mckee ML, Shevlin PB (2002) Tetrahedron Lett 43:735–736
30. Gimarc BM, Zhao M (1996) Inorg Chem 35:825–834
31. Johnson BFG (1986) Chem Commun 1:27–30
32. Roberts YV, Johnson BFG (1994) J Chem Soc Dalton Trans 5:759–766
33. Wales DJ, Mingos DMP, Zhenyang L (1989) Inorg Chem 28:2754–2764
34. Wales DJ, Mingos DMP (1989) Polyhedron 8:1933–1938
35. Wales DJ, Stone AJ (1987) Inorg Chem 26:3845–3850
36. Wu S, Jones M Jr (1989) J Am Chem Soc 111:5373–5384
37. Edvenson GM, Gaines DF (1990) Inorg Chem 29:1210–1216
38. Frisch MJ, Trucks GW, Schlegel HB, Scuseria GE, Robb MA, Cheeseman JR, Montgomery JA Jr, Vreven T, Kudin KN, Burant JC, Millam JM, Iyengar SS, Tomasi J, Barone V, Mennucci B, Cossi M, Scalmani G, Rega N, Petersson GA, Nakatsuji H, Hada M, Ehara M, Toyota K, Fukuda R, Hasegawa J, Ishida M, Nakajima T, Honda Y, Kitao O, Nakai H, Klene M, Li X, Knox JE, Hratchian HP, Cross JB, Bakken V, Adamo C, Jaramillo J, Gomperts R, Stratmann RE, Yazyev O, Austin AJ, Cammi R, Pomelli C, Ochterski JW, Ayala PY, Morokuma K, Voth GA, Salvador P, Dannenberg JJ, Zakrzewski VG, Dapprich S, Daniels AD, Strain MC, Farkas O, Malick DK, Rabuck AD, Raghavachari K, Foresman JB, Ortiz JV, Cui Q, Baboul AG, Clifford S, Cioslowski J, Stefanov BB, Liu G, Liashenko A, Piskorz P, Komaromi I, Martin RL, Fox DJ, Keith T, *Al*-Laham MA, Peng CY, Nanayakkara A, Challacombe M, Gill PMW, Johnson B, Chen W, Wong MW, Gonzalez C, Pople JA (2004) Gaussian, Wallingford, CT
39. Mutterties EL (1975) Boron hydride chemistry. Academic, New York
40. Grafstein D, Dvorak J (1963) Inorg Chem 2:1128–1133
41. Papetti S, Heying TL (1964) J Am Chem Soc 86:2295
42. Turner AR, Robertson HE, Borisenko KB, Rankin DWH, Fox MA (2005) J Chem Soc Dalton Trans 7:1310–1318
43. Bohn RK, Bohn MD (1971) Inorg Chem 10:350–355

J Mol Model (2006) 12: 665–672
DOI 10.1007/s00894-005-0021-y

ORIGINAL PAPER

Yirong Mo

Probing the nature of hydrogen bonds in DNA base pairs

Received: 2 May 2005 / Accepted: 27 June 2005 / Published online: 6 May 2006

Abstract Energy decomposition analyses based on the block-localized wave-function (BLW-ED) method are conducted to explore the nature of the hydrogen bonds in DNA base pairs in terms of deformation, Heitler–London, polarization, electron-transfer and dispersion-energy terms, where the Heitler–London energy term is composed of electrostatic and Pauli-exchange interactions. A modest electron-transfer effect is found in the Watson–Crick adenine–thymine (AT), guanine–cytosine (GC) and Hoogsteen adenine-thymine (H-AT) pairs, confirming the weak covalence in the hydrogen bonds. The electrostatic attraction and polarization effects account for most of the binding energies, particularly in the GC pair. Both theoretical and experimental data show that the GC pair has a binding energy (−25.4 kcal mol^{-1} at the MP2/6-31G** level) twice that of the AT (−12.4 kcal mol^{-1}) and H-AT (−12.8 kcal mol^{-1}) pairs, compared with three conventional N-H···O(N) hydrogen bonds in the GC pair and two in the AT or H-AT pair. Although the remarkably strong binding between the guanine and cytosine bases benefits from the opposite orientations of the dipole moments in these two bases assisted by the π-electron delocalization from the amine groups to the carbonyl groups, model calculations demonstrate that π-resonance has very limited influence on the covalence of the hydrogen bonds. Thus, the often adopted terminology "resonance-assisted hydrogen bonding (RHAB)" may be replaced with "resonance-assisted binding" which highlights the electrostatic rather than electron-transfer nature of the enhanced stabilization, as hydrogen bonds are usually regarded as weak covalent bonds.

Dedicated to Professor Paul von Ragué Schleyer on the occasion of his 75th birthday

Y. Mo
Department of Chemistry, Western Michigan University, Kalamazoo, MI, 49008 USA
E-mail: yirong.mo@wmich.edu

Keywords Hydrogen bond · DNA base pair · Charge transfer · Electrostatic interaction · Resonance-assisted hydrogen bonding (RAHB)

Introduction

Molecular simulations of biological systems and computational high-throughput screening of drug candidates rely heavily on the quality of force fields, where the potential-energy function is generally expressed as the summation of various bonded and nonbonded energy terms [1–3]. However, there are few straightforward experimental proofs to justify the formulations and magnitudes of these energy terms. Only theoretical studies can provide valuable insights into the interatomic and intermolecular interactions pivotal for the development of next-generation force fields. In particular, force-field formulations and parameterizations may be monitored and refined by intermolecular interaction-energy decomposition analyses [4–11]. For instance, it has been shown that a simple electrostatic model works impressively well for strong noncovalent cation–π interactions [12–15], which play a key role in biological recognition. On the other hand, recent work shows that the polarization effect is comparable to the magnitude of electrostatic interactions [16] and consequently a polarizable force field is necessary to model cation–π interactions [17]. This dilemma was resolved by our recent energy-decomposition analyses on a series of cation–π complexes, where the results show that electrostatic, polarization and charge-transfer energy terms have good linear relationships with the total interaction energies [18]. This finding indicates that if the electrostatic energy is scaled in a way that the final interaction energies are close to the experimental or high-level quantum mechanical results, even simple force fields without explicit polarization and charge-transfer terms can model cation–π complexes very successfully.

Similarly, it has also been conceived for a long time that the hydrogen bonds in the DNA base pairs are

Table 1 Comparison of the hydrogen-bond lengths (Å) in the DNA base pairs

Level of theory	WC–AT		WC–GC			H–AT	
	$R(N_1\text{–}O_1)$	$R(N_2\text{–}N_3)$	$R(O_6\text{–}N_4)$	$R(N_1\text{–}N_3)$	$R(N_2\text{–}O_2)$	$R(N_1\text{–}O_1)$	$R(N_5N_3)$–
HF/6-31G**[a]	3.09	2.99	2.92	3.04	3.02	–	–
HF/cc-pVTZ(-f)[b]	3.06	2.92	2.83	2.95	2.92	–	–
BP86/TZ2P[c]	2.85	2.81	2.73	2.88	2.87	–	–
B3LYP/6-31G**[d]	2.94	2.84	2.79	2.93	2.92	–	–
MP2/6-31G*	2.99	2.88	2.84	2.97	2.95	3.03	2.84
MP2/6-31G**	2.97	2.84	2.81	2.94	2.93	3.01	2.81
RI-MP2/cc-pVTZ[e]	2.86	2.83	2.75	2.90	2.89	–	–
X-ray[f]	2.95	2.82	2.91	2.95	2.86	–	–

[a]Šponer et al. [53, 54].
[b]Brameld et al. [37].
[c]Guerra et al. [25].
[d]Bertran et al. [36].
[e]Šponer et al. [21].
[f]Saenger [55].

H-AT pairs have similar binding energies. Since there are three regular N–H···O(N) hydrogen bonds in the GC pair, compared with two in the AT and H-AT pairs, it is intriguing to explore the nature of the very high-binding energy in the GC pair. Examination of individual energy contributions reveals that the correlation energy term is comparable in all systems, indicating that the dispersion force plays the same role in all base pairs. Moreover, although both the deformation and charge-transfer energy terms do show differences between the GC and AT or H-AT pairs, their variations are small. Remarkable discrepancies are found in the Heitler–London and polarization-energy terms. Strong electrostatic attraction and polarization effects are observed in the GC pair. Apparently, this stems from the favorable dipole–dipole electrostatic interaction between the guanine and cytosine bases, as widely recognized [21]. As the amine group is a π electron-donor group while the carbonyl group is a π electron-acceptor group, guanine and cytosine exhibit opposite dipole moments in the interfacial region, which as a result significantly stabilize the complex via classical electrostatic interaction. In contrast, such a favorable dipole–dipole interaction does not exist in the AT and H-AT pairs. The enhanced binding in the GC pair due to the stabilizing electrostatic interaction and polarization induced by π-electron resonance seems in accord with the definition of resonance-assisted hydrogen bonding (RAHB). RAHB was proposed to explain some abnormally strong intra- or intermolecular hydrogen bonds in π-conjugated systems [38–43]. In both guanine and cytosine, the π-electron resonance from the amine groups to the carbonyl groups may make the proton acceptor (carbonyl oxygen) more negative and the proton donor (amine nitrogen) more positive. On one hand, this can potentially reinforce the N–H···O hydrogen bonds as the proton acceptor can donate more electrons to the proton donor. On the other hand, since the π-electron delocalization in the guanine and cytosine bases occurs in anti-parallel directions, the binding between the guanine and cytosine is significantly enhanced due to the resonance-induced dipole moments.

Similar to Lewis acid–base complexes, the covalence in hydrogen bonds can be measured by the magnitude of the charge-transfer effect as the bond is formed by the interaction between an occupied donor orbital and virtual acceptor orbital. However, unlike strong Lewis acid–base complexes such as BH_3NH_3, where the charge-transfer energy is comparable to the overall interaction energy [30, 31, 44], in DNA base pairs I observed a relatively weak charge-transfer interaction, which accounts for around 20–35% of the overall interaction (excluding the structural deformation energy and note the Pauli repulsion largely offset the electrostatic attraction). This indicates a low covalence in the hydrogen bonds. Interestingly, as there are two hydrogen bonds in the AT and H-AT pairs but three hydrogen bonds in the GC pair, the charge-transfer stabilization energy is roughly proportional to the number of hydrogen bonds. This linear correlation will be improved further if the very weak C–H···O=C hydrogen bonding is taken into account (see below). It is also noted that in Guerra et al.'s analyses, the orbital-interaction energy term (ΔE_{oi}) is essentially a sum of the polarization, charge-transfer and dispersion energies in our approach [24–26]. Among these three components, only the charge-transfer term reflects the strength of covalence in hydrogen bonds. The polarization effect, which is induced by the mutual influence of two monomers, essentially has an electrostatic origin. Thus, our current analyses show that, although there is indeed some covalence nature in the hydrogen bonds in DNA base pairs, much of the binding energies come from the electrostatic attraction, particularly in the GC pair.

Conventionally, the bonding nature in DNA base pairs can be illustrated by the frontier orbitals (HOMO's and LUMO's) of individual bases [25]. Alternatively, we can plot electron density difference (EDD) maps, which can intuitively demonstrate the electron movement due to either the polarization or charge-transfer effects. Since the BLW defines a diabatic state where the electron flow among monomers is forbidden, the difference between the electron densities of the delocalized HF wavefunction Ψ_{AB} and optimal BLW Ψ_{AB}^{BLW} uniquely highlights the electron transfer between the monomers A and B. On the other hand, the polarization effect can be visualized by the electron-density difference between Ψ_{AB}^{BLW} and Ψ_{AB}^{BLW0}.

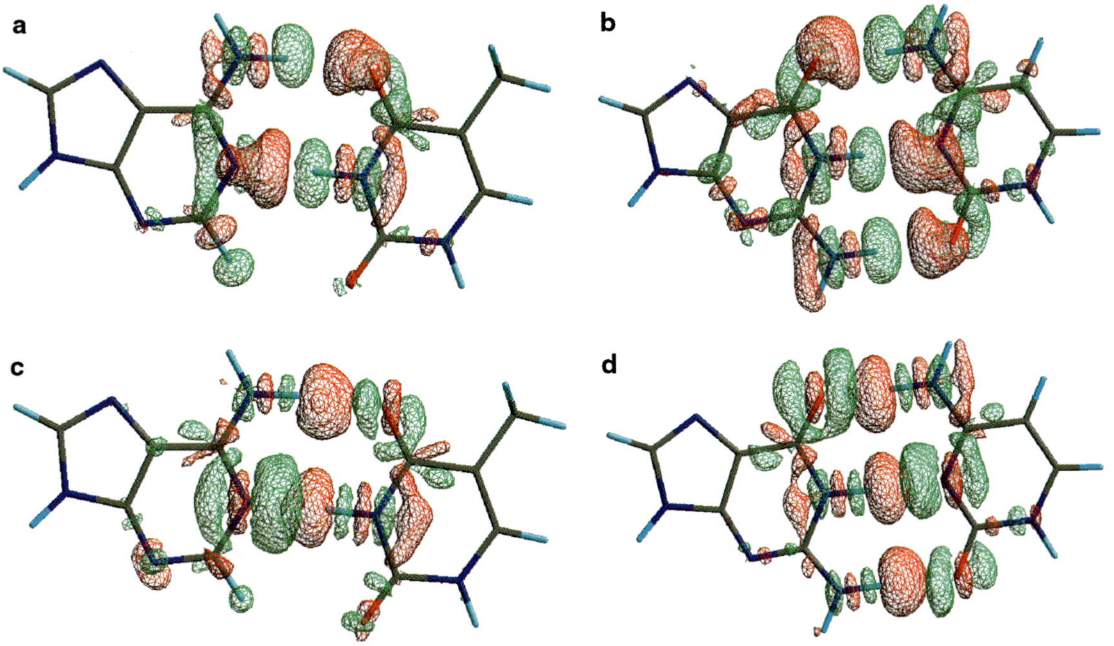

Fig. 2 Electron density difference (EDD) maps: **a** and **b** show the polarization effect in the AT and GC pairs (isodensity $1.2{\times}10^{-3}$ a.u.); **c** and **d** show the charge transfer effect in AT and GC (isodensity $2{\times}10^{-4}$ a.u.)

Figure 2 shows the polarization and charge-transfer effects in AT and GC pairs, where the red denotes the gain of electron density while green means the loss of electron density. In the interface between adenine and thymine or guanine and cytosine, N–H bond polarization from hydrogen to nitrogen in all monomers is observed (Fig. 2a, b), whereas hydrogen-bond acceptors gain electron density from adjacent atoms to prepare for the formation of hydrogen bonds, which makes these acceptors lose electron density to the proton donors (Fig. 2c, d). However, I find that both the polarization and charge-transfer effects are mostly local, indicating the secondary role of the π-aromatic rings in the hydrogen bonding in the DNA base pairs. The only exception is the polarization in the GC pair (Fig. 2b), where I see a small polarization involving the aromatic rings in the guanine and cytosine bases. With the formation of hydrogen bonds, proton acceptors lose electron density to the opposite protons. This confirms the sort of covalent nature of the hydrogen bonds in the DNA base pairs.

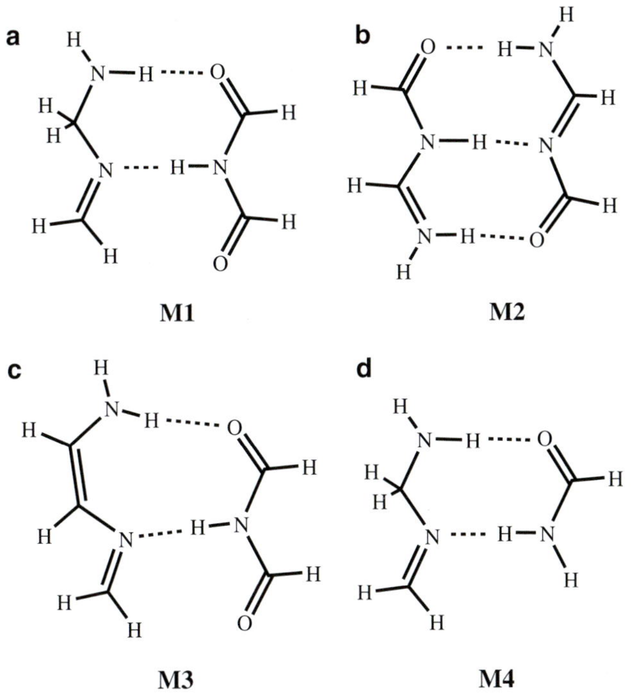

Fig. 3 Geometries of simplified models to simulate the interactions in, WC–AT **a** Watson–Crick guanine–cytosine, **b** and Hoogsteen adenine–thymine, **c** base pairs. **d** is a further simplified model of **a** by removing the carbonyl group in the right side

To verify our above claim that the hydrogen-bonding in the DNA base pairs is mostly due to the local polarization as well as favorable electrostatic interactions further, I designed three simplified models, M1, M2 and M3, to mimic the interactions in the AT, GC and H-AT pairs by retaining only the interfacial parts with unchanged structural parameters as shown in Fig. 3. Similarly, BLW-ED analyses on these systems were conducted and the results are listed in Table 3 (note here the total interaction energy ΔE_{int} does not consider the deformation energy). Comparison of the energy terms in these simplified models with those of complete DNA pairs reveals interesting information. Firstly, it is found that the dispersion energy loses about 1 kcal mol^{-1} in all simplified models. Secondly, the charge transfer effect has little change in the AT and H-AT pairs, but reduces by about 0.5 kcal mol^{-1} in the GC pair. This indicates that the covalence in the hydrogen

Table 2 Energy contributions to the binding energies in nucleic acid base pairs (kcal mol^{-1})[a]

Energy term	6-31G*			6-31G**			6-311+G**		
	WC–AT	WC–GC	H–AT	WC–AT	WC–GC	H–AT	WC–AT	WC–GC	H–AT
Deformation energy (ΔE_{def})	2.5	4.4	2.3	2.5	4.4	2.2	2.6	4.4	2.2
Heitler–London energy (ΔE_{HL})	−1.4	−10.9	−2.4	−1.7	−11.2	−2.7	0.0	-9.0	−1.0
Polarization energy (ΔE_{pol})	−5.2	−10.4	−4.9	−5.2	−10.4	−4.9	−5.9	−11.1	−5.7
Charge-transfer energy (ΔE_{ct})	−4.1	−5.5	−3.9	−4.3	−5.6	−4.1	−4.8	−6.1	−4.6
Dispersion energy (ΔE_{corr})	−3.8	−3.1	−3.7	−3.7	−2.6	−3.3	−4.1	−3.4	−3.3
Binding energy (ΔE_{B})	−12.0	−25.4	−12.6	−12.4	−25.4	−12.8	−12.2	−25.2	−12.4

[a]Structural parameters are based on the optimal geometries at the MP2/6-31G(d,p) level.

Table 3 Energy contributions to the interaction energy in reduced model systems (kcal mol^{-1})[a]

Energy term	6-31G*				6-31G**				6-311+G**			
	M1	M2	M3	M4	M1	M2	M3	M4	M1	M2	M3	M4
ΔE_{HL}	1.0	-3.5	0.4	3.4	0.7	-3.7	0.2	3.2	2.4	-1.5	1.8	5.1
ΔE_{pol}	−4.6	−8.8	−4.8	−4.3	−4.6	−8.8	−4.8	−4.3	−5.2	−9.4	−5.4	−4.9
ΔE_{ct}	−4.1	−5.1	−4.0	−3.6	−4.2	−5.2	−4.2	−3.8	−4.8	−5.6	−4.8	−4.3
ΔE_{corr}	−2.5	−2.3	−2.9	−2.5	−2.6	−2.4	−3.0	−2.7	−2.7	−2.5	−3.1	−3.0
ΔE_{int}	−10.2	−19.7	−11.3	−7.0	−10.7	−20.1	−11.8	−7.6	−10.3	−19.0	−11.6	−7.1

[a]Structural parameters are based on the optimal geometries of DNA base pairs at the MP2/6-31G** level.

bonds changes only slightly. Thirdly, the polarization effect reduces slightly in the AT and H-AT pairs (less than 0.6 kcal mol^{-1}), but a modest reduction is observed in the GC pair (about 2 kcal mol^{-1}). This echoes the finding in the EDD map (Fig. 2b), which suggests a slight involvement of aromatic rings in the polarization of the GC pair. Finally, the biggest impact by removing the aromatic rings in the DNA pairs is the reduction of the electrostatic attractions. Table. 2 and 3 show that there are 2.4 and 2.8 kcal mol^{-1} reductions in the Heitler–London energies for the AT and H-AT pairs. In contrast, for the GC pair, the reduction is as high as 7.5 kcal mol^{-1}. Since Pauli-exchange repulsion decays more quickly (in exponential function [31]) than the electrostatic interaction and the binding regions are unchanged in the simplified model, I reckon that these energy changes are completely due to the reduction of electrostatic attractions. In this sense, I feel that the nature of RAHB is actually the resonance-enhanced electrostatic attraction and polarization effect, as the covalence in hydrogen bonds has only slight changes. To confirm this claim, we can further analyze individual hydrogen bonds in DNA base pairs with simple models, such as those shown in Scheme 1 for the AT pair, where the hydrogen bond lengths and orientations retain as in the base pair. With the 6-311+G** basis set, I computed the charge-transfer energies in the N–H···O and N···H–N hydrogen bonds which are −0.7 and −2.8 kcal mol^{-1}, respectively. The low value in the N-H···O bond may originate from the longer bond distance (2.97 Å) than the N···H–N bond (2.84 Å). Nevertheless, the sum of the charge-transfer energies in these two simple models without any resonance assistance (−3.5 kcal mol^{-1}) accounts for most of the charge-transfer stabilization energy in the AT base pair (−4.8 kcal mol^{-1}), indicating the limited enhancement of the hydrogen bond strength due to the resonance (Scheme 1).

The possible role of the C–H···O=C hydrogen bond in biological structures and functions has been widely discussed [45–52], primarily due to the large number of this kind of bonds in biosystems. To investigate the C–H···O=C hydrogen bond in the AT pair, I further removed the carbonyl group in model M1 to generate model M4, as shown in Fig. 3d. The results of the subsequent energy decomposition analysis of M4 are also listed in Table 3. Compared with model M1, once again the largest energetic change comes from the electrostatic interaction, which accounts for around 2.5 kcal mol^{-1} of 3.2 kcal mol^{-1} reduction in the total interaction. The removal of the carbonyl group reduces the electron-transfer stabilization by 10%, suggesting extremely weak covalence in C–H···O=C hydrogen bonding, whose interaction is dominated by the electrostatic forces. Here the allocation of a small fraction of the charge-transfer energy to the C–H···O=C hydrogen bond further legi-

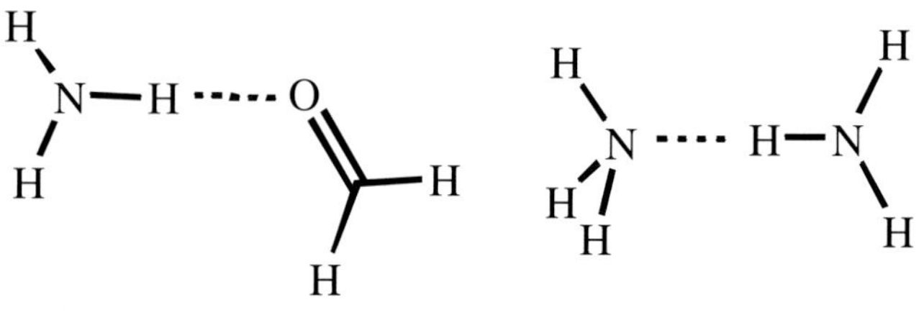

Scheme 1 Individual models for the H-bonds in the AT pair

timates our claim that the RAHB is of the electrostatic rather than charge-transfer nature.

Conclusion

There have been controversies regarding the nature of hydrogen bonds. While all experimental and theoretical studies confirm the directionality of hydrogen bonds, in accord with the definition of covalent bonds, a widespread belief is that hydrogen bonds are predominantly electrostatic in nature. The present studies of the DNA base pairs show a modest charge-transfer effect between pairing bases, and the charge-transfer stabilization accounts for only 20–35% of the overall interaction energy, while both the electrostatic (which is largely cancelled out by the Pauli-exchange interaction) and polarization energies play larger roles than the charge-transfer effect in the binding of base pairs, notably in the GC pair. This suggests a low covalence in the hydrogen bonds. Guerra et al.'s recent work, however, concluded a much more significant role of the electron-transfer effect in the DNA base pair than previous and current analyses and they found that donor-acceptor orbital interactions (i.e. charge transfer) is of the same order of magnitude as the electrostatic interaction term [24, 25, 26]. However, it should be noted that the orbital-interaction energy term in their analyses is composed of both the polarization and charge-transfer effects, and the former is of electrostatic origin.

Further analyses show that the strong binding in the GC results from favorable dipole-dipole electrostatic interaction between the guanine and cytosine bases, which is assisted by the π-electron delocalization from the amine groups to the carbonyl groups in both bases (but in reversal directions). I find that the π-electron delocalization in DNA bases strengthens the hydrogen bonds very moderately, and most of the enhanced stabilization due to π-resonance comes from the electrostatic and polarization effects. As a consequence, I recommend the terminology "resonance-assisted binding" instead of "resonance-assisted hydrogen bonding (RHAB)" to emphasize the electrostatic rather than electron-transfer nature of the enhanced stabilization in DNA base pairs.

Acknowledgement This work was supported by Western Michigan University.

References

1. Burkert U, Allinger NL (1982) Molecular mechanics. American Chemical Society, Washington DC
2. Brooks BR, Bruccoleri RE, Olafson BD, States DJ, Swaminathan S, Karplus M (1983) J Comp Chem 4:187–217
3. Reindl B, Clark T, Schleyer PvR (1996) J Comp Chem 17:1406–1430
4. Kitaura K, Morokuma K (1976) Int J Quantum Chem 10:325–340
5. Stevens WJ, Fink WH (1987) Chem Phys Lett 139:15–22
6. Gutowski M, Piela L (1988) Mol Phys 64:337–355
7. Cybulski SM, Scheiner S (1990) Chem Phys Lett 166:57–64
8. Moszynski R, Heijmen TGA, Jeziorski B (1994) Mol Phys 88:741–758
9. Glendening ED, Streitwieser A (1994) J Chem Phys 100:2900–2909
10. van der Vaart A, Merz KM Jr (1999) J Phys Chem A 103:3321–3329
11. Mo Y, Gao J, Peyerimhoff SD (2000) J Chem Phys 112:5530–5538
12. Gentle IR, Ritchie GLD (1989) J Phys Chem 93:7740–7744
13. Craven IE, Hesling MR, Laver DR, Lukins PB, Ritchie GLD, Vrbancich J (1989) J Phys Chem 93:627–631
14. Dougherty DA (1996) Science 271:163–168
15. Mecozzi S, West AP, Dougherty DA (1996) J Am Chem Soc 118:2307–2308
16. Cubero E, Luque FJ, Orozco M (1998) Proc Natl Acad Sci USA 95:5976–5980
17. Caldwell JW, Kollman PA (1995) J Am Soc Chem 117:4177–4178
18. Mo Y, Subramanian G, Ferguson DM, Gao J (2002) J Am Chem Soc 124:4832–4837
19. Morokuma K (1977) Acc Chem Res 10:294–300
20. Jeffrey GA (1997) An introduction to hydrogen bonding. Oxford University Press, New York
21. Šponer J, Jurecka P, Hobza P (2004) J Am Chem Soc 126:10142–10151
22. Cheatham TE III, Kollman PA (2000) Ann Rev Phys Chem 51:435–471
23. Kratochvíl M, Šponer J, Hobza P (2000) J Am Chem Soc 122:3495–3499
24. Guerra CF, Bickelhaupt FM (1999) Angew Chem Int Ed 38:2942–2945
25. Guerra CF, Bickelhaupt FM, Snijders JG, Baerends EJ (1999) Chem Eur J 5:3581–3594
26. Guerra CF, Bickelhaupt FM, Snijders JG, Baerends EJ (2000) J Am Chem Soc 122:4117–4128
27. Mo Y, Peyerimhoff SD (1998) J Chem Phys 109:1687–1697
28. Cubero E, Luque FJ, Orozco M (2001) J Am Chem Soc 123:12018–12025
29. Boys SF, Bernardi F (1970) Mol Phys 19:553–566
30. Mo Y, Gao J (2001) J Phys Chem A 105:6530–6536
31. Mo Y, Song L, Wu W, Zhang Q (2004) J Am Chem Soc 126:3974–3982
32. Gianinetti E, Raimondi, Tornaghi E (1996) Int J Quantum Chem 60:157–166
33. Gianinetti E, Vandoni I, Famulari A, Raimondi M (1998) Adv Quantum Chem 31:251–266
34. Schmidt MW, Baldridge KK, Boatz JA, Elbert ST, Gordon MS, Jensen JJ, Koseki S, Matsunaga N, Nguyen KA, Su S, Windus TL, Dupuis M, Montgomery JA (1993) J Comput Chem 14:1347–1363
35. Frisch MJ, Trucks GW, Schlegel HB, Scuseria GE, Robb MA, Cheeseman JR, Zakrzewski VG, Montgomery JAJ, Stratmann RE, Burant JC, Dapprich S, Millam JM, Daniels AD, Kudin KN, Strain MC, Farkas O, Tomasi J, Barone V, Cossi M, Cammi R, Mennucci B, Pomelli C, Adamo C, Clifford S, Ochterski J, Petersson GA, Ayala PY, Cui Q, Morokuma K, Malick DK, Rabuck AD, Raghavachari K, Foresman JB, Cioslowski J, Ortiz JV, Baboul AG, Stefanov BB, Liu G, Liashenko A, Piskorz P, Komaromi I, Gomperts R, Martin RL, Fox DJ, Keith T, Al-Laham MA, Peng CY, Nanayakkara A, Challacombe M, Gill PMW, Johnson B, Chen W, Wong MW, Andres JL, Gonzalez C, Head-Gordon M, Replogle ES, Pople JA (1998) Gaussian 98, Ed. A.9. Gaussian Inc, Pittsburgh PA
36. Bertran J, Oliva A, Rodríguez-Santiago L, Sodupe M (1998) J Am Chem Soc 120:8159–8167

37. Brameld K, Dasgupta S, Goddard WA III (1997) J Phys Chem B 101:4851–4859
38. Gilli G, Bellucci F, Ferretti V, Bertolasi V (1989) J Am Chem Soc 111:1023–1028
39. Bertolasi V, Gilli P, Ferretti V, Gilli G (1991) J Am Chem Soc 113:4917–4925
40. Steiner T (1998) Chem Commun 411–412
41. Gilli P, Bertolasi V, Ferretti V, Gilli G (2000) J Am Chem Soc 122:10405–10417
42. Munn RW, Eckhardt CJ (2001) J Phys Chem A 105:6938–6942
43. Gilli P, Bertolasi V, Pretto L, Ferretti V, Gilli G (2004) J Am Chem Soc 126:3845–3855
44. Fiacco DL, Mo Y, Hunt SW, Ott ME, Roberts A, Leopold KR (2000) J Phys Chem A 105:484–493
45. Sutor DJ (1962) Nature 195:68–69
46. Desiraju GR, Steiner T (2001) The weak hydrogen bond in structural chemistry and biology. Oxford University Press, New York
47. Meadows ES, De Wall SL, Barbour LJ, Fronczek FR, Kim M-S, Gokel GW (2000) J Am Chem Soc 122:3325–3335
48. Senes A, Ubarretxena-Belandia I, Engelman DM (2001) Proc Natl Acad Sci USA 98:9056–9061
49. Aravinda S, Shamala N, Bandyopadhyay A, Balaram P (2003) J Am Chem Soc 125:15065–15075
50. Manikandan K, Ramakumar S (2004) Proteins 56:768–781
51. Petrella RJ, Karplus M (2004). Proteins 54:716–726
52. Guo H, Beahm RF, Guo H (2004) J Phys Chem B 108:18065–18072
53. Šponer J, Leszczynski J, Hobza P (1996) J Phys Chem 100:1965–1974
54. Šponer J, Hobza P (1998) In: Schleyer PvR (ed) Encyclopedia of computational chemistry. Wiley, New York, pp 777–789
55. Saenger W (1984) Principles of nucleic acid structure (and references therein). Springer, Berlin Heidelberg New York, pp 123–124

J Mol Model (2006) 12: 673–680
DOI 10.1007/s00894-005-0045-3

ORIGINAL PAPER

Andrew Streitwieser

Ion pair aggregates and reactions; experiment and theory

Received: 10 May 2005 / Accepted: 28 September 2005 / Published online: 9 December 2005

Abstract Two methods have been developed for determining the aggregation equilibrium constants of lithium enolates based on the change in UV-vis spectrum with concentration and the effect of aggregation on proton-transfer equilibria. Dimers and tetramers are common. Substitution α to the carbonyl group generally reduces aggregation. Kinetic studies show that S_N2 alkylations generally occur with the monomers, even in the presence of large amounts of aggregate. The qualitative chemistry is well modeled by ab initio computations at the HF level with modest basis sets; in these studies solvation is modeled by a combination of coordination of lithium cation with an ether oxygen and the electrostatic interaction of the resulting dipoles and quadrupoles with the solvent dielectric continuum.

Keywords Aggregation · Alkylation · Equilibrium constant · Acidity · Molecular orbital

Introduction

Modern organic syntheses frequently take place in organic solvents in which the active reagents are often present as ion pairs and aggregates. Relatively little research has been done on the equilibrium constants for aggregate formation and their reactivities. Much of my research during the past decade or so has been devoted to the study of aggregates of lithium and cesium enolates. In this review, I will emphasize lithium enolates and our methods for determining aggregation equilibrium constants and reactivities, the experimental results and our initial approaches in modeling this chemistry with ab initio molecular orbital theory at a modest level, RHF with the 6-31+G* basis set.

Preformed lithium enolates are important reagents in organic synthesis because they can be alkylated by S_N2 reactions with alkyl halides and sulfonates and they undergo aldol addition reactions with carbonyl compounds; both are useful routes to generating new carbon-carbon bonds [1–5]. These enolates have long been known to be aggregated in ethereal solutions as well as in the solid state [6], and such aggregates have been suggested to be involved in reactions. [7] Actual evidence for the involvement of lithium enolate aggregates, however, is sparse. The problem is that at normal temperatures the equilibria among aggregates are rapid, giving rise to a classic Curtin–Hammett situation (Scheme 1) [8].

The total rate of reaction is given by Eq. 1 in which the aggregation equilibrium constants are defined relative to the ion-pair monomer whose concentration is $[M]$. It is possible that even though an aggregate dominates the equilibrium, the rate constant of a monomer could be sufficiently high that most of the flux of reaction occurs through the monomer. Indeed, this is what we have generally found.

$$\text{Rate} = \Sigma k_n K_n [M]^n \tag{1}$$

Methods and results

In order to establish the relative roles of monomer and aggregate in reactivity, it is necessary to determine the individual aggregation equilibrium constants and rate constants. Determining the equilibrium constants involves measurements over a concentration range in which the

Dedicated to Professor Dr. Paul von Ragué Schleyer on the occasion of his 75th birthday

This review is based on the fifth annual Paul vR Schleyer Lecture presented at Professor Schleyer's 75th birthday symposium, Athens GA, Feb 26, 2005. Carbon Acidity 123. For no. 122 see Streitwieser A, Kaufman MJ, Bors DA, MacArthur CA, Murphy JT, Guibe F, Arkivoc (2005) (vi) 200–210

A. Streitwieser (✉)
Department of Chemistry, University of California,
Berkeley, CA 94720-1460, USA
e-mail: astreit@berkeley.edu
Tel.: +1-510-6422205
Fax: +1-510-8416072

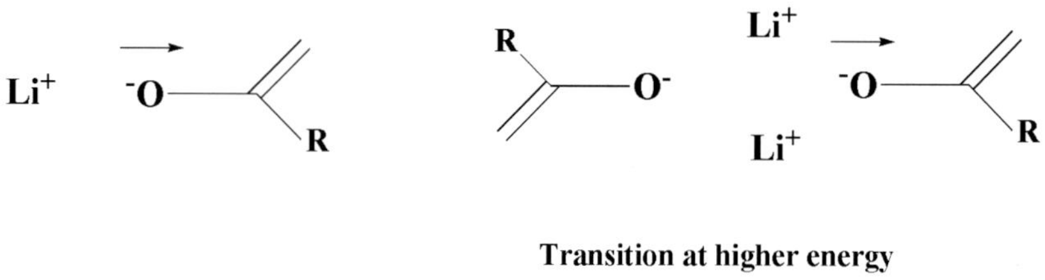

Fig. 1 The transition moment of an aggregate is at higher energy because of the electrostatic repulsion to additional cations

relative aggregation concentrations vary. In practice, this concentration range is of the order of magnitude of 10^{-5} to 10^{-3} M, a concentration so dilute that the only practical analytical tool is UV-vis spectroscopy. Accordingly, the enolates studied generally contain a phenyl or biphenylyl ring to provide a useful chromophore for measurements. Two general tools were discovered to be effective in this study:

1. The UV-vis spectrum of some enolates was found to vary as a function of concentration, indicating that different aggregates have different spectra [9].
2. Coupled equilibria: the proton-transfer equilibria to monomeric indicators are skewed by aggregation [10].

In applying the first method, a series of spectra were taken over a more than tenfold concentration range and the digitized data were subjected to Singular Value Decomposition, SVD, a method that involves a matrix diagonalization to give a series of vectors and associated coefficients [11]. The first vector represents the mean of the spectra and has a large coefficient. The second is the first order deviation from the mean and has a significant but smaller value. The remaining coefficients are much smaller and represent noise. The two significant vectors suggest two components. In many cases the existence of an isospestic point also points to two components. From further matrix manipulation, the spectra of the two components can be determined. When each experimental spectrum is analyzed in terms of these component spectra, the composition is given. For a monomer–dimer mixture a plot of [dimer] vs [monomer]^2 gives a straight line whose slope is $K_{1,2}$; for a monomer–tetramer mixture the corresponding plot is of [tetramer] vs [mono-

Table 1 Aggregation of some lithium enolates in THF

Entry	Li Enolate	$K_{1,2}$	$K_{1,4}$	[M] in 1 M	pK
1	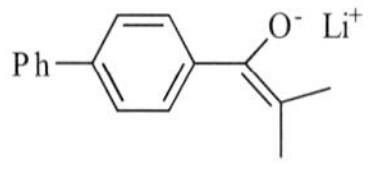 LiPhIBP		5.0E+8	0.0047	15.9
2	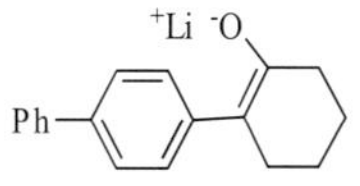 LiBPCH	4300		0.011	12.6
3	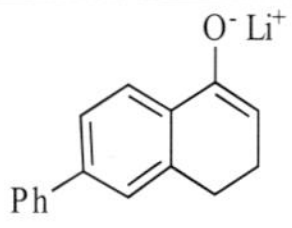 LiPAT		4.7E+10	0.0015	14.2
4	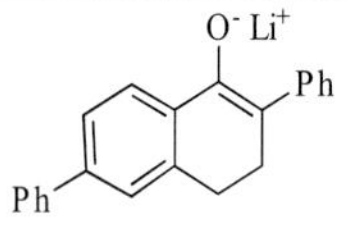 LiPhPAT	2650		0.014	11.1
5	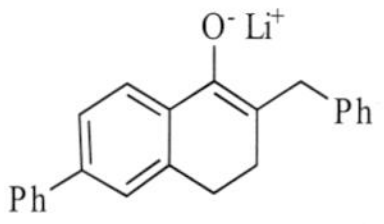 LiBnPAT	3800		0.011	14.0
6	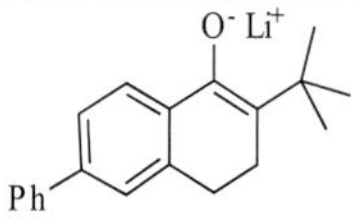 LiBuPAT	1400		0.019	16.7

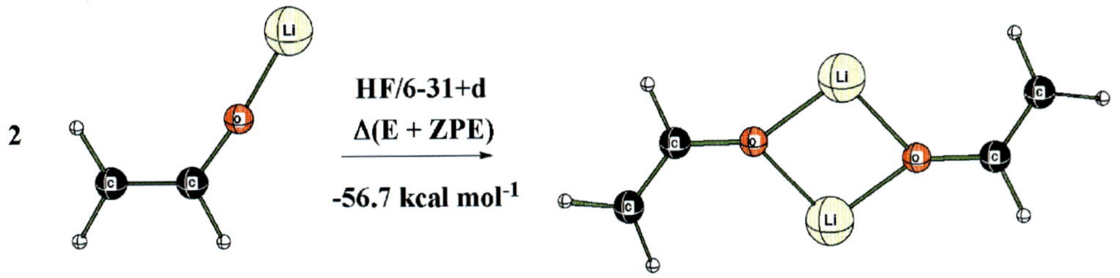

Fig. 2 Dimerization of lithium vinyloxide at HF/6-31+G* as ΔE plus unscaled ZPE

mer][4]. The corresponding linear relation establishes the nature of the equilibrium (Eqs. 2 and 3).

$$2\,\text{Monomer} \overset{K_{1,2}}{\rightleftharpoons} \text{Dimer} \tag{2}$$

$$4\,\text{Monomer} \overset{K_{1,4}}{\rightleftharpoons} \text{Tetramer} \tag{3}$$

We have found the spectrum of the aggregate always to have λ_{max} at shorter wavelength than the monomer. We interpret this effect simply on the basis of Fig. 1. The transition involves movement of electron density from the oxide anion to the organic moiety. Such movement is electrostatically more difficult for an aggregate with more Li^+ cations close to the negative charge.

The second method is based on the proton-transfer equilibrium in Eq. 4 in which Ind-H is an indicator whose lithium salt is known to be monomeric. This equilibrium is skewed to the right by aggregation of $R^- Li^+$ (Eq. 5).

$$\text{R}-\text{H} + \text{Ind}^-\text{Li}^+ \overset{K_o}{\rightleftharpoons} \text{Ind}-\text{H} + \text{R}^-\text{Li}^+ \tag{4}$$

$$\text{n}\,\text{R}^-\text{Li}^+ \overset{K_{1,\text{n}}}{\rightleftharpoons} (\text{R}^-\text{Li}^+)_\text{n} \tag{5}$$

The equilibrium constant defined by Eq. 6 is equivalent to the relative ion pair p*K* (Eq. 7). The symbol $\{R^- Li^+\}$ is used to denote the formal concentration of $R^- Li^+$ as given, for example, by spectroscopy.

$$K_{\text{obs}} = \{\text{R}^-\text{Li}^+\}[\text{Ind}-\text{H}]/[\text{R}-\text{H}][\text{Ind}^-\text{Li}^+] \tag{6}$$

$$\text{p}K_{\text{obs}}(\text{RH}) = \text{p}K(\text{Ind}-\text{H}) - \log K_{\text{obs}} \tag{7}$$

We have determined the p*K*'s of a number of such indicators relative to the assumed value of 22.9 for fluorene, the per-hydrogen p*K* in DMSO [12].

Summed over the full set of aggregates in a lithium enolate equilibrium, K_{obs} is given by the expression in Eq. 8 [11]. In turn, this equation reduces to Eq. 9 for a simple monomer–dimer equilibrium and to Eq. 10 for a monomer–tetramer equilibrium. In Eq. 9, for example, a plot of K_{obs} vs $\{R^- Li^+\}/K_{obs}$ gives a straight line from which K_o and $K_{1,2}$ can be derived from the intercept and slope.

$$K_{\text{obs}} = \sum \text{n}K_{1,\text{n}}K_\text{o}^\text{n}\left(\{\text{R}^-\text{Li}^+\}/K_{\text{obs}}\right)^{\text{n}-1} \tag{8}$$

$$K_{\text{obs}} = K_o + 2K_{1,2}K_o^2\{\text{R}^-\text{Li}^+\}/K_{\text{obs}} \tag{9}$$

$$K_{\text{obs}} = K_o + 4K_{1,2}K_o^4\left(\{\text{R}^-\text{Li}^+\}/K_{obs}\right)^3 \tag{10}$$

For enolates in which both methods could be applied, the results are generally in good agreement. These methods have been applied to a range of lithium enolates [13–21]. A few of the values obtained are summarized in Table 1. One important generalization from these results is that substitution in the α-position results in reduced aggregation. This generalization can be seen by comparison of entries 1 with 2 in Table 1 and of entry 3 with 4–6.

A simple ab initio model of enolate dimerization is shown in Fig. 2.

The reaction is strongly exothermic, as expected electrostatically for the union of two ion pairs, but this picture is misleading because no role is assigned to solvent. The aggregation of the lithium salt of *p*-phenylisobutyrophenone (LiPhIBP) is almost independent of temperature (Table 2), indicating that the primary driving force is entropic rather than enthalpic [16]. This had been shown previously by others, notably Jackman for lithium phenolates [22] and Chabanal for other 1:1 salts [23]. When the coordinated solvent is included with tetracoordinated lithium cations, the aggregation equilibria become Eqs. 11 and 12 (in which S is the solvent, generally THF).

$$2\text{ROLiS}_3 \rightleftharpoons (\text{ROLiS}_2)_2 + 2\text{S} \tag{11}$$

$$2(\text{ROLiS}_2)_2 \rightleftharpoons (\text{ROLiS})_4 + 4\text{S} \tag{12}$$

Table 2 Effect of temperature on the aggregation of the lithium enolate of *p*-phenylisobutyrophenone (LiPhIBP); ΔH°=1.5±1 kcal mol^{-1}, ΔS°=40±5 e.u. [16]

Temp. °C	$K_{1,4}M^{-3}$
25.0	5.2×10^7
0.0	4.4×10^7
−20.0	3.2×10^7

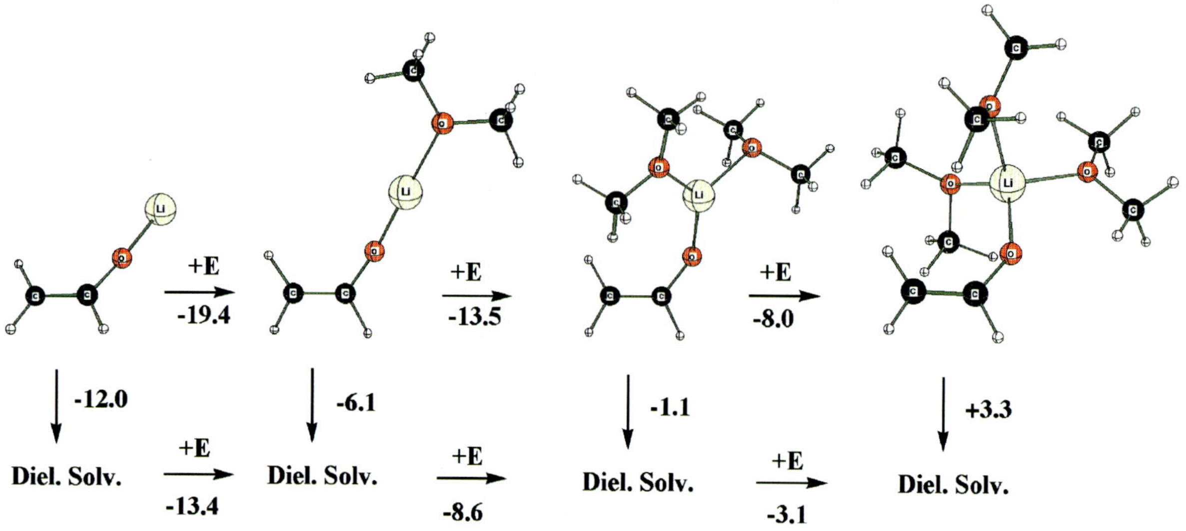

Fig. 3 Effect of successive solvation of lithium vinyloxide by dimethyl ether (*E*), HF/6-311+G*, ΔE + unscaled ZPE (kcal mol^{-1}), as a combination of coordination and electrostatic interaction with the dielectric continuum using the dielectric constant of THF [25]

In each case lithium remains coordinated to four oxygens, giving only a small enthalpic change but aggregation liberates solvent molecules with their increased entropy.

In computations, dimethyl ether has frequently been used in place of THF because of its smaller size [24, 25]. The effect of solvation by dimethyl ether is summarized in Fig. 3.

We considered solvation as two specific components. One is the specific coordination of solvent with lithium cation and computed as the "supermolecule" with the coordinated solvent specifically included in the molecule. The second component is the "dielectric solvation" of the coordinated species, the electrostatic interaction of the supermolecule with a "Polarized Continuum" dielectric as included in the Gaussian program using the dielectric constant of THF [26]. Coordination is a significant exothermic term but is smaller with successive coordinations. This effect is consistent with a simple electrostatic model of interaction of the solvent dipole with the cationic charge. The successive coordinations involve the additional electrostatic interactions with the opposing dipoles of the preceding solvents. The dielectric term is initially important but rapidly diminishes with successive solvation. The first complex is a dipole whose electrostatic interaction with the dielectric is significant. As successive solvent is added, the resulting dipole moments are smaller. Further-

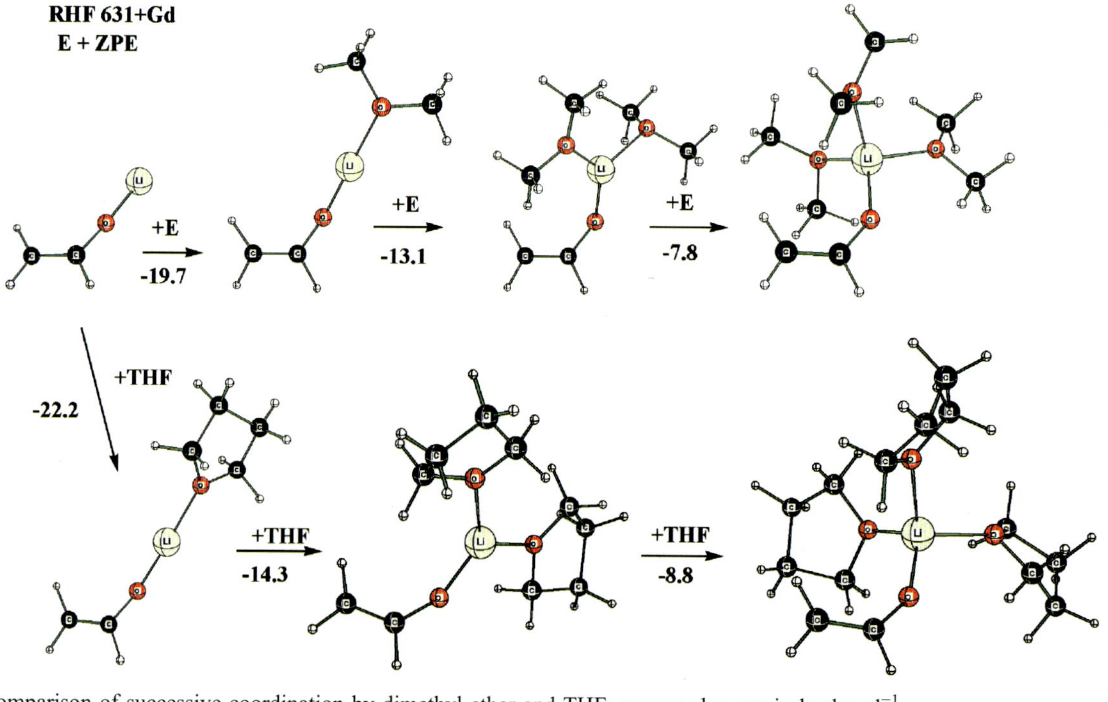

Fig. 4 Comparison of successive coordination by dimethyl ether and THF; energy changes in kcal mol^{-1}

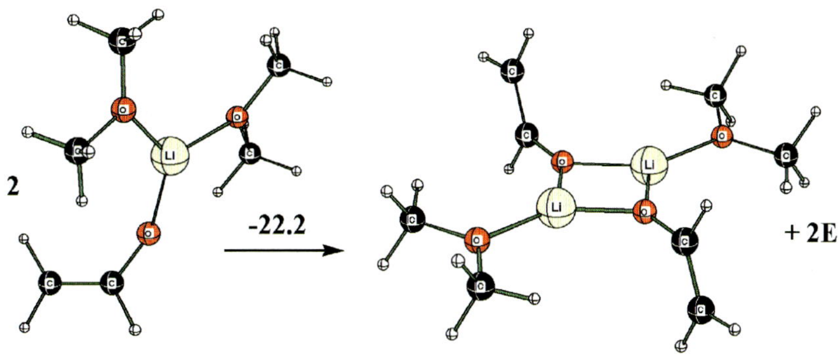

Fig. 5 Effect of solvation on dimerization of lithium vinyloxide at HF/6-31+G* + unscaled ZPE, kcal mol^{-1}

more, the increasing size of the solvated moiety requires a greater cavity in the continuous dielectric until at the final structure the electrostatic term is insufficient to compensate for the cavity required. The net energy change for the final coordination of solvent, −3.1 kcal mol^{-1}, is small and interesting. Chabanal [23] has suggested that the entropy change of such a reaction be approximated by the entropy of freezing of the solvent; for THF this quantity is −12.4 e.u. which is equivalent at room temperature to an enthalpy change of 3.7 kcal mol^{-1}; that is, the last coordination in Fig. 2 has a $\Delta G°$ of about 0 or $K \cong 1$. This means that the lithium enolate is not fully coordinated in solution—substantial populations are apparently tricoordinated. Because of the relatively small role of dielectric solvation in these systems we have not generally computed this term and have focused the computed models on the role of coordination solvation.

We tested how well dimethyl ether works as a computational model for THF by calculating the energetics of successive THF coordination of lithium vinyloxide compared to Me_2O at the smaller basis set level of 6-31+G*; we had shown previously that this is as effective a level as the larger 6-311+G* for many purposes [25]. The comparison shown in Fig. 4 shows that THF is generally more coordinating than Me_2O but only by about 1 kcal mol^{-1}. Even this difference is expected to be reduced by the lower dielectric solvation of THF's larger cavity size.

The effect of solvation on dimerization of lithium vinyloxide is summarized in Fig. 5. Coordination about lithium has been kept constant in the process shown. Dimerization is now much less exothermic than the uncoordinated model in Fig. 2 and would be less exothermic if the monomer has greater coordinative solvation than the dimer.

Having determined a number of aggregation equilibrium constants, we can now look at the kinetics. Kinetic measurements were made using the UV spectra of the enolates, which were therefore at low concentration, typically 10^{-3}–10^{-4} M, much less than the concentrations of the alkyl halides used, typically about 0.02 M. The second-order reactions are thus of pseudo-first order. Moreover, the reactions were typically allowed to run to only about 10% in order to avoid possible complications with mixed aggregates involving the reaction products. Under these conditions the ln (fraction reaction)≅(fraction reaction) and it was only necessary to plot the extent reaction *vs* time to determine the pseudo-first-order rate constants. These numbers were divided by the concentration of the alkyl halide to obtain the corresponding second-order rate constant. These rate constants are composites since the alkylation reaction is an S_N2 process with all of the various aggregates present. For example, for a monomer–dimer system, the total rate is given by Eq. 13, which can be rearranged to Eq. 14. In these equations the *k*'s are the appropriate second-order rate constants.

$$\mathrm{Rate}/[\mathrm{RX}] = k_M[\mathrm{Monomer}] + k_D[\mathrm{Dimer}] \tag{13}$$

$$\mathrm{Rate}/[\mathrm{RX}][\mathrm{Dimer}] = k_M[\mathrm{Monomer}]/[\mathrm{Dimer}] + k_D \tag{14}$$

Since the aggregation equilibrium constant is known, we can calculate [Monomer] and [Dimer] at any given concentration of {Li enolate}. Eq. 14 has the convenient form that a plot of the rate quantity on the left *vs* [Monomer]/[Dimer] should give a straight line with slope k_M and intercept k_D. Many examples of such treatments have given normal values of k_M and values for k_D (or k_T) of essentially zero. For these cases, a simple plot of Rate/[RX] vs [Monomer] gives a normal straight line without any upward curvature indicative of a role for k_D or k_T.

For the alkylation of the lithium enolate of *p*-phenylsulfonylisobutyrophenone (LiSIBP) with *p*-*t*-butylbenzyl bromide, a complete analysis was possible to give k_M=0.33 $M^{-1}s^{-1}$ and k_D=1.1×10^{-4} $M^{-1}s^{-1}$ or k_M/k_D=3,000

S_N2 Monomer

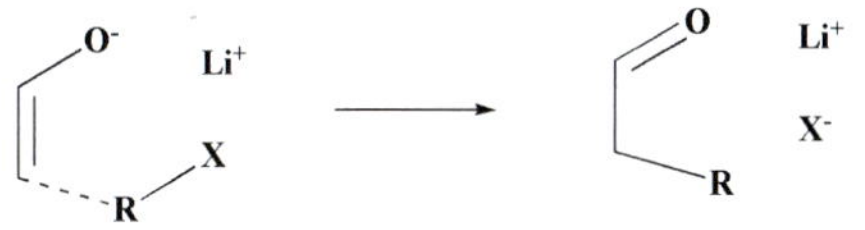

S_N2 Dimer

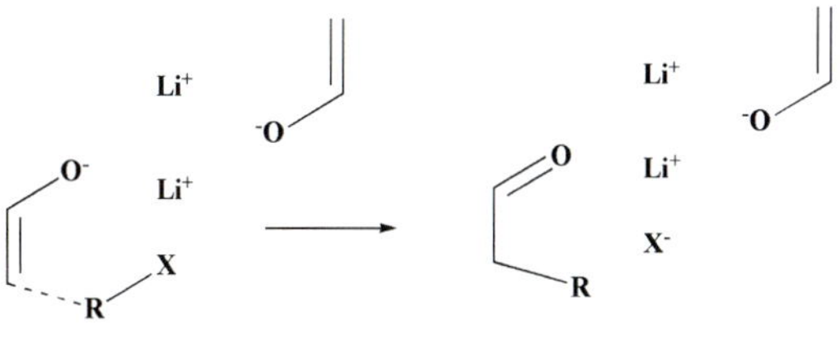

Fig. 6 Assumed cyclic transition structure models for alkylation of lithium enolates

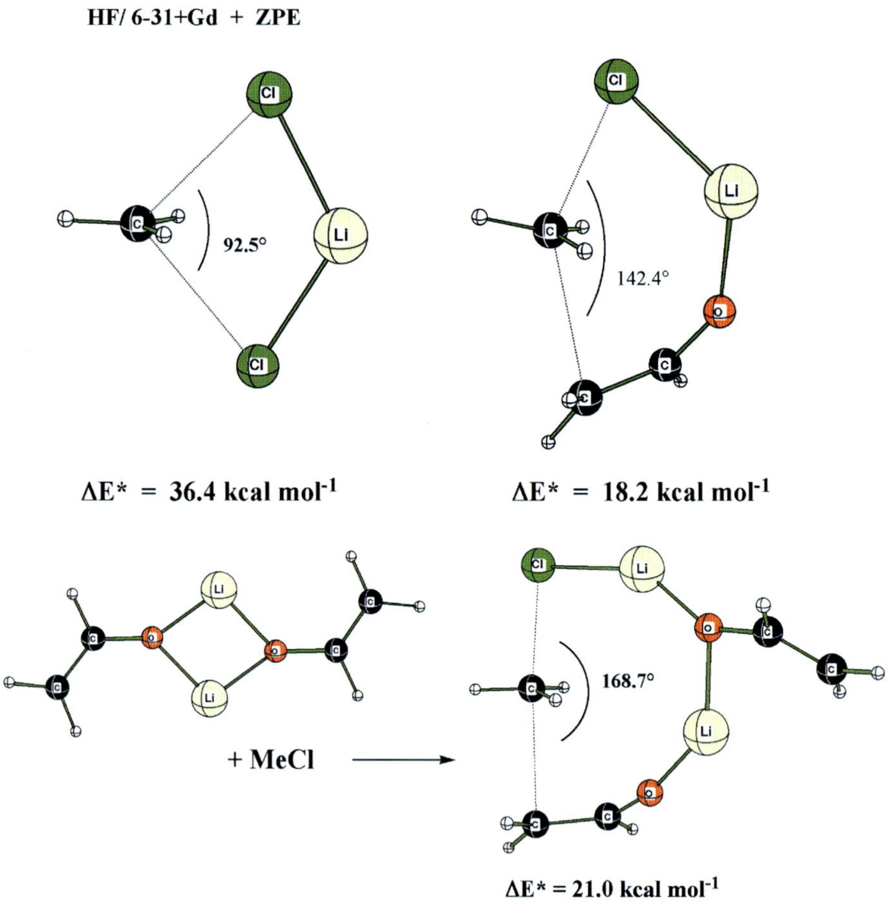

Fig. 7 Computational models for some ion pair S_N2 reactions

[14]. The general finding is that the monomers are much more reactive in alkylation reactions than the aggregates. This behavior was rationalized on the basis of the assumed cyclic transition structures in Fig. 6.

The S_N2 reaction of the dimer is then expected to be slower because the oxide ion is neighbor to two lithium cations and is therefore less nucleophilic and the lithium cation is close to two oxide ions and is therefore less electrophilic. This hypothesis was tested by *ab initio* computations, as summarized in Fig. 7.

These are ion-pair S_N2 reactions and simple such reactions of methyl halides with alkali halides were shown previously to have strongly bent reaction angles and consequently high reaction barriers [27]. The computed transition structure for reaction of lithium vinyloxide with methyl chloride is that of a six-membered ring with an expanded reaction bond angle and a much lower and more accessible reaction barrier. The corresponding reaction of the dimer has broken one Li–O coordination bond with a correspondingly larger and almost normal reaction bond angle; nevertheless, the loss of coordination results in a substantially higher reaction barrier.

Reactions of dilithiated carbonyl compounds have achieved great importance in synthetic chemistry. For example, substituted acetic acids can usually be dimetallated to the dilithium enediolates, which can then be alkylated to give the corresponding disubstituted acetic acids cleanly (Eq. 15).

$$RCH_2COOLi \rightarrow RCH = C(OLi)_2 \xrightarrow{R'X} RR'CHCOOLi \tag{15}$$

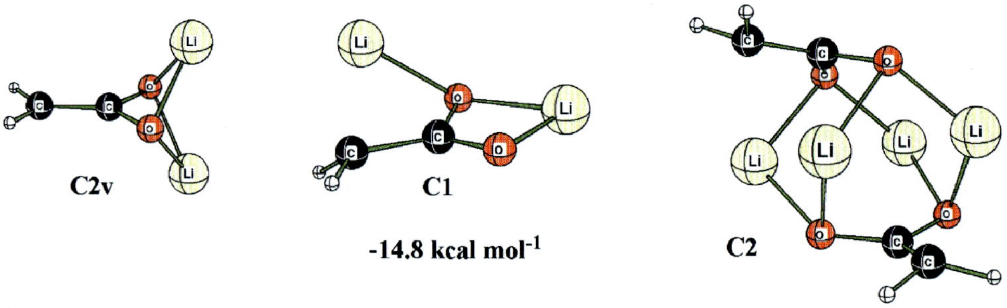

Fig. 8 Computed structures for unsolvated monomer and dimer of $CH_2{=}C(OLi)_2$ at HF/6-31+G*. The C1 structure of the monomer is 14.8 kcal mol^{-1} more stable than the C2v structure

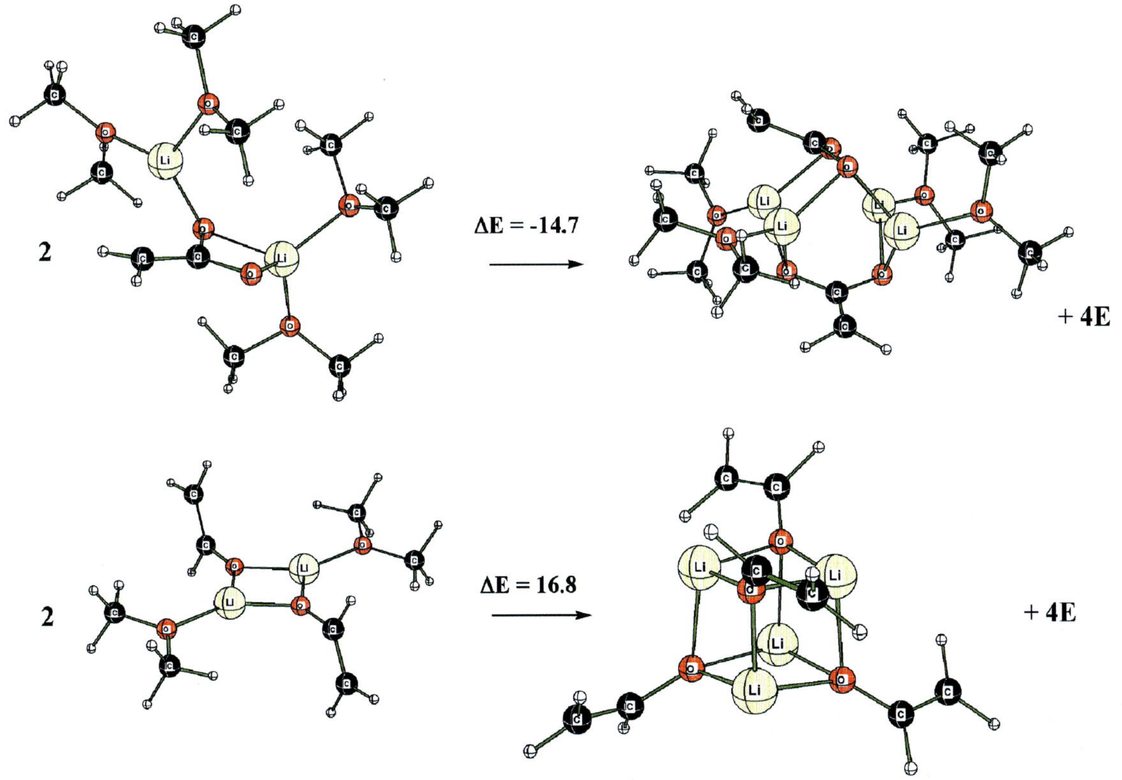

Fig. 9 Computed dimerization reactions at HF/6-31+G* with lithium solvated such that the same numbers of dimethyl ethers are liberated. Energy changes shown are kcal mol^{-1}

In earlier work, the dilithium salt of α-naphthylacetic acid was found to be aggregated and to have a p*K* of about 23 [28]. The spectrum does not change with concentration and lithium carboxylates themselves are aggregated, so neither of our two general methods could be used to determine the extent of aggregation. The aggregates are broken up with hexamethylphosphoramide (HMPA), however, and the stoichiometry indicates the dilithium salt to be a dimer [29]. This dimer is exceptionally tight and the amount of monomer present is too small to show up in alkylation kinetics.

Ab initio modeling at the HF/6-31+G* level showed two minima on the potential energy surface for the unsolvated dilithium salt of acetic acid, a C2v structure and a

Fig. 10 Computed energetics of some equilibria with dilithium enediolate to give mixed aggregates with lithium amide. Energy changes in kcal mol^{-1} indicate that lithium amide forms 2:1 mixed aggregates

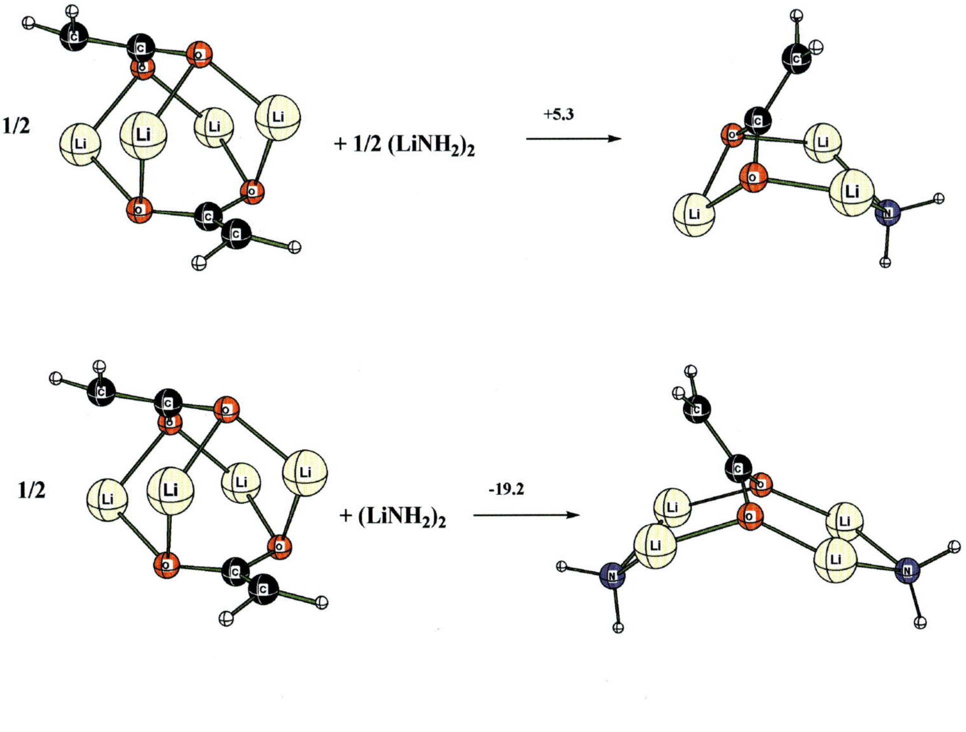

lower energy C1 structure as shown in Fig. 8. The corresponding dimer has a structure that resembles a distorted cubic structure of a lithium enolate tetramer. The dimerization of CH_2=C(OLi)$_2$ was found to be much more exothermic than that of CH_2=CHOLi dimer going to tetramer (Fig. 9). In both cases the lithiums were coordinatively solvated to the same degree such that the same numbers of ethers were liberated in the reaction. The computed energetics are in good qualitative agreement with the experimental chemistry.

Dilithium α-naphthylacetate was found to form mixed aggregates with some lithium amides but the stoichiometry was not determined. Computed structures and reactions summarized in Fig. 10 suggest that the amides form 2:1 complexes with the dilithium enediolate.

The molecular orbital computations were accomplished with various versions of the Gaussian program up to Gaussian 03 [26].

Acknowledgements In the period since I became Emeritus in 1993, a large group of undergraduate and graduate students and postdoctoral coworkers have done experimental and computational research in this project: (in alphabetical order) Alessandro Abbotto, Godwin S-C Choy, Jayasree Elambalassery, Antonio Facchetti, Rustram Gareyev, Faraj Hasanayn, Marc Husemann, Kathleen V. Kilway, Yeong-Joon Kim, James A. Krom, Simon S-W Leung, Joe C-Y Liang, Arlene McKeown, James Pugh, Manolis Stratakis, and Daniel Z-R. Wang. Two contributing visitors were Prof. Eusabio Juaristi and Prof. Larry M. Pratt. This work was assisted by grants from the National Science Foundation and the National Institutes of Health.

References

1. Evans DA, Nelson JV, Taber TR (1982) Topics in Stereochemistry 13:1–115
2. Mukaiyama T (1982) Org React 28:203–331
3. Heathcock CH (1991) In: Comprehensive organic synthesis, vol 1. Pergamon, Oxford, pp 181–238
4. Braun M (1992) Advances in carbanion chemistry, vol 1. Jai, Greenwich CT, pp 177–247
5. Arya P, Qin H (2000) Tetrahedron 56:917–947
6. Wen JQ, Grutzner JB (1986) J Org Chem 51:4220, Seebach D (1988) Angew Chem Int Ed Engl 27:1624-1654, Boche G (1989) Angew Chem Int Ed Engl 28:277, Arnett EM, Moe KD (1991) J Am Chem Soc 113:7288–7293, Jackman LM, Bortiatynski J (1992) Adv Carbanion Chem 1:45–87
7. Seebach D, Amstutz R, Dunitz JD (1981) Helv Chim Acta 64:2622-2626, Williard PG, Hintze MJ (1990) J Am Chem Soc 112:8602-8604, Bach RD, Andres JL, Davis FA (1992) J Org Chem 57:613, Juaristi E, Beck AK, Hansen J, Matt T, Mukhopadhyay T, Simson M, Seebach D (1993) Synthesis 1271, Wei Y, Bakthavatchalam R (1993) Tetrahedron 49:2373–2390, Solladie-Cavallo A, Csaky AG, Gantz I, Suffert J (1994) J Org Chem 59:5343–5346
8. Seeman JI (1983) Chem Rev 83:83–134
9. Ciula JC, Streitwieser A (1992) J Org Chem 57:431–432, correction p 668
10. Kaufman MJ, Streitwieser A (1987) J Am Chem Soc 109: 6092–6097
11. Krom JA, Petty JT, Streitwieser A (1993) J Am Chem Soc 115: 8024–8030
12. Streitwieser A, Wang DZ, Stratakis M, Facchetti A, Gareyev R, Abbotto A, Krom JA, Kilway KV (1998) Can J Chem 76:765–769
13. Abbotto A, Streitwieser A (1995) J Am Chem Soc 117:6358–6359
14. Abu-Hasanayn F, Stratakis M, Streitwieser A (1995) J Org Chem 60:4688–4689
15. Gareyev R, Ciula JC, Streitwieser A (1996) J Org Chem 61: 4589–4593
16. Abbotto A, Leung S S-W, Streitwieser A, Kilway KV (1998) J Am Chem Soc 120:10807–10813
17. Streitwieser A, Wang DZ-R (1999) J Am Chem Soc 121:6213–6219
18. Wang DZ-R, Streitwieser A (1999) Can J Chem 77:654–658
19. Wang DZ-R, Kim Y-J, Streitwieser A (2000) J Am Chem Soc 122:10754–10760
20. Streitwieser A, Kim Y-J, Wang DZ-R (2001) Org Lett 3:2599–2601
21. Wang DZ, Streitwieser A (2003) J Org Chem 68:8936–8942
22. Jackman LM, DeBrosse CW (1983) J Am Chem Soc 105: 4177–4184
23. Chabanel M (1990) Pure & Appl Chem 62:35–46
24. Abbotto A, Streitwieser A, Schleyer PvR (1997) J Am Chem Soc 119:11255–11268
25. Pratt LM, Streitwieser A (2003) J Org Chem 68:2830–2838
26. Frisch MJ, Trucks GW, Schlegel HB, Scuseria GE, Robb MA, Cheeseman JR, Montgomery JA, Vreven, T, Kudin KN, Burant JC, Millam JM, Iyengar SS, Tomasi J, Barone V, Mennucci B, Cossi M, Scalmani G, Rega N, Petersson GA, Nakatsuji H, Hada M, Ehara M, Toyota K, Fukuda R, Hasegawa J, Ishida M, Nakajima T, Honda Y, Kitao O, Nakai H, Klene M, Li X, Knox JE, Hratchian HP, Cross JB, Adamo C, Jaramillo J, Gomperts R, Stratmann RE, Yazyev O, Austin AJ, Cammi R, Pomelli C, Ochterski JW, Ayala PY, Morokuma K, Voth GA, Salvador P, Dannenberg JJ, Zakrzewski VG, Dapprich S, Daniels AD, Strain MC, Farkas O, Malick DK, Rabuck AD, Raghavachari K, Foresman JB, Ortiz JV, Cui Q, Baboul AG, Clifford S, Cioslowski J, Stefanov BB, Liu G, Liashenko A, Piskorz P, Komaromi I, Martin RL, Fox DJ, Keith T, Al-Laham MA, Peng CY, Nanayakkara A, Challacombe M, Gill PMW, Johnson B, Chen W, Wong MW, Gonzalez C, Pople JA (2003) Gaussian 03, Revision B.05. Gaussian Inc, Pittsburgh PA
27. Harder S, Streitwieser A, Petty JT, Schleyer PvR (1995) J Am Chem Soc 117:3253–3259
28. Gronert S, Streitwieser A (1988) J Am Chem Soc 110:4418–4419
29. Streitwieser A, Husemann M, Kim Y-J (2003) J Org Chem 68:7937–7942

J Mol Model (2006) 12: 681–686
DOI 10.1007/s00894-006-0114-2

ORIGINAL PAPER

Ying Xiao · Dawoon Jung · Tamara Gund ·
Sanjay V. Malhotra

An experimental and theoretical study of the enantioselective deprotonation of cyclohexene oxide with isopinocampheyl-based chiral lithium amides

Received: 9 May 2005 / Accepted: 2 March 2006 / Published online: 17 May 2006

Abstract The mechanism of the enantioselective deprotonation of cyclohexene oxide with isopinocampheyl-based chiral lithium amide was studied by quantum chemical calculations. The transition states of eight molecules were fully optimized at the ab initio HF/3-21G and density functional B3LYP/3-21G levels with Gaussian 98. The activation energies were calculated at the B3LYP/6-31+G (3df,2p)//B3LYP/3-21G level. We found the theoretical evaluation to be consistent with the experimental data. At the best case, an enantiomeric excess of up to 95% for (*R*)-2-scyclohexen-1-ol was achieved with (−)-*N, N*-diisopinocampheyl lithium amide.

Keywords Theoretical study · Enantioselective deprotonation · Cyclohexene oxide · Chiral lithium amides · Density functional activation energy · B3LYP/6-31+G(3df,2p)

Experimental data

Lithium amides are an important class of reagents in organic synthesis and have been used extensively as strong bases for various reactions [1–3]. The deprotonation of an epoxide with a lithium amide to obtain an allylic alcohol was first reported in 1970 in a deuterium-labeling study by Thummel and Rickborn [4]. The reaction is thought to proceed via a cyclic six-membered transition state, formed by a 1:1 epoxide to base complex, where the base coordinates to the lone pair of electrons on oxygen, thereby facilitating the β-hydrogen removal. On the other hand, if a prochiral epoxide is deprotonated with a chiral lithium amide, it could give an optically active product. Such a rationale was first proposed over two decades ago, for a nonenzymatic "asymmetric deprotonation" of cyclohexene oxide to obtain 2-cyclohexen-1-ol (Scheme 1) [5].

Over the years, a number of different chiral lithium amides have been designed and used for various deprotonation reactions [6]. As a result, asymmetric synthesis using chiral lithium amides has emerged as a useful method for the preparation of nonracemic compounds [7–11]. The methodology offers the advantage that the chiral auxiliaries can be recycled easily, thereby making the process effective and cost efficient. These chiral lithium bases have been exploited by a variety of efficient enantioselective reactions. Considering the remarkable success of pinene-based reagents for various organic transformations [12, 13], we synthesized a set of chiral secondary amines derived from α-pinene (Scheme 2) (Malhotra and Brown 1998, unpublished results).

The lithium salts of these amines were tested in the deprotonation of *meso*-epoxides [14]. Diisopinocampheylamine (DIPAM) prepared from (+)-α-pinene has the advantage of C_2-symmetry. *N*-cyclohexyl-*N*-isopinocampheylamine (ChxIPAM), *N*-benzyl-*N*-isopinocampheylamine (BzIPAM) and *N*-isopropyl-*N*-isopinocampheylamine (^{i}PIPAM) were chosen to study the steric effect of the isopinocampheyl moiety in the deprotonation of a *meso*-epoxide to allylic alcohol. Reaction with a stoichiometric amount of DIPAM gave the product with >99% enantiomeric excess (EE) on deprotonation of cyclohexene oxide.

To generate a catalytic cycle [15, 16] (Scheme 3), numbers of achiral lithium amides were compared with chiral amines for the opening of *meso*-epoxides. Lithium diisopropylamide (LDA) emerged as the achiral amide of choice because first, the chiral amine could be regenerated easily, and secondly, due to the fact that there was a significant difference in the rate of competitive reactions

Dedicated to Professor Dr. Paul von Ragué Schleyer on the occasion of his 75th birthday.

Electronic Supplementary Material Supplementary material is available for this article at http://dx.doi.org/10.1007/s00894-006-0114-2

Y. Xiao · D. Jung · T. Gund · S. V. Malhotra (✉)
Department of Chemistry and Environmental Science,
New Jersey Institute of Technology, University Heights,
Newark, NJ 07102, USA
e-mail: malhotra@njit.edu

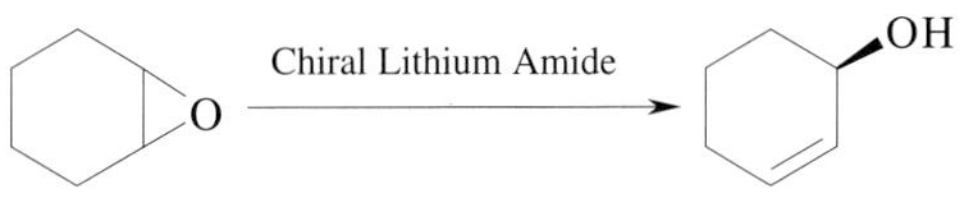

Scheme 1 Asymmetric deprotonation of cyclohexene oxide

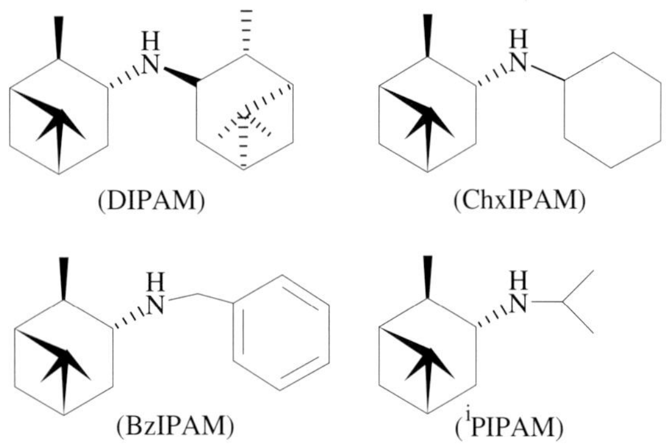

Scheme 2 Four kinds of lithium amides

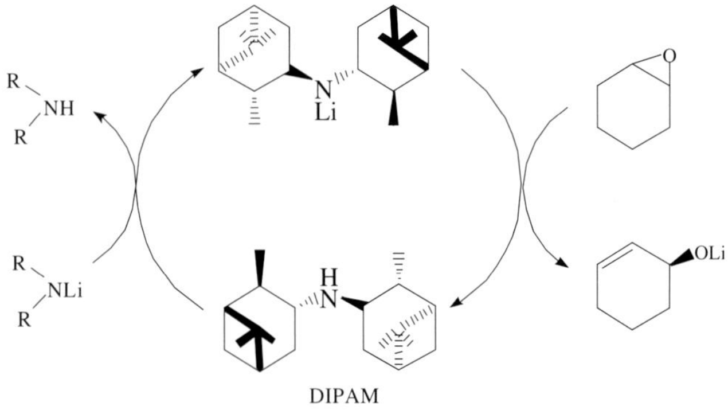

Scheme 3 Reaction flow chart

with the chiral amides, i.e., when two amides (achiral and chiral) were present together, the deprotonation of the epoxide was mainly due to the chiral amide. Also, a systematic investigation with varied amounts of chiral amine DIPAM, and at different temperatures revealed that the best results could be obtained at 0 °C using 20 mol% of the chiral and 125% of the achiral amines.

The results of the deprotonation of *meso*-cyclohexene oxide with various chiral amides are shown in Table 1.

The experimental data clearly indicate that the isopinocampheyl moiety is the key to obtaining high yields and EEs.

Theoretical calculations

Successful rationalizations of enantioselectivity depend on comprehensive knowledge about the reaction mechanism [17, 18]. To verify the experimental results shown in Table 1, we embarked on a theoretical calculation of the deprotonation reaction of cyclohexene oxide using Gaussian 98. The possible reaction routes for deprotonation of cyclohexene oxide with various chiral amides are shown in Scheme 4.

First, all transition states for reactions with various chiral amines were located and optimized by ab initio HF/3-21G and density functional B3LYP/3-21G calculations with Gaussian 98. The deprotonation of the epoxide with lithium amide occurs via a six-membered ring transition state, where lithium approaches the lone pair of electrons on oxygen, thus facilitating the removal of the β hydrogen. As shown in Fig. 1a, in the case of deprotonation with DIPAM-Li (R1 optimized with B3LYP/3-21G), when

Table 1 Catalytic enantioselective deprotonation of cyclohexene oxide 1

Entry	Amine	Mol (%)	Temperature	Yield (%)[a]	EE (%)[b]	Config[c]
1	DIPAM	5	0 °C(rt)	59 (62)	80 (77)	*R*
2	DIPAM	10	0 °C(rt)	66 (68)	81 (74)	*R*
3	DIPAM	15	0 °C(rt)	69 (73)	83 (79)	*R*
4	DIPAM	20	0 °C(rt)	77 (82)	95 (87)	*R*
5	DIPAM	25	0 °C(rt)	78 (79)	89 (83)	*R*
6	ChxIPAM	5	0 °C	55	65	*R*
7	ChxIPAM	10	0 °C	57	74	*R*
8	ChxIPAM	15	0 °C	66	76	*R*
9	ChxIPAM	20	0 °C	70	78	*R*
10	BzIPAM	5	0 °C	51	34	*R*
11	BzIPAM	10	0 °C	53	39	*R*
12	BzIPAM	15	0 °C	58	46	*R*
13	BzIPAM	20	0 °C	57	48	*R*
14	^{i}PIPAM	5	0 °C	50	43	*R*
15	^{i}PIPAM	10	0 °C	51	47	*R*
16	^{i}PIPAM	15	0 °C	55	51	*R*
17	^{i}PIPAM	20	0 °C	58	52	*R*

[a]Isolated yield
[b]EE of the (*R*)-MTPA ester derivative [20, 21]
[c]Assignment based on the sign of optical rotation

Scheme 4 The deprotonation reaction equations of cyclohexene oxide

DIPAM-Li + Cyclohexene Oxide —R1→ DIPAM + Lithium R-2-Cyclohexen-1-ol
DIPAM-Li + Cyclohexene Oxide —S1→ DIPAM + Lithium S-2-Cyclohexen-1-ol

ChxIPAM-Li + Cyclohexene Oxide —R2→ ChxIPAM + Lithium R-2-Cyclohexen-1-ol
ChxIPAM-Li + Cyclohexene Oxide —S2→ ChxIPAM + Lithium S-2-Cyclohexen-1-ol

BzIPAM-Li + Cyclohexene Oxide —R3→ BzIPAM + Lithium R-2-Cyclohexen-1-ol
BzIPAM-Li + Cyclohexene Oxide —S3→ BzIPAM + Lithium S-2-Cyclohexen-1-ol

^{i}PIPAM-Li + Cyclohexene Oxide —R4→ ^{i}PIPAM + Lithium R-2-Cyclohexen-1-ol
^{i}PIPAM-Li + Cyclohexene Oxide —S4→ ^{i}PIPAM + Lithium S-2-Cyclohexen-1-ol

* R1, R2, R3, R4 and S1, S2, S3, S4 are signals of transition states

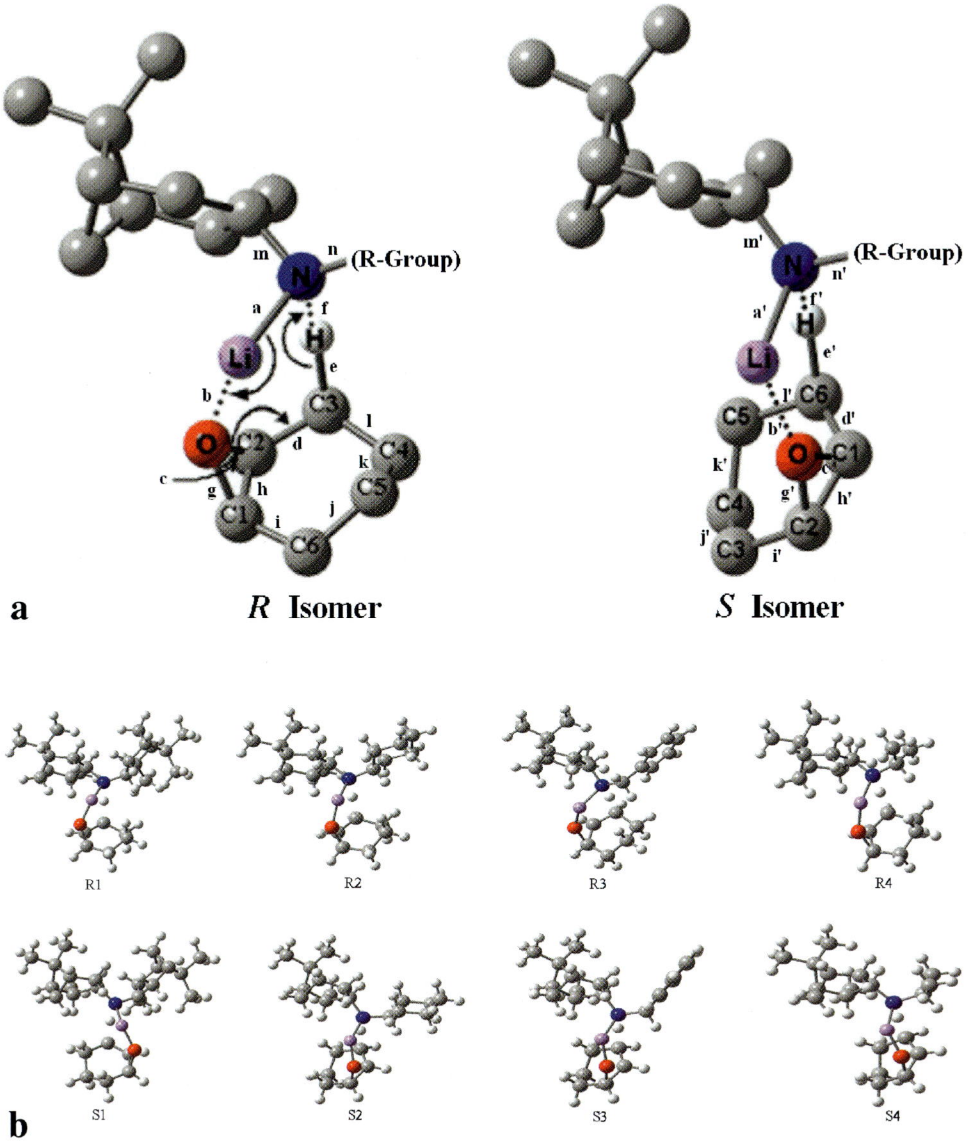

Fig. 1 **a** Six-membered ring transtion states. **b** The (*R*) and (*S*) transition states for the deprotonation reactions shown in Scheme 4

Table 2 Important bond distances in optimized transition states

Bond	HF/3-21G				B3LYP/3-21G			
	T.S. of DIPAM Distance (Å)	T. S. of ChxIPAM Distance (Å)	T. S. of BzIPAM Distance (Å)	T. S. of ^{i}PIPAM Distance (Å)	T.S. of DIPAM Distance (Å)	T. S. of ChxIPAM Distance (Å)	T. S. of BzIPAM Distance (Å)	T. S. of ^{i}PIPAM Distance (Å)
a (a′)	1.89(1.90)	1.89(1.89)	1.89(1.90)	1.89(1.89)	1.87(1.87)	1.86 (1.87)	1.87 (1.87)	1.86 (1.86)
b (b′)	1.74(1.76)	1.73(1.73)	1.73(1.73)	1.73(1.73)	1.74(1.73)	1.73 (1.73)	1.72 (1.72)	1.73 (1.73)
c (c′)	1.64(1.58)	1.66(1.62)	1.65(1.60)	1.66(1.62)	1.84(1.83)	1.85 (1.83)	1.86 (1.84)	1.85 (1.84)
d (d′)	1.46(1.47)	1.45(1.46)	1.46(1.47)	1.45(1.46)	1.44(1.44)	1.44 (1.44)	1.44 (1.44)	1.44 (1.44)
e (e′)	1.44(1.48)	1.43(1.46)	1.45(1.48)	1.43(1.46)	1.24(1.25)	1.24 (1.25)	1.24 (1.25)	1.24 (1.24)
f (f′)	1.32(1.29)	1.33(1.30)	1.31(1.29)	1.32(1.30)	1.59(1.57)	1.62 (1.57)	1.60 (1.56)	1.62 (1.59)
g (g′)	1.47(1.49)	1.46(1.47)	1.47(1.48)	1.46(1.47)	1.47(1.47)	1.47 (1.47)	1.47 (1.47)	1.47 (1.47)
h (h′)	1.46(1.47)	1.46(1.46)	1.46(1.46)	1.46(1.46)	1.48(1.48)	1.48 (1.48)	1.48 (1.48)	1.48 (1.48)
i (i′)	1.51(1.51)	1.51(1.51)	1.51(1.51)	1.51(1.51)	1.52(1.53)	1.52 (1.53)	1.52 (1.53)	1.52 (1.53)
j (j′)	1.54(1.54)	1.53(1.54)	1.53(1.54)	1.53(1.54)	1.54(1.54)	1.54 (1.54)	1.54 (1.54)	1.54 (1.54)
k (k′)	1.54(1.54)	1.54(1.54)	1.54(1.54)	1.54(1.54)	1.54(1.54)	1.54 (1.54)	1.54 (1.54)	1.54 (1.54)
l (l′)	1.54(1.54)	1.54(1.55)	1.54(1.54)	1.54(1.55)	1.54(1.55)	1.54 (1.54)	1.54 (1.55)	1.54 (1.55)
m (m′)	1.49(1.48)	1.49(1.48)	1.49(1.48)	1.49(1.48)	1.50(1.49)	1.50 (1.49)	1.50 (1.49)	1.50 (1.49)
n (n′)	1.48(1.49)	1.48(1.48)	1.48(1.48)	1.48(1.48)	1.49(1.50)	1.49 (1.50)	1.48 (1.49)	1.49 (1.49)

lithium approaches the lone pair electrons on oxygen, the N–Li bond is stretched to 1.870 Å (from 1.750 Å). As a result, the C–O bond (1.840 Å) is broken. Furthermore, the β C–H bond is stretched from 1.080 to 1.240 Å and then broken. The transition states for (*R*) and (*S*) isomers obtained with different ligands are shown in Fig. 1b. When the "β hydrogen" is deprotonated from the C-3 atom of cyclohexene oxide, the transition states result in the *R* configuration, and the product is (*R*)-2-cyclohexen-1-ol. On the other hand, if the "β hydrogen" is deprotonated from the C-6 atom of cyclohexene oxide, the transition states give the *S* configuration, thereby (*S*)-2-cyclohexen-1-ol is obtained. The selected bond distances of the transition state (*R* and *S*) structures were optimized using ab initio HF/3-21G and density-functional B3LYP/3-21G calculations. The changed bond distances are shown in Table 2. The first imaginary frequency shown in the Appendix explains that ab initio HF and density-functional B3LYP methods with the 3-21G-basis set were successful in finding transition states in this study.

As Fig. 1 and Table 2 show, the bond distances varied with different *R*-groups. *R* isomers have absolutely the same atoms as *S* isomers, but the bond distances are different. Bonds a, b, c, d, e, f or a′, b′, c′, d′, e′, and f′ are most important for locating the transition states. Bond g–n or g′–n′ did not change much in various transition state structures.

The total energies in Hartree at 0 K for the reactants and transition states obtained by way of single-point calculations on the B3LYP/3-21G optimized geometries using a modern SCRF/PCM [19] implementation are shown in Table 3. The solvent in the SCRF calculations is tetrahydrofolate (THF), which is the same as used in the experiments.

Table 3 The energies for all transition states and reactants in Hartree at 0 K

Amine	B3LYP/6-31+G(3df,2p)//B3LYP/3-21G (Hartree)			
	R figuration	S figuration	(S)-(R)	(S)-(R) kcal mol^{-1}
DIPAM	(R1) −1152.031094	(S1) −1152.024281	0.006813	4.275
ChxIPAM	(R2) −996.847617	(S2) −996.840595	0.007022	4.406
BzlPAM	(R3) −1031.810112	(S3) −1031.802679	0.007433	4.664
^{i}PIPAM	(R4) −879.958785	(S4) −879.951436	0.007349	4.612
DIPAM-Li	−842.197713			
ChxIPAM-Li	−687.015267			
BzlPAM-Li	−721.984766			
^{i}PIPAM-Li	−570.132208			
Cyclohexene Oxide	−309.889039			

Table 4 Activation energies calculated by B3LYP/6-31+G(3df,2p)// B3LYP/3-21G

Amine	Activation energy	
	Hartree	kcal mol^{-1}
DIPAM	0.055658	34.926
ChxIPAM	0.056689	35.573
BzIPAM	0.063693	39.968
^{i}PIPAM	0.062462	39.196

All transition states correspond to *R* configuration products

As the calculation and data in Table 3 show, the transition state energy for *R* isomer is lower than the *S* isomer in each case. Hence, the transition states for the *R* products are more stable and therefore easier to obtain. The energy differences between the transition states for the *R* and *S* structures were also calculated. However, all computed (*S*)-(*R*) energies in Table 3 correspond to >99% ee, which did not reflect the observed experimental selectivities. The activation energies can be calculated by subtracting the sum of the energies of reactants from the energy of the corresponding transition state. For example:

$$\begin{aligned}\text{Activation Energy of DIPAM} &= \text{R1} - (\text{DIPAM} - \text{Li} + \text{Cyclohexene Oxide}) \\ &= -1152.031094 - (-842.197713) + (-309.889039) \\ &= 0.055658(\text{Hartree}) \\ 0.055658 * 627.5095 &= 34.926\left(\text{kcal mol}^{-1}\right)\end{aligned}$$

The activation energies were calculated at the density functional B3LYP level with the 6-31+G(3df,2p) basis set. We only calculated the *R* configuration structures and the data for *R* configuration are shown in Table 4. Similarly, the calculations for the *S* configuration structures could use the same method. One thing that should be pointed out in this study is that the monomeric Li-amide may be a poor model for the substrates in solution because of its tendency to form higher aggregates.

In the case of reaction with the chiral amine DIPAM, the activation energy obtained is the lowest. This reaction can occur selectively and rapidly, resulting in high EE and yield of product. Substituting one isopinocampheyl group in DIPAM with cyclohexyl leads to higher activation energy for the reaction to occur. Similarly, substitution with a benzyl or isopropyl group also gives increased reaction energy, resulting in lower product yield and EE. We also note that with the benzyl group, the relative activation energies are highest, which means that this kind of reaction is the most difficult. Therefore, lower yields and EE for this product should be obtained, as is the case in Table 1. These theoretical calculations confirm the experimental findings that the isopinocampheyl moiety plays an important role in orienting the epoxide and the lithium salt in such way that the deprotonation takes place very enantioselectively.

Conclusion

The mechanism of the enantioselective deprotonation of cyclohexene oxide with isopinocampheyl-based chiral lithium amide was examined at the B3LYP/6-31+G (3df,2p)//B3LYP/3-21G levels using Gaussian 98. We uncovered the mechanistic details of the catalytic cycle, which are supported well by the experimental results. The highest EE (95%) for the conversion of cyclohexene oxide to 2-cyclohexen-1-ol has been achieved with lithium DIPAM through the catalytic system. These theoretical and experimental results certainly promise the catalytic application of DIPAM and similar pinene-based ligands in the complex synthesis of natural products and biologically important compounds.

Acknowledgements We thank Professor Joseph Bozzelli for letting us use the cluster PCs and quantum mechanical programs. The Molecular Modeling Laboratory used in this study was funded jointly by the NSF Grant No 9750809 and New Jersey Institute of Technology.

References

1. Koga K (1995) New Methods Drug Res 4:25–27
2. O'Brien P (1998) J Chem Soc Perkin Trans 1:1439–1457
3. Simpkins NS (1996) Asymmetric deprotonation reactions using enantiopure lithium amide bases. In: Stephenson GR (ed) Advanced asymmetric synthesis. Blackie, Glasgow, Scotland, pp 111–125
4. Thummel RP, Rickborn B (1970) J Am Chem Soc 92: 2064–2067
5. Whitesell JK, Felman SW (1980) J Org Chem 45:755–756
6. Simpkins NS (2000) Chimia 54:53–54
7. Cox PJ, Simpkins NS (1991) Tetrahedron: Asymmetry 2:1–26
8. Simpkins NS (1996) Pure Appl Chem 68:691–694
9. Jones S (2002) J Chem Soc Perkin Trans 1:1–21
10. O'Brien P (2001) J Chem Soc Perkin Trans 1:95–113
11. Koga K (1994) Pure Appl Chem 66:1487–1492
12. Brown HC, Kramer GW, Levy AB, Midland MM (1975) Organic synthesis via Boranes. Wiley, New York
13. Brown HC, Ramachandran PV (1996) Sixty years of hydride reductions. In: Abdel-Magid AF (ed) Reductions in organic synthesis. American Chemical Society, Washington, DC, pp 1–30
14. Malhotra SV (2003) Tetrahedron: Asymmetry 14:645–647

15. Asami M (1985) Tetrahedron Lett 26:5803–5806
16. Asami M, Ishizaki T, Inoue S (1994) Tetrahedron: Asymmetry 5:793–796
17. Toukoniitty E, Nieminen V, Taskinen A, Paivarinta J, Hotokka M, Murzin DY (2004) J Catal 224:326–339
18. Brandt P, Hedberg C, Lawonn K, Pinho P, Andersson PG (1999) Chem Eur J 5:1692–1699
19. Foresman JB, Keith TA, Wiberg KB, Snoonian J, Frisch MJ (1996) J Phys Chem 100:16098–16104
20. Sodergren MJ, Andersson PG (1998) J Am Chem Soc 120: 10760–10761
21. Gayet A, Bertilsson S, Andersson PG (2002) Org Lett 4: 3777–3779

J Mol Model (2006) 12: 687–701
DOI 10.1007/s00894-005-0046-2

ORIGINAL PAPER

Ulrike Salzner · Ozan Karaltı · Serdar Durdaği

Does the donor–acceptor concept work for designing synthetic metals?

III. Theoretical investigation of copolymers between quinoid acceptors and aromatic donors

Received: 12 May 2005 / Accepted: 27 September 2005 / Published online: 13 January 2006

Abstract Homopolymers of quinoxaline (QX), benzothiadiazole (BT), benzobisthiadiazole (BBT), thienopyrazine (TP), thienothiadiazole (TT), and thienopyrazinothiadiazole (TTP) and copolymers of these acceptors with thiophene (TH) and pyrrole (PY) were investigated with density functional theory. Theoretical band-gap predictions reproduce experimental data well. For all but six copolymers, band-gap reductions with respect to either homopolymer are obtained. Four of the acceptors, BBT, TP, TT, and TTP, give rise to copolymers with band gaps that are smaller than that of polyacetylene. BBT and TTP copolymers with PY in 1:2 stoichiometry are predicted to be synthetic metals. Band-gap reductions result from upshifts of HOMO energies and much smaller upshifts of LUMO values. The smallest band gaps are predicted with TTP, since changes in LUMO energies upon copolymerization are particularly small. The consequence of the small interactions between LUMO levels of donor and acceptor are vanishingly small conduction bandwidths.

Keywords Conducting polymers · Donor–acceptor concept · Band gap · Band width · Conductivity · DFT

Introduction

Since the discovery of conducting organic polymers (COPs) in 1977, [1] enormous progress has been made in tailoring properties of COPs through chemical modification [2, 3]. One of the goals is decreasing band gaps, since zero-band-gap systems would be synthetic metals with intrinsic conductivity. Most COPs have band gaps between 1 and 4 eV and are semiconductors [4, 5]. Systems with band gaps above about 0.5 eV are insulating in their neutral ground state and require doping to conduct electricity. In general, oxidizing or p-doping is easier than reducing or n-doping. Thus, stable p-doped COPs with high conductivity are common. In contrast, n-doped polymers tend to loose electrons and show lower conductivities [6, 7]. To make organic transistors, materials that can be p- and n-doped are desirable. Therefore, there is interest in lowering conduction band energies of COPs, since n-type conductors as well as synthetic metals could be produced in this way. Tanaka, Yamashita, Yammamoto, and their coworkers [8–17] have succeeded in designing a series of low-band-gap materials with large electron affinities (EAs) by copolymerizing quinoxaline (QX), [8, 11, 14] benzothiadiazole (BT), [12, 14] benzobisthiadiazole (BBT), [12, 14] thienopyrazine (TP), [10, 14] thienothiadiazole (TT), [9, 12] and thienopyrazinothiadiazole (TTP) [17] (Scheme 1) with thiophene (TH) and pyrrole (PY) (Schemes 2 and 3). The smallest band gaps were achieved with BBT and TTP in combination with PY. Poly-PY-TTP-PY (Scheme 3) has no electrochemical band gap [17].

All acceptors shown in Scheme 1 have small HOMO-LUMO gaps, even as monomers. Their ionization potentials (IPs) are relatively close to that of thiophene but their EAs are much higher than those of thiophene or pyrrole. Therefore copolymers with thiophene and pyrrole were classified as donor–acceptor polymers [18]. The six acceptors can be divided into two groups. QX and BT adopt aromatic structures. Homopolymers of TP, TT, BBT, and TTP prefer quinoid conformations. In the quinoid form, rearrangement of the double bonds allows the sulfur atoms to form two single bonds with nitrogen and avoid the unfavorable non-classical structure with two double bonds between S and N. In this way the thiadiazole group forces the ring attached to it out of its preferred aromatic conformation. The strong preference for quinoid structures leads to a geometrical mismatch in copolymers with aromatic systems like thiophene and pyrrole. Since quinoid and aromatic polymers have their positions of single and double bonds reversed, copolymerization of quinoid and

Dedicated to Professor Dr. Paul von Ragué Schleyer on the occasion of his 75th birthday

U. Salzner (✉) · O. Karaltı · S. Durdaği
Department of Chemistry, Bilkent University,
800 Bilkent, Ankara, Turkey
e-mail: salzner@fen.bilkent.edu.tr
Tel.: +91-312-2902122
Fax: +90-312-2664579

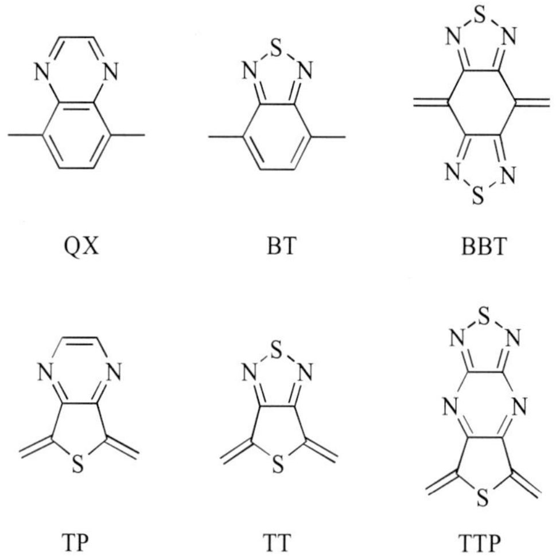

Scheme 1 Quinoxaline (QX), benzothiadiazole (BT), benzobisthiadiazole (BBT), thienopyrazine (TP), thienothiadiazole (TT), and thienopyrazinothiadiazole (TTP) monomers, shown in the conformation that is preferred in the polymers

aromatic units tends to reduce bond length alternation (BLA). Since BLA is one of the factors that increase band gaps, [2, 3] reduction of BLA could also contribute to the band-gap reductions observed for the copolymers [19]. The present investigation was carried out with the aim to analyze the electronic structures of these interesting systems and to evaluate the factors responsible for the band-gap reductions.

Materials and Methods

Theoretical level

Homo and cooligomers with between 1 and 8 repeat units shown in Schemes 2 and 3 were investigated. The necessary chain lengths were determined by the convergence properties. Cooligomers are named by using abbreviations of their names in the sequence of their connection. The calculations were carried out with density functional theory as implemented in Gaussian 98, revision A11 [20] and Gaussian 03, revision C.02 [21]. We used Becke's three-parameter hybrid functional [22] with Hartree-Fock exchange increased to 30% and Perdew's 86 [23] correlation functional. (The keywords to create this functional are: "BP86 IOP (5/45=10000300, 5/46=07200700, 5/47= 08101000, 5/42=406)" in G98 and "BP86 IOP (3/76=1000003000, 3/77=0720007000, 3/78=0810010000, 3/ 74=406)" in G03). Stevens Basch Krauss pseudopotentials [24] combined with polarized split-valence basis sets were employed throughout for all atoms (except hydrogen).

Structure determination

The thiadiazole unit is sometimes described as "non-classical" employing two double bonds between N and S. Such structures have been shown to be very unfavorable in a recent experimental and theoretical investigation [25]. In agreement with this conclusion, we find a high zwitterionic character in monomers of TT, BBT, and TTP. Adopting quinoid structures in the polymers relieves the unfavorable bonding situation between N and S that exists in the monomers. Therefore, oligomers that are terminated with H may be singlet biradicals and closed-shell calculations may lead to severe underestimation of the energy gaps. For such systems we performed closed- and open-shell calculations on oligomers end-capped with H, and closed-shell calculations on oligomers end-capped with CH2 groups. Open-shell calculations were carried out with the same functional as specified above in combination with the "guess=mix" keyword in G98 and G03.

To determine whether quinoid or aromatic structures are preferred, energy differences were evaluated for fully optimized monomers and dimers end-capped with H and with CH_2. The energy of the inner fragment was estimated by subtracting energies of monomers from those of dimers. The procedure is shown in Scheme 4 for quinoxaline. In this way, the energies of the different end groups are removed. The lower-energy inner fragment (quinoid or aromatic) was used as repeat unit and the oligomers were terminated with the appropriate end-groups. Additional indications of the correct geometry are the bond lengths in the inner part of longer oligomers and speed of convergence of properties. Since end effects do not reach very far, inner fragments switch to the preferred conformation upon geometry optimization as the chain length increases. This is reflected in bond-length changes with increasing chain lengths if the wrong form is used. Switching to a different geometry is accompanied by discontinuities and slow convergence of IPs, EAs, and energy gaps.

In copolymers with pyrrole, the orientation of the pyrrole rings is fixed into the *trans* position by hydrogen bridges between the pyrrole N-H group and the nitrogen atoms of the acceptors. With thiophene, *cis* or *trans* forms are possible. We have checked the preference for the monomers and used the more stable conformation throughout. As a result, 1:1 copolymers with a *cis*-preference are non-linear and polymers would be helical. The relatively short oligomers considered here are planar. In 1:2 stoichiometry, the copolymers are linear since the adjacent thiophene rings adopt the *trans* conformation. Band gaps of *cis*- and *trans* forms are almost identical. The more stable conformers have slightly higher IPs and EAs. The differences are in the range of 0.1 eV.

Extrapolation

Polymer properties were obtained by extrapolation of oligomer data for increasing chain lengths. Geometries of all oligomers were fully optimized with appropriate symmetry constraints. It is common to employ linear fits using a couple of data points plotted vs inverse chain length [26]. Fig. 1. shows plots of HOMO-LUMO gaps and of time-dependent Hartree-Fock (TDHF) excitation energies vs

Scheme 2 Copolymers of QX, BT, and BBT with thiophene (TH) and pyrrole (PY) in 1:1 and 1:2 stoichiometry

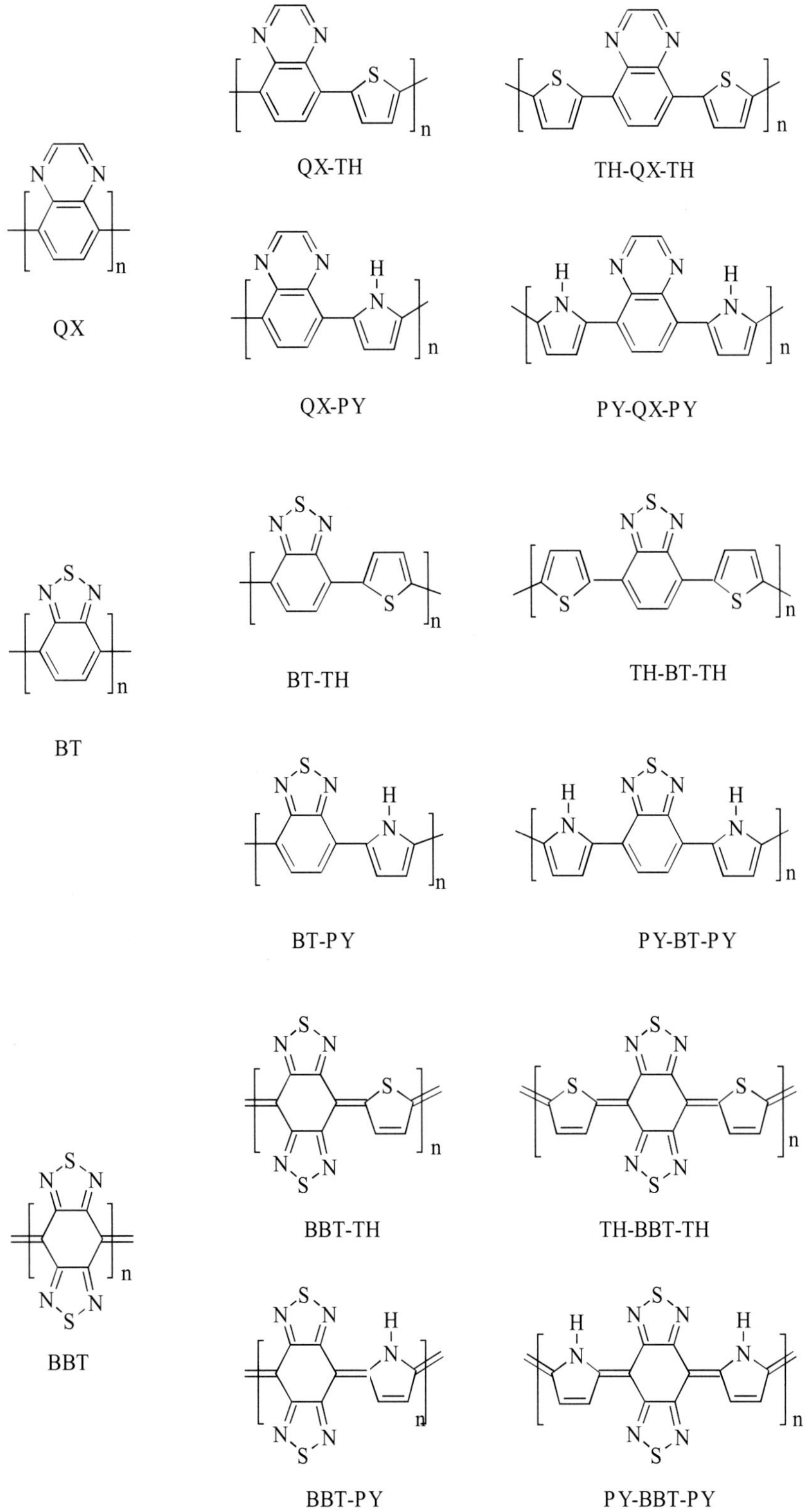

inverse chain lengths for oligomers of acetylene, thiophene, and pyrrole with up to 80 double bonds. The dependence is non-linear at either level of theory and saturation is reached for long oligomers. It is therefore evident that linear fitting will lead to different results depending on the number of data points and on the lengths of the oligomers. Therefore, it appears better to calculate oligomers of increasing size until the onset of saturation is reached and to fit only the data for the longest oligomers. We employ polynomial fitting to account for the saturation in the infinite-chain-length limit. In general, our longest oligomers have HOMO-LUMO gaps that are only 0.2–

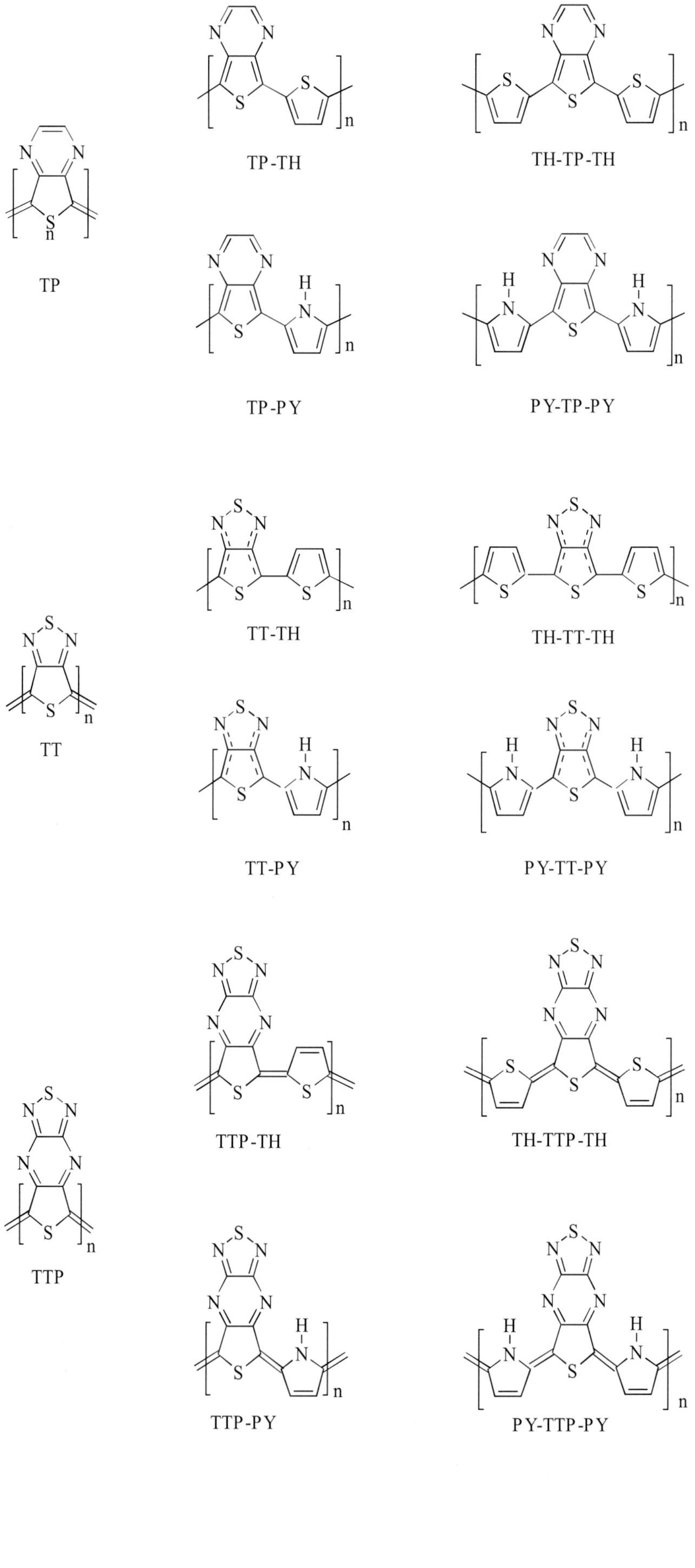

Scheme 3 Copolymers of TP, TT, and TTP with thiophene (TH) and pyrrole (PY) in 1:1 and 1:2 stoichiometry

Scheme 4 Determination of the preferred conformation of a polymer from oligomer calculations. Monomers and dimers are calculated in aromatic and quinoid conformation. Energies of end groups are removed by subtracting total energies of monomers from those of dimers. The form with the lower energy difference is the preferred conformation

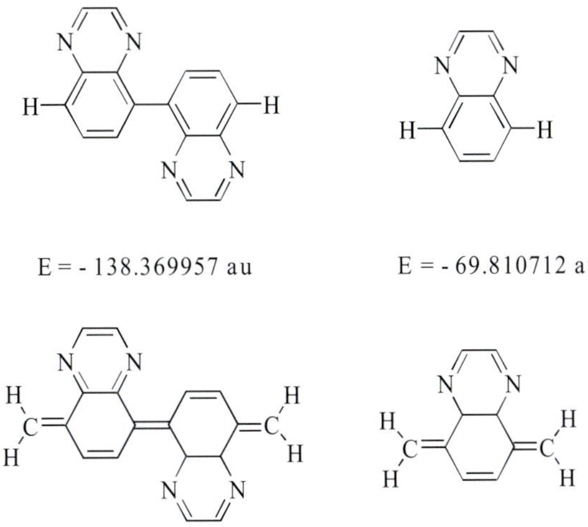

E = - 138.369957 au   E = - 69.810712 au   ΔE = -68.559245 au

E = -151.063819 au   E = - 82.523041 au   ΔE = -68.540778 au

ΔΔE = -0.018467 au = -11.59 kcal mol^{-1}

0.4 eV larger than the extrapolated values. Therefore, little error is introduced due to extrapolation.

Accuracy of the band gaps

The accuracy of DFT HOMO-LUMO gaps for predicting band gaps of conducting polymers has been studied in detail by one of us [27–30]. With the above functional, HOMO-LUMO gaps give reasonable estimates for λ_{max} in gas-phase spectra of medium-sized polyenes. HOMO-LUMO gaps converge to a limiting value of 1.7 eV for polyacetylene in the gas phase. The same value is obtained with one-dimensional periodic boundary calculations. More accurate time-dependent Hartree-Fock (TDHF) calculations lead to a limiting value of 2.2 eV. Polyenes are the only systems for which experimental gas phase data are available. In solution the excitation energies are about 0.3–0.4 eV lower [31]. In the solid-state inter-chain interactions lower the lowest excitation energy due to Davydov splitting compared to solution data. The exact amount of Davydov splitting in the long-chain limit is unknown but it is predicted [32] be smaller than for medium-sized thiophene oligomers, which show Davydov splittings of about 0.4 eV and lowering of the lowest excitation energies of 0.1–0.2 eV [33]. Since the lowering of excitation energies due to Davydov splitting is small, we do not consider it here. Thus, from experiment we deduce that gas-phase values should be around 0.4 eV higher than solution data. λ_{max} of PA films occurs at 1.9 eV [34, 35]. TDHF, which gives a gas-phase value of 2.2 eV, is therefore quite accurate but HOMO-LUMO gaps underestimate λ_{max} of PA by about 0.7 eV.

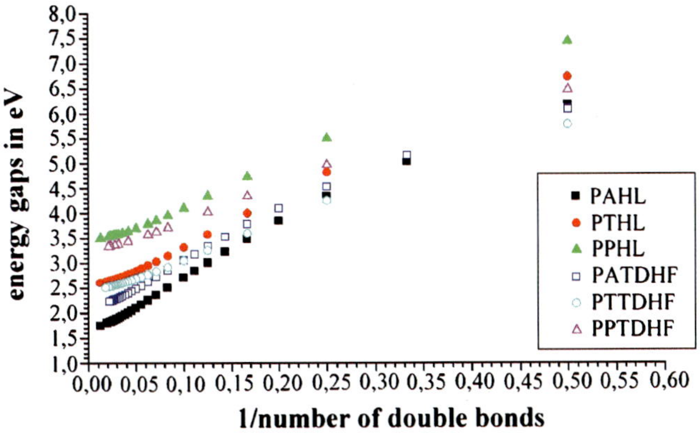

Fig. 1 DFT HOMO-LUMO gaps (*filled symbols*) and TDHF excitation energies (*open symbols*) in electron volts (eV) for acetylene, thiophene and pyrrole oligomers with up to 80 double bonds plotted vs. inverse number of double bonds

There are no gas-phase data for thiophene and pyrrole oligomers and for the systems shown in Schemes 2 and 3. We assume therefore that solvent effects are the same as for polyenes, ~0.4 eV. Comparison between RT solutions and low-temperature solid solutions of thiophene oligomers [36–38] showed that thiophene oligomers are non-planar in liquid solution. Cooling leads to a lowering of the excitation energies by about 0.4 eV. Cooling has no such effect on polyene spectra. Thus, at room temperature, effects due to solvent and non-planarity cancel roughly for thiophene oligomers. This is reflected in the fact that TDHF reproduces excitation energies of thiophene oligomers in RT solution very accurately. The limiting values in the long-chain limit are 2.5 eV at the TDHF level and 2.6 eV with DFT HOMO-LUMO gaps. λ_{max} of PT occurs at 2.5–2.6 eV [39, 40]. Thus, both levels of theory agree nicely for PT and are close to experiment. The error is thus much smaller than for PA. λ_{max} of polypyrrole (PPy) converges to 3.3 eV with TDHF and to 3.5 eV with DFT HOMO-LUMO gaps. Since pyrrole oligomers are non-planar in the gas phase, effects of non-planarity should cancel solvent effects roughly. Therefore, HOMO-LUMO gaps overestimate λ_{max} by 0.3 eV (3.5 vs. 2.9–3.2 eV [40, 41]). Thus, the overall accuracy of DFT HOMO-LUMO gaps is about ±0.3 eV for aromatic polymers. Since the systems investigated here are ring systems and since they are planar, we expect that HOMO-LUMO gaps overestimate condensed-phase values of oligomers by about 0.4 eV. This correction is applied when theory and experiment are compared.

Experimental band gaps are obtained as differences between oxidation and reduction potentials or as the onset of optical absorptions. For many polymers, both coincide. [42] Onset of absorption and λ_{max} usually differ by about 0.5 eV [34, 39, 43]. Adding the correction for the solvent effect of 0.4 eV, our HOMO-LUMO gaps should be about 0.9 eV higher than experimental band gaps. We will refer to the theoretical estimates as HOMO-LUMO or *energy* gaps in order to distinguish them from experimental *band* gaps. When a comparison is made, 0.9 eV is subtracted from the theoretical gas-phase values.

IPs and EAs

IPs and EAs are obtained as negative HOMO and LUMO energies, respectively. Using Koopmans' theorem [44] in connection with density functional theory (DFT) is controversial [29, 45–50] but experience shows that LUMO energy levels are much better approximations to EAs than virtual orbital energies obtained with the HF approximation [48]. In contrast HOMO energies are too high to give good IPs. The error is about 1 eV and similar for HOMO and LUMO, which is the reason for the good energy gaps [29]. The error is also roughly the same for different systems. Therefore, we will use HOMO and LUMO energies chiefly to compare IPs and EAs of homo- and copolymers. We will not try to produce absolute values.

Results

Quinoxaline QX

Quinoxaline oligomers prefer aromatic structures. The energy difference ΔΔE obtained by subtracting the energy differences between dimer and monomer in aromatic conformations end-capped with H and in quinoid conformations end-capped with CH_2 (Scheme 4), is 11.59 kcal mol^{-1} in favor of the aromatic form. Since TH and PY oligomers prefer aromatic structures as well, all copolymers are assumed to be aromatic and only closed-shell calculations were performed. With pyrrole, hydrogen bridges between pyrrole hydrogen and quinoxaline nitrogen atoms determine the orientation of the rings. With thiophene, the opposite orientation of the rings is conceivable. Geometry optimizations show, however, that the structures in Scheme 2 are preferred by 0.35 kcal mol^{-1} for the 1:1 and by 1.1 kcal mol^{-1} for the 1:2 monomer. The effect on the energy gap is below 0.1 eV.

Energy gaps, IPs and EAs for the oligomers of the QX series are summarized in Table 1. The extrapolated energy gap of p-QX is the same as that of p-PT, 2.6 eV. IP and EA are 0.3 eV higher than those of p-PT. Copolymerization with PT and PY in 1:1 and 1:2 ratio leads to band-gap reductions of up to 0.67 eV, which are due to lowering of the IPs. Experimentally, λ_{max} of the QX homopolymer was determined to be 3.81 eV, the onset of absorption was estimated to occur at 2.6 eV [8]. Correcting our data as described in the methods section would predict a value of 2.2 eV for λ_{max} and 1.7 eV for the onset of absorption. These are unusually large discrepancies. The experimental difference of 1.2 eV between λ_{max} and the onset of absorption is also unusually large.

The copolymer with one TH unit has a band gap of 1.7 eV; with two TH units the band gap is 1.4 eV [51, 52]. Our predictions of 1.2 eV are in much better agreement with experimental values than those for the homopolymer. For pyrrole systems, the theoretical energy gaps are about 0.2 eV lower than with thiophene. Although the band gaps

Table 1 Energy gaps, IPs, and EAs according to Koopmans' theorem for oligomers of the QX system

Number of units		QX	QX-TH	TH-QX-TH	QX-PY	PY-QX-PY
1	Eg	5.40	4.06	3.45 (3.06) [14]	3.64 (2.95) [51]	3.00 (2.47) [51]
	IP	7.79	6.70	6.31	6.24	5.78
	EA	2.39	2.64	2.86	2.60	2.79
2	Eg	4.24	3.13	2.76	2.83 (2.32) [51]	2.43
	IP	7.00	6.09	5.82	5.68	5.31
	EA	2.76	2.96	3.07	2.85	2.89
3	Eg	3.72	2.76	2.52	2.51	2.24
	IP	6.67	5.86	5.69	5.47	5.18
	EA	2.95	3.10	3.16	2.96	2.94
4	Eg	3.43		2.42		2.15
	IP	6.49		5.63		5.12
	EA	3.06		3.21		2.96
inf	Eg	2.63	2.08 (1.7) [52]	2.14 (1.4) [52]	1.88	2.13
	IP	5.98	5.47	5.48	5.10	4.96
	EA	3.35	3.28	3.34	3.13	3.03

Experimental values are quoted in brackets. For oligomers these are λ_{max} values obtained in solution, for polymers they are onsets of absorption. All values are given in electron volts (eV)

are lowered upon copolymerization, all QX systems are predicted to have larger energy gaps than polyacetylene (PA).

Benzothiadiazole BT

Due to the influence of the thiadiazole side group, the benzene ring is strongly distorted. It contains a butadiene fragment and a relatively long bond between the carbon atoms attached to the thiadiazole group. In long oligomers, the inter-ring bond has a similar bond length as in p-TH and p-PY. Therefore, p-BT is aromatic. There are no major geometry changes upon copolymerization with TH and PY. Although BT is structurally very similar to QX, the thiophene rings prefer opposite orientations in the copolymers. The 1:2 monomer, for instance, is 1.92 kcal mol^{-1} more stable in the orientation shown in Scheme 2 than in the conformation analogous to TH-QX-TH. With pyrrole the hydrogen bridges require the opposite orientation.

The agreement between theory and experiment is very good for the BT systems, as can be seen from Table 2. The BT homopolymer has a very small band gap but there are only small band-gap reductions upon copolymerization. With one or two thiophene rings, there is practically no change in band gaps. With pyrrole, the band gap decreases by 0.5 with 1:1 stoichiometry and 0.5 eV with 1:2 stoichiometry. Table 2 reveals that HOMO and LUMO are affected by copolymerization with thiophene but the upshift of the HOMO is stronger than that of the LUMO with pyrrole. While copolymers with thiophene are predicted to have similar energy gaps to those of QX, the pyrrole-BT copolymers have about 0.3 and 0.5 eV smaller energy gaps than those with QX.

Benzobisthiadiazole BBT

With two thiadiazole groups, the benzene ring cannot form double bonds at four of its six carbon atoms. Therefore, a quinoid structure with an inter-ring double bond is favorable. However, a planar BBT polymer would be a highly strained because of repulsions between the nitrogen lone pairs in the thiadiazole units. Frequency calculations show that aromatic and quinoid forms are twisted around the inter-ring bonds. The optimized dihedral angles are 60° in the aromatic and 40° in the quinoid form. Since the inter-ring double bond is weakened in the quinoid form, the aromatic form turns out to be 3.1 kcal mol^{-1} more stable when evaluated between dimer and monomer. The preference for the aromatic form, however, decreases when longer oligomers are considered (Scheme 5). For oligomers

Table 2 Energy gaps, IPs, and EAs according to Koopmans' theorem for oligomers of the BT system

Number of repeat units		BT	BT-TH	TH-BT-TH	BT-PY	PY-BT-PY
1	Eg	4.92	3.84 (3.18) [51]	3.27 (2.77) [51] (2.66) [14]	3.38 (2.81) [51]	2.73 (2.33) [51]
	IP	7.76	6.85	6.44	6.35	5.86
	EA	2.84	3.01	3.16	2.97	3.12
2	Eg	3.86	2.90	2.66 (2.38) [51]	2.59	2.17
	IP	7.11	6.24	6.05	5.82	5.41
	EA	3.25	3.34	3.39	3.23	3.24
3	Eg	3.33	2.61	2.44	2.26	1.98
	IP	6.81	6.15	5.94	5.62	5.27
	EA	3.48	3.54	3.50	3.36	3.29
4	Eg	3.03	2.44	2.34	2.08	1.89
	IP	6.65	6.07	5.89	5.52	5.21
	EA	3.61	3.63	3.55	3.44	3.32
5	Eg	2.85	2.34		1.97	
	IP	6.54	6.02		5.46	
	EA	3.70	3.68		3.49	
6	Eg	2.72	2.28		1.91	
	IP	6.48	6.00		5.43	
	EA	3.76	3.72		3.52	
8	Eg	2.57				
	IP	6.40				
	EA	3.83				
inf	Eg	2.11	2.05	2.07 (1.2) [14]	1.61	1.63
	IP	6.15	5.93	5.85	5.31	5.05
	EA	4.05	3.91	3.71	3.65	3.40

Experimental values are quoted in brackets. For oligomers these are λ_{max} values obtained in solution, for polymers they are onsets of absorption. All values are given in electron volts (eV)

Scheme 5 Determination of the preferred conformation of poly-BBT from oligomer calculations. Although BBT is inherently quinoid, steric strain prevents formation of a planar structure. As a result, the preference for the aromatic conformation decreases on chain length increase and converges to about 1.2 kcal/mol per BBT unit for longer oligomers

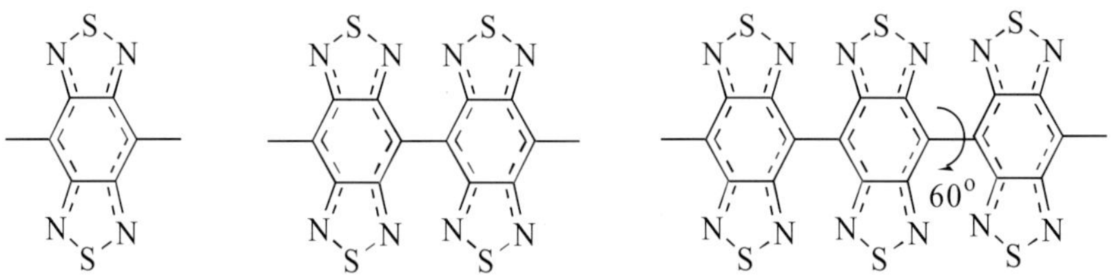

E = - 96.404798 E = - 191.569682 E = - 286.734484

ΔE_{1-2} = - 95.164884 ΔE_{2-3} = - 95.164802 ΔE_{3-4} = - 95.164766

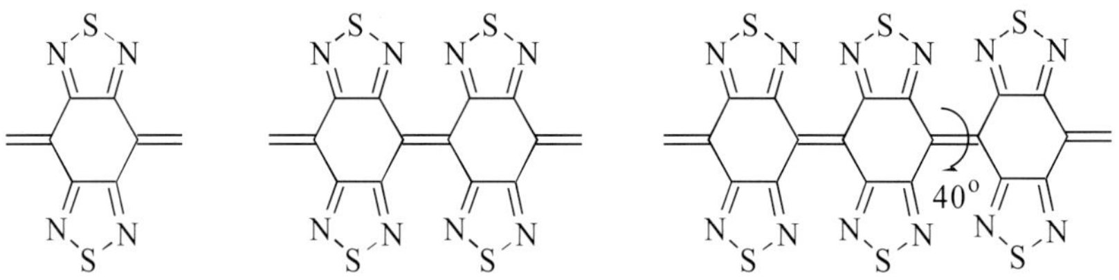

E = - 109.161519 E = - 204.321409 E = - 299.484237

ΔE_{1-2} = - 95.159890 ΔE_{2-3}= - 95.162828 ΔE_{3-4} = - 95.162824

$\Delta\Delta E_{1-2}$ = - 3.13 kcal/mol $\Delta\Delta E_{2-3}$ = - 1.24 kcal/mol $\Delta\Delta E_{3-4}$ = - 1.22 kcal/mol

Table 3 Energy gaps, IPs, and EAs according to Koopmans' theorem for oligomers of the BBT system

Number of repeat units		BBT	BBT-TH	TH-BBT-TH	BBT-PY	PY-BBT-PY
1	Eg	3.24	3.32	q 2.99/a 2.01 (1.77) [14]	3.27	q 2.84/a 1.73
	IP	7.34	6.51	6.10	6.29	5.75
	EA	4.11	3.19	3.11	3.02	2.91
2	Eg	2.76	2.42	1.89	2.32	1.66
	IP	6.99	6.19	5.77	5.93	5.38
	EA	4.23	3.77	3.89	3.61	3.71
3	Eg	2.50	2.16	1.65	2.08	1.43
	IP	6.80	6.07	5.68	5.81	5.27
	EA	4.30	3.91	4.03	3.73	3.84
4	Eg	2.19	2.03	1.54	1.96	1.32
	IP	6.43	6.02	5.64	5.76	5.23
	EA	4.23	3.99	4.10	3.80	3.91
5	Eg	2.02				
	IP	6.34				
	EA	4.32				
6	Eg	1.90				1.22
	IP	6.28				5.20
	EA	4.38				3.98
8	Eg					1.18
	IP					5.19
	EA					4.01
inf	Eg	1.30	1.66	1.36 (0.5) [14]	1.60	1.05
	IP	5.98	5.83	5.57	5.65	5.17
	EA	4.70	4.23	4.26	4.01	4.12

Experimental values are quoted in brackets. For oligomers these are λ_{max} values obtained in solution, for polymers they are onsets of absorption. All values are given in electron volts (eV)

longer than trimer, the quinoid form is more stable. In Table 3, data are given accordingly. For extrapolation, tetramers and longer oligomers of the quinoid form are used. Extrapolated energy gaps for p-BBT are as follows: aromatic closed-shell: 1.9 eV, aromatic open-shell: 2.4 eV, quinoid 1.3 eV. Applying the correction for using gas-phase λ_{max} values, the band gap of the quinoid form is 0.4 eV.

The inherent preference for the quinoid form comes to the fore upon copolymerization, since this increases the separation of the nitrogen lone pairs and allows formation of strong double bonds between the rings. The preference for the quinoid form is 7.4 kcal mol^{-1} for BBT-TH, 3.6 kcal mol^{-1} for TH-BBT-TH, and 8.0 kcal mol^{-1} for BBT-PY. The TH-BBT-TH monomer has a λ_{max} value of 1.77 eV in $CHCl_3$. Our value of 2.99 eV for the quinoid monomer, even after correction for the solvent effect, is 0.8 eV too high. We therefore checked the energy gap of the aromatic form and found it to be in much better agreement with experiment. The energy gap is 2.01 eV and after correction for the solvent effect our prediction is 1.6 eV. This suggests that quinoid structures develop only at longer chain lengths.

All copolymers of BBT are quinoid and energy gaps were evaluated for oligomers end-capped with CH_2. For extrapolation, the aromatic monomers are not included. Copolymerization tends to increase the energy gaps. 1:2 Stoichiometry is more favorable with respect to small energy gaps than 1:1 stoichiometry. Poly-TH-BT-TH is predicted to have a 0.06 eV larger band gap than p-BBT. Only with PY in 1:2 stoichiometry does the energy gap decrease (by 0.25 eV) compared to p-BBT. The band-gap reduction is due to lowering of IP and EA, but the latter is smaller. Correcting the energy gap as described in the methods section suggests that p-PY-BBT-PY has a band gap of about 0.15 eV. There is no direct experimental comparison since BBT was copolymerized only with N-methylpyrrole. It turned out that the methyl groups lead to steric hindrance and non-planarity. The band gap was 0.6, 0.1 eV larger than that of p-TH-BBT-TH [14].

Thienopyrazine TP

In TP, pyrazine and thiophene rings compete for aromaticity. The energy difference between quinoid and aromatic forms is evaluated in the same way as shown for QX in Scheme 4. For the homopolymer, there is a slight preference (3.5 kcal mol^{-1}) for the quinoid structure. Therefore, all data for the homopolymer are obtained for the quinoid oligomers, terminated with CH_2 groups. Copolymerization with aromatic TH and PY induces a change from quinoid to aromatic structures. For p-TP-TH, the preference for the aromatic conformation amounts to 2.1 kcal mol^{-1}, for p-TP-PY it is 4.12 kcal mol^{-1}. Copolymers with 1:2 stoichiometry are therefore assumed to be aromatic as well. The orientation of the pyrrole and thiophene rings is *trans* with respect to the central unit. The agreement with experimental data is good (compare Table 4).

Table 4 Energy gaps, IPs, and EAs according to Koopmans' theorem for oligomers of the TP system

Number of repeat units		TP	TP-TH	TH-TP-TH	TP-PY	PY-TP-PY
1	Eg	4.37	3.54	2.93 (2.35) [14] (2.79) [12]	3.27	2.62
	IP	6.78	6.50	6.07	6.14	5.63
	EA	2.41	2.96	3.14	2.88	3.00
2	Eg	3.31	2.54	2.30	2.38	2.10
	IP	6.18	5.86	5.68	5.54	5.23
	EA	2.86	3.32	3.38	3.15	3.13
3	Eg	2.78	2.13	2.06	2.03	1.91
	IP	5.89	5.62	5.55	5.31	5.11
	EA	3.12	3.49	3.48	3.27	3.19
4	Eg	2.47	1.92	1.95	1.85	1.83
	IP	5.73	5.50	5.49	5.20	5.05
	EA	3.26	3.58	3.53	3.34	3.22
5	Eg		1.80			
	IP		5.43			
	EA		3.63			
6	Eg	2.15				
	IP	5.56				
	EA	3.41				
8	Eg	1.99				
	IP	5.47				
	EA	3.48				
inf	Eg	1.55	1.41	1.66 (1.0) [14]	1.32	1.58
	IP	5.23	5.14	5.29	4.85	4.84
	EA	3.68	3.82	3.69	3.53	3.32

Experimental values are quoted in brackets. For oligomers these are λ_{max} values obtained in solution, for polymers they are onsets of absorption. All values are given in electron volts (eV)

Table 5 Energy gaps, IPs, and EAs according to Koopmans' theorem for oligomers of the TT system

Number of repeat units		TT	TT-TH	TH-TT-TH	TT-PY	PY-TT-PY
1	Eg	4.38	3.38	2.72 (2.00) [12]	3.18	2.47 (1.74) [52]
	IP	7.06	6.28	6.12	5.93	5.68
	EA	2.68	2.90	3.40	2.75	3.21
2	Eg	3.54	2.56	2.33	2.25	2.07
	IP	6.60	5.92	5.87	5.46	5.39
	EA	3.06	3.36	3.54	3.20	3.32
3	Eg		2.27	2.19	1.94	1.95
	IP		5.80	5.79	5.27	5.29
	EA		3.53	3.60	3.34	3.34
4	Eg	3.01	2.11	2.13	1.76	1.89
	IP	6.30	5.74	5.76	5.18	5.25
	EA	3.29	3.63	3.63	3.42	3.36
6	Eg	2.82				
	IP	6.20				
	EA	3.39				
inf	Eg	2.40	1.61	1.97 (0.9) [12]	1.17	1.73
	IP	5.97	5.59	5.71	4.90	5.14
	EA	3.57	3.92	3.73	3.63	3.40

Experimental values are quoted in brackets. For oligomers these are λ_{max} values obtained in solution, for polymers they are onsets of absorption. All values are given in electron volts (eV)

P-TP has a small theoretical energy gap as a homopolymer, 1.6 eV. Therefore, the band gap would be around 0.7 eV. Copolymerization in 1:1 stoichiometry with TH and with PY leads to reductions of the energy gap. As always, the effect of pyrrole is stronger than that of thiophene. In contrast to most of the other systems, increasing the content of TH and PY increases the energy gaps.

TP is the only acceptor for which the homopolymer has a lower IP (5.23 eV) than p-TH (5.66 eV at the same level of theory). Therefore, the valence band is pushed up compared to p-TH, not down as for the other systems. With one TH ring and with one and two PY rings, the IPs of the copolymers are smaller than those of either homopolymer. The small energy gap of p-TH-TP is due to a slight increase in EA and a slight decrease in IP. For p-TP-PY the reason is the stronger decrease of the IP compared to that of the EA.

Table 6 Energy gaps, IPs, and EAs according to Koopmans' theorem for oligomers of the TTP system

Number of repeat units		TTP	TTP -TH	TH- TTP -TH	TTP -PY	PY- TTP -PY
1	Eg	q 3.29/a 3.01 (2.13) [17]	2.63	q 2.28/a 1.71 (1.25) [17]	2.52	q 2.20/a 1.31 (0.92) [17]
	IP	q 7.19/a 7.32	6.45	q 6.03/a 6.09	6.16	q 5.66/a 5.68
	EA	q 3.90/a 4.31	3.82	q 3.66/a 4.38	3.64	q 3.46/a 4.37
2	Eg	2.59	1.98	1.58	1.78	1.27
	IP	6.81	6.19	5.73	5.76	5.33
	EA	4.22	4.21	4.15	3.98	4.07
3	Eg	2.41	1.82	1.45	1.63	1.14
	IP	6.64	6.12	5.66	5.62	5.27
	EA	4.23	4.29	4.20	3.99	4.13
4	Eg	2.31	1.75	1.40	1.56	1.10
	IP	6.56	6.10	5.63	5.56	5.26
	EA	4.26	4.35	4.23	4.01	4.16
5	Eg	2.26				
	IP	6.52				
	EA	4.26				
6	Eg	2.22				
	IP	6.49				
	EA	4.27				
inf	Eg	2.11	1.57	1.29 (0.3) [52]	1.30	1.02 (0) [52]
	IP	6.39	6.07	5.57	5.43	5.22
	EA	4.34	4.50	4.32	4.07	4.25

Experimental values are quoted in brackets. For oligomers these are λ_{max} values obtained in solution, for polymers they are onsets of absorption. All values are given in electron volts (eV)

Thienothiadiazole TT

P-TT prefers the quinoid conformation by 11.2 kcal mol^{-1}. In 1:1 stoichiometry, copolymers are quinoid. Copolymers with thiophene prefer the *cis*-conformation. The band gaps are lowered by 0.8 eV with thiophene and by 1.2 eV with pyrrole (Table 5). In the case of thiophene, this is due to a decrease of the IP and increase of the EA. Pyrrole has only a small influence on the EA of TT but lowers the IP by 1 eV. It is remarkable that such a strong mixing of the EAs occurs with TT since TT has the second smallest EA of all the acceptors in this investigation. The increase of the EAs of the copolymers compared to p-TT of up to 0.35 eV (with TH) seems to be possible since the difference between the EAs of p-TT and p-TH is only 0.5 eV. According to the donor–acceptor concept, the smallest band gaps are expected when the energy differences between the energy levels are largest.

With two thiophene rings, there is only a very small energy difference between quinoid and aromatic structures, 0.24 kcal mol^{-1} in favor of the quinoid form. For PY-TT-PY, the energy difference is 1.23 kcal mol^{-1}. When open-shell (unrestricted) calculations are used for comparison, aromatic forms are preferred but there is increasing spin contamination with increasing chain length, which may artificially lower the energies. It is therefore difficult to decide from a theoretical point of view which structure is correct. Since energy gaps depend strongly on the choice of structure, we evaluated aromatic, quinoid and open-shell aromatic forms for the 1:2 oligomers. For quinoid as well as for closed-shell aromatic forms, HOMO-LUMO gaps extrapolate to very small energy gaps, around 1 eV. Quinoid p-PY-TT-PY has an energy gap of 0.34 eV. The band gaps of the 1:2 copolymers would therefore be close to zero. Poly-PY-TT-PY, in particular, would be a metal. For p-TH-TT-TH, the experimental band gap is known, 0.9 eV [12]. Thus, the theoretical energy gap of 1.1 eV is too small since it leads to a predicted band gap of 0.2 eV. With open-shell calculations the energy gaps increase. For p-TH-TT-TH we obtain a value of 1.97 eV, so that the predicted band gap, 1.07 eV, agrees with experiment. Thus, it seems that symmetry breaking occurs when the mismatch between quinoid and aromatic structure leads to no clear structural preference.

Thienopyrazinothiadiazole TTP

The homopolymer is planar and has a strong preference (12.2 kcal mol^{-1}) for the quinoid conformation. Even if copolymerized with two TH and two PY rings, the preference for the quinoid form, 4.0 and 5.8 kcal mol^{-1}, respectively, remains and TH and PY are forced to adopt quinoid structures as well. As a result of the geometrical mismatch, bond-length alternation in the TH or PY fragments is very small. The longest bond along the conjugated pathway is the C–C bond attached to the pyrazine group in the thiophene ring of the TTP fragment. With one thiophene ring, the *cis*-conformation is more stable. With two thiophene rings, the monomer prefers the *cis*-conformation but longer oligomers are more stable in the *trans* form. The energetic preferences are subtle and the effect on the energy gaps is small.

As for monomers in the BBT system, the energy gap of the quinoid PY-TTP-PY monomer is much larger than the experimental value, 2.20 vs 0.92 eV (Table 6). Using the aromatic form leads to good agreement. The HOMO-LUMO gap is 1.31 eV. The TDHF excitation energy is 1.28 eV. If 0.4 eV is subtracted to account for the solvent effect, the agreement is almost perfect. To obtain polymer values, the aromatic monomers were excluded from the extrapolation.

TTP copolymers have very small band gaps and p-PY-TTP-PY is predicted to have the smallest energy gap of all systems in this investigation, 1.02 eV. After correction of 0.9 eV, the experimental zero band gap is confirmed. This is interesting since residual doping was suggested as a possible reason for the absence of an electrochemical band gap [51, 52].

In the TTP system (Table 6), small energy gaps are due to large gap reductions upon copolymerization. Copolymerization decreases the relatively high IP of the homopolymer without lowering the EA. This is due to the fact that the LUMO of TTP is localized on the pyrazino thiadiazole side group. The thiophene rings contribute little and the overlap between the rings upon polymerization is small. Therefore, the EA of p-TTP, 4.34 eV, is not much larger than that of the monomer, 3.90 eV, and the conduction band is flat. The EAs of the copolymers are close to that of p-TTP, independent of whether TH or PY is used and independent of the stoichiometry.

Discussion

There are three main questions we wish to answer with this investigation. Firstly, do theoretical gas-phase calculations reproduce the experimental findings of dramatic band-gap lowering upon copolymerization of donor and "non-classical" acceptor units? Secondly, is the donor–acceptor concept as proposed by Havinga [18] responsible? Thirdly, what is the influence of the geometrical mismatch when quinoid and aromatic units are copolymerized?

In Fig. 2, our best estimates as discussed in the previous sections for λ_{max} values (HOMO-LUMO gaps −0.4 eV) and band gaps (extrapolated HOMO-LUMO gaps −0.9 eV) are plotted vs experimental values. The straight line corresponds to perfect agreement between theory and experiment. Open circles are oligomer data; closed circles are polymer band gaps. The overall accuracy is good. Thus, theory reproduces the small experimental band gaps. The good agreement justifies using DFT calculations to analyze electronic structures of these systems.

In Table 7, the band-gap changes upon copolymerization of the acceptors with thiophene and pyrrole in 1:1 and in 1:2 stoichiometry are summarized. We note large band-gap reductions with QX, TT, and with TTP. With BT there is no effect with thiophene but substantial lowering with pyrrole.

The band gaps of TP and BBT are affected much less and may decrease or increase depending on donor and stoichiometry. In general the band-gap reductions are larger with pyrrole than with thiophene. The general trend is that band gaps are smaller in 1:2 copolymers, only with TP and TT, the 1:1 copolymers have smaller band gaps.

The acceptors can be divided into three groups, QX and BT homopolymers are aromatic, p-TP and p-TT are quinoid but switch to aromatic structures when copolymerized with aromatic donors, TTP and BBT are quinoid and force thiophene and pyrrole units to become quinoid upon copolymerization. In each of the three groups, one system exhibits strong band-gap reduction upon copolymerization (QX, TTP, and TT) and one does not (BT, BBT, and TP). There is thus no clear indication that quinoid acceptors are required for band-gap reductions of copolymers with aromatic donors. It will be shown below that small band gaps are obtained, however, when the energetic preference for either form is small. Neither the pyrazino group nor the thiadiazole group is crucial for band-gap reduction, as the comparison between QX, BT, TP, and TT shows.

Changes in IPs and EAs upon copolymerization are listed in Tables 8 and 9. The general trend is that copolymerization decreases IPs and EAs. This can be attributed to the fact that pyrrole and thiophene have lower IPs and EAs than the acceptors. The changes range from +0.06 to −1.17 eV for IPs and from +0.35 to −0.69 eV for EAs. Smaller reductions of EAs are responsible for the band-gap reductions. The effects are larger with pyrrole than with thiophene, especially on the IPs. The exceptionally large band-gap reductions in the TTP system are due to the nearly constant EAs of all of the polymers involving TTP. Only two acceptors with relatively low EAs interact strongly with unoccupied levels of TH and PY. The corresponding copolymers may even have increased EAs compared to the homopolymers. In summary, strong interactions arise only between similar energy levels. Interactions between levels of very different energy are small as predicted by perturbation theory.

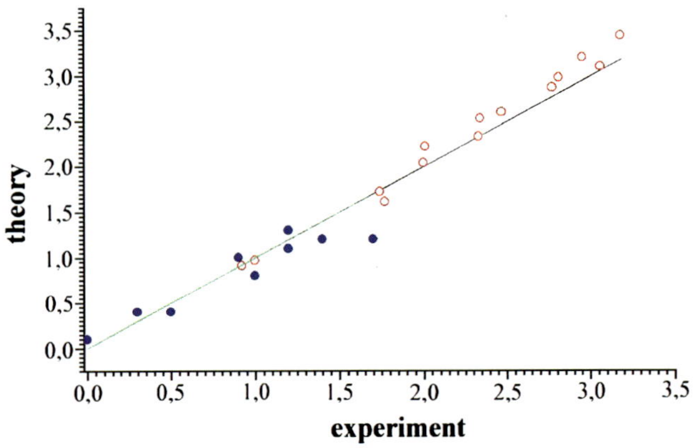

Fig. 2 Theoretical λ_{max} and band gap estimates plotted vs. measured values. *Open circles* are oligomer data, *closed circles* are extrapolated polymer band gaps. All energies are in electron volts (eV)

Table 7 Band gap changes of copolymers with respect to the corresponding acceptor homopolymers in electron volt (eV)

	TH		PY	
	1:1	1:2	1:1	1:2
QX	−0.44	−0.49	−0.67	−0.50
BT	−0.06	−0.04	−0.50	−0.48
BBT	+0.36	+0.06	+0.30	−0.25
TP	−0.14	+0.11	−0.23	+0.03
TT	−0.79	−0.47	−1.23	−0.68
TTP	−0.54	−0.82	−0.81	−1.09

The smallest band gaps are predicted to be those of p-PY-TTP-PY ~ p-PY-BBT-PY< p-TT-PY<TH-TTP-TH ~ TTP-PY ~ TP-PY< TH-BBT-TH< TP-TH. With TP and BBT the small band gaps do not result from band-gap reductions but from small band gaps of the homopolymers. BBT is unusual since the small band gap of the homopolymer is probably caused by steric strain. It is noteworthy that small band gaps result for copolymers of acceptors that strongly prefer quinoid structures in 1:2 stoichiometry and of acceptors that weakly prefer quinoid structures in 1:1 stoichiometry with aromatic donors. Thus, the smallest band gaps are found when energetic differences between quinoid and aromatic forms are small. The trend is continued with the 1:2 polymers of the TT system, for which we obtained extremely small energy gaps with closed-shell calculations. However, in these systems symmetry breaking increases the energy gap.

Since the original prediction of the donor–acceptor concept is that band gaps are smallest when electronegativity differences are largest, we tried to find correlations between band-gap reductions and energy differences between IPs and EAs of donors and acceptors. No such correlation could be established in a variety of plotting schemes. In contrast, the crucial factor for band-gap lowering is that the mixing between HOMO levels must be large, so that the IPs are lowered, and mixing between LUMO levels must be small, so that the large EAs of the homopolymers are preserved.

Small mixing between energy levels has a negative effect on band width. In Table 10 we show estimates for band widths. The numbers are not true band widths but energy differences between the four highest occupied orbitals and between the four lowest unoccupied orbitals

Table 8 IP changes of copolymers with respect to the corresponding acceptor homopolymers in electron volts (eV)

	TH		PY	
	1:1	1:2	1:1	1:2
QX	−0.51	−0.50	−0.88	−1.02
BT	−0.22	−0.30	−0.84	−1.10
BBT	−0.15	−0.41	−0.33	−0.81
TP	−0.09	+0.06	−0.38	−0.39
TT	−0.38	−0.32	−1.07	−0.85
TTP	−0.32	−0.82	−0.96	−1.17

Table 9 EA changes of copolymers with respect to the corresponding acceptor homopolymers in electron volts (eV)

	TH		PY	
	1:1	1:2	1:1	1:2
QX	−0.07	−0.01	−0.22	−0.32
BT	−0.14	−0.34	−0.40	−0.65
BBT	−0.47	−0.44	−0.69	−0.58
TP	+0.15	+0.02	−0.14	−0.35
TT	+0.35	+0.16	+0.06	−0.16
TTP	+0.16	−0.02	−0.27	−0.09

for tetramers. For a valid comparison, the same four levels are taken for thiophene and pyrrole octamers and dodecamers. Octamers can be considered to be tetramers of dimeric, dodecamers of trimeric repeat units. Dimers have the same number of double bonds as 1:1 and trimers as 1:2 comonomers. Since the repeat units of 1:2 are larger than those of 1:1 polymers, their "band widths" are smaller. The valence "band widths" of cooligomers are comparable to those of TH and PY systems. The conduction "band widths" tend to be smaller, about half of those of the TH and PY tetramers. Aromatic polymers have wider valence and conduction bands than quinoid polymers.

According to the donor–acceptor concept, maximizing the electronegativity difference decreases band gaps and increases band widths. Thus copolymers with pyrrole should have wider valence and conduction bands than with thiophene. The opposite is the case for the conduction bands of the QX, BT, and TP systems. Only the quinoid polymers seem to have wide conduction bands, even wider than those of pyrrole and thiophene oligomers. Closer inspection of the energy levels shows that for the quinoid tetramers, the first three unoccupied orbitals are virtually degenerate. The fourth level has a substantially higher energy and is localized at the ends of the chains. This is probably an end-effect due to the CH_2 groups, which are

Table 10 Valence and conduction "band" width of tetramers in eV

	TH		PY	
	1:1	1:2	1:1	1:2
	1.70	1.02	1.72	0.99
	1.44	0.90	1.15	0.70
QX	1.62	1.13	1.60	1.17
	0.75	0.47	0.56	0.25
BT	1.56	1.03	1.65	1.18
	0.70	0.44	0.57	0.26
BBT	1.31	1.10	1.41	1.21
	0.72 (0.19)	0.88	0.72	0.92 (0.22)
TP	1.84	1.07	1.86	1.11
	0.81	0.49	0.62	0.29
TT	1.58	0.92	1.82	0.98
	0.60 (0.33)	0.32	0.69 (0.30)	0.19
TTP	1.38	1.10	1.55	1.15
	0.60 (0.13)	0.64 (0.05)	0.69 (0.11)	0.83 (0.08)

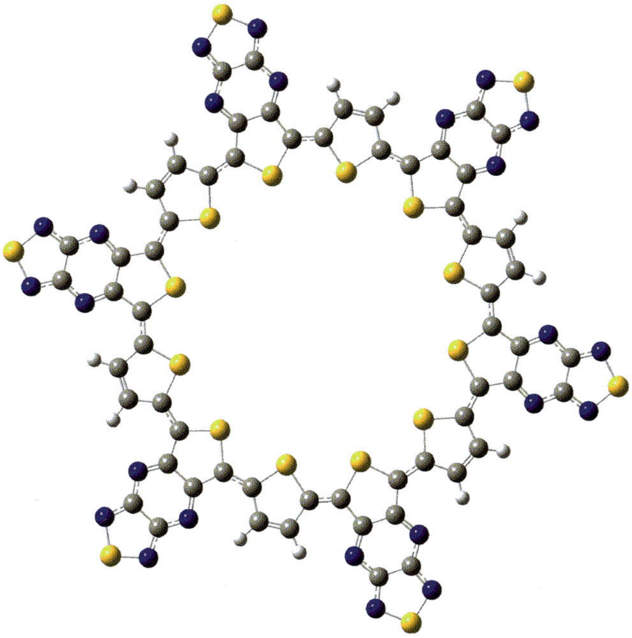

Fig. 3 Six-membered ring of *cis*-TTP-TH used as a model for p-TTP-TH avoiding end groups

needed to terminate the oligomers properly. For aromatic tetramers, the four levels are equally spaced. Without this last level, homo- and copolymers of BBT and TTP have negligible conduction-band dispersions. The corresponding values are given in parentheses for some of the systems in Table 10. To check whether this approach is valid, a test calculation on a symmetric 6-TTP-TH ring (Fig. 3) was performed. The TTP-TH system prefers the *cis*-conformation and leads to an unstrained six-membered ring. IP (6.12 eV), EA (4.46 eV), and band gap (1.66 eV) are close to those of the tetramer and of the polymer. Valence and conduction band widths estimated as the energy range of the six highest occupied MOs and the six lowest unoccupied MOs are 1.74 and 0.24 eV, respectively. This confirms that the conduction band widths of the TTP system is very small. Without end groups, no high-lying unoccupied level exists.

A totally flat conduction band was observed also for the only other zero-band-gap conducting polymer ever synthesized, p-CDM-EDOT [6]. This polymer has high intrinsic conductivity but the ratio between the mobilities of p- and n-type charge carriers is 500. Theoretical investigation revealed that the low mobility of the electrons is due to the lack of interaction between the high-lying unoccupied energy levels of EDOT and the low-lying unoccupied levels of CDM [53]. Because of their narrow conduction bands, we expect copolymers of BBT and TTP to have poor n-conductivities as well.

Conclusions

Estimates of λ_{max} values of oligomers and of band gaps of polymers obtained from DFT HOMO-LUMO gaps reproduce experimental values reliably if corrections for solvent effects and for the difference between E_g and λ_{max} are applied. Care must be taken to consider the correct conformations (aromatic vs. quinoid) since the band gaps are dramatically underestimated when the wrong geometry is used. We used comparison with experiment for several systems to select the proper structures. Purely theoretical predictions require great care.

Poly-PY-TTP-PY is predicted to have no band gap. This supports the experimental finding and indicates that this system is a zero-band-gap material in its neutral form, without residual doping. We predict p-PY-BBT-PY and p-TT-PY to have almost the same low band gaps. Thus, synthesizing p-PY-BBT-PY without methyl groups on the pyrrole rings could be worthwhile. For p-TH-TTP-TH, p-TTP-PY, p-TH-BBT-TH, p-TP-PY, and p-TP-TH, we predict band gaps of about 0.4–0.5 eV.

Band-gap reductions are a result of smaller upshifts of EAs as compared to IPs. The smaller general effect of copolymerization on EAs is due to reduced interactions between donor and acceptor because of large LUMO energy differences. This leads to smaller conduction-band widths compared to homopolymers for aromatic polymers and vanishingly small conduction-band widths of polymers involving TTP and BBT. This makes the low-band-gap quinoid polymers like p-PY-TTP-PY good candidates for intrinsic conductors. We predict their p-type conductivities to be comparable to those of homopolymers and n-type conductivities to be very low.

Since band-gap reductions do not correlate with IPs or EAs of monomer or homopolymers, we believe that the origin of the band-gap reductions is not due to the donor–acceptor concept. On the contrary, the reduced interactions between the energetically different LUMOs of donors and acceptors prevent averaging of the EAs. At the same time, the HOMOs of more similar energies interact and average their values. Thus, the two predictions of the donor–acceptor concept, namely that band gaps are smaller the larger the electronegativity difference and that polymerizing donor with acceptors leads to increase in band widths, are both wrong.

Geometrical mismatch between quinoid and aromatic form seems to be an important factor when the structural preferences of donor and acceptor are opposite and about of equal size. This is shown when comparing TP, TT, and TTP. TP and TT have a weak preference for the quinoid form and the band gap is smaller in 1:1 stoichiometry than in 1:2 stoichiometry with TH and PY. TTP has a strong preference for the quinoid form. Here, increasing the amount of donor additionally lowers the energy gap. When there is no clear preference for either structure as with 1:2 copolymers involving TT, closed-shell calculations predict vanishingly small energy gaps. However, symmetry breaking may occur, which increases the band gap.

Acknowledgements We thank TÜBITAK and Bilkent University for financial support. We also thank Burcin Temel for carrying out part of the calculations on the BT system during her senior project at Bilkent University.

References

1. Chiang CK, Fincher CR, Park YW, Heeger AJ, Shirakawa H, Louis EJ, Gau SC, MacDiarmid AG (1977) Phys Rev Lett 39:1098–1101
2. Roncali J (1992) Chem Rev 92:711–738
3. Roncali J (1997) Chem Rev 97:173–205
4. Skotheim TA (1986) Handbook of Conducting Polymers. Marcel Dekker, New York
5. Skotheim TA, Elsenbaumer RL, Reynolds JR (1997) Handbook of Conducting Polymers. In: Marcel Dekker Inc, New York
6. Huang H, Pickup PG (1998) Chem Mater 10:2212–2216
7. Lambert TL, Ferraris JP (1991) J Chem Soc, Chem Commun 752–754
8. Kanbara T, Yamamoto T (1993) Macromolecules 26:3464–3466
9. Tanaka S, Yamashita Y (1993) Synth Met 55-57:1251–1254
10. Kitamura C, Tanaka S, Yamashita Y (1994) J Chem Soc, Chem Commun 1585–1586
11. Kanbara T, Miyazaki Y, Yamamoto T (1995) J Polym Sci A 33: 999–1003
12. Karikomi M, Kitamura C, Tanaka S, Yamashita Y (1995) J Am Chem Soc 117:6791–6792
13. Tanaka S, Yamashita Y (1995) Synth Met 69:599–600
14. Kitamura C, Tanaka S, Yamashita Y (1996) Chem Mater 8: 570–578
15. Kitamura C, Tanaka S, Yamashita Y (1996) Chem Lett 63–64
16. Yamamoto T, Zhou Z-h, Kanbara T, Shimura M, Kizu K, Maruyama T, Nakamura Y, Fukuda T, Lee B-L, Ooba N, Tomaru S, Kurihara T, Kaino T, Kubota K, Sasaki S (1996) J Am Chem Soc 118:10389–10399
17. Tanaka S, Yamashita Y (1997) Synth Met 84:229–230
18. Havinga EE, ten Hoeve W, Wynberg H (1993) Synth Met 55-57:299–306
19. Kürti J, Surján PR, Kertesz M, Frapper G (1993) Synth Met 55-57:4338–4343
20. Frisch MJ, Trucks GW, Schlegel HB, Scuseria GE, Robb MA, Cheeseman JR, Zakrzewski VG, Montgomery JJA, Stratmann RE, Burant JC, Dapprich S, Millam JM, Daniels AD, Kudin KN, Strain MC, Farkas O, Tomasi J, Barone V, Cossi M, Cammi R, Mennucci B, Pomelli C, Adamo C, Clifford S, Ochterski J, Petersson GA, Ayala PY, Cui Q, Morokuma K, Rega N, Salvador P, Dannenberg JJ, Malick DK, Rabuck AD, Raghavachari K, Foresman JB, Cioslowski J, Ortiz JV, Baboul AG, Stefanov BB, Liu G, Liashenko A, Piskorz P, Komaromi I, Gomperts R, Martin RL, Fox DJ, Keith T, Al-Laham MA, Peng CY, Nanayakkara A, Challacombe M, Gill PMW, Johnson B, Chen W, Wong MW, Andres JL, Gonzalez C, Head-Gordon M, Replogle ES, Pople JA (1998) Gaussian 98, revision A.11.2. Gaussian Inc, Pittsburgh PA
21. Frisch MJ, Trucks GW, Schlegel HB, Scuseria GE, Robb MA, Cheeseman JR, Montgomery JJA, Vreven T, Kudin KN, Burant JC, Millam JM, Iyengar SS, Tomasi J, Barone V, Mennucci B, Cossi M, Scalmani G, Rega N, Petersson GA, Nakatsuji H, Hada M, Ehara M, Toyota K, Fukuda R, Hasegawa J, Ishida M, Nakajima T, Honda Y, Kitao O, Nakai H, Klene M, Li X, Knox JE, Hratchian HP, Cross JB, Adamo C, Jaramillo J, Gomperts R, Stratmann RE, Yazyev O, Austin AJ, Cammi R, Pomelli C, Ochterski JW, Ayala PY, Morokuma K, Voth GA, Salvador P, Dannenberg JJ, Zakrzewski VG, Dapprich S, Daniels AD, Strain MC, Farkas O, Malick DK, Rabuck AD, Raghavachari K, Foresman JB, Ortiz JV, Cui Q, Baboul AG, Clifford S, Cioslowski J, Stefanov BB, Liu G, Liashenko A, Piskorz P, Komaromi I, Martin RL, Fox DJ, Keith T, Al-Laham MA, Peng CY, Nanayakkara A, Challacombe M, Gill PMW, Johnson B, Chen W, Wong MW, Gonzales C, Pople JA (2003) Gaussian 03, Revision C. 02. Gaussian Inc, Pittsburgh PA
22. Becke AD (1993) J Chem Phys 98:5648–5652
23. Perdew JP (1986) Phys Rev B 33:8822–8824
24. Stevens W, Basch H, Krauss M (1984) J Chem Phys 81:6026
25. Leusser D, Henn J, Kocher N, Engels B, Stalke D (2004) J Am Chem Soc 126:1781–1793

26. Brédas L-L, Silbey R, Boudreaux DS, Chance RR (1983) J Am Chem Soc 105:6555–6559
27. Salzner U (2004) Curr Org Chem 8:569–590
28. Salzner U (2005) Conjugated Organic Polymers: From Bulk to Molecular Wire. In: Rieth M, Schommers W (eds) Handbook of Theoretical and Computational Nanotechnology, ASP (accepted)
29. Salzner U, Lagowski JB, Pickup PG, Poirier RA (1997) J Comput Chem 18:1943–1953
30. Salzner U, Lagowski JB, Pickup PG, Poirier RA (1998) J Phys Chem 102:2572–2578
31. D'Amico KL, Manos C, Christensen RL (1980) J Am Chem Soc 102:1777–1782
32. Cornil J, Beljonne D, Calbert JP, Brédas JL (2001) Adv Mater 13:1053–1067
33. Taliani C, Blinov LM (1996) Adv Mater 8:353–359
34. Fincher JCR, Ozaki M, Tanaka M, Peebles D, Lauchlan L, Heeger AJ (1979) Phys Rev B 20:1589–1602
35. Leising G (1988) Phys Rev B 38:10313–10322
36. Birnbaum D, Fichou D, Kohler BE (1992) J Chem Phys 96:165–169
37. Birnbaum D, Kohler BE (1989) J Chem Phys 90:3506–3510
38. Birnbaum D, Kohler BE (1991) J Chem Phys 95:4783–4789
39. Chung T-C, Kaufman JH, Heeger AJ, Wudl F (1984) Phys Rev B 30:702–710
40. Tourillon G, Garnier F (1983) J Phys Chem 87:2289–2292
41. Street GB (1986) Polypyrrole from powders to plastics. In: Skotheim TA (ed) Handbook of conducting polymers, vol 1. Dekker, New York
42. Pei Q, Zuccarello G, Ahlskog M, Inganäs O (1994) Polymer 35:1347–1351
43. Sato M, Tanaka S, Kaeriyama K (1987) J Chem Soc, Chem Commun 1725–1726
44. Koopmans TA (1933) Physica 1:104–113
45. Perdew JP, Levy M (1983) Phys Rev Lett 51:1884–1887
46. Williams AR, von Barth U (1983) Applications of Density Functional Theory to Atoms, Molecules, and Solids. In: Lundqvist S, March NH (eds) Theory of the Inhomogeneous Electron Gas. Plenum, London
47. Godby RW, Schlüter M, Sham LJ (1988) Phys Rev B 37: 10159–10175
48. Baerends EJ, Gritsenko OV (1997) J Phys Chem A 101:5383–5403
49. Stowasser R, Hoffmann R (1999) J Am Chem Soc 121:3414–3420
50. Gritsenko OV, Baerends EJ (2002) J Chem Phys 117:9154–9159
51. van Mullekom HAM, Vekemans JAJM, Meijer EW (1998) Chem Eur J 4:1235–1243
52. van Mullekom HAM (2000) The chemistry of high and low band gap π-conjugated polymers. University of Technology, Eindhoven
53. Salzner U, Köse ME (2002) J Phys Chem B 106:9221–9226

J Mol Model (2006) 12: 703–711
DOI 10.1007/s00894-005-0047-1

ORIGINAL PAPER

William N. Setzer · Glenn F. Rozmus · Mary C. Setzer ·
Jennifer M. Schmidt · Bernhard Vogler · Sabine Reeb ·
Betsy R. Jackes · Anthony K. Irvine

Bioactive principles in the bark of *Pilidiostigma tropicum*

Received: 13 May 2005 / Accepted: 27 September 2005 / Published online: 7 April 2006

Abstract The crude dichloromethane bark extract of *Pilidiostigma tropicum* (Myrtaceae) from north Queensland, Australia, shows antibacterial and cytotoxic activity. Bioactivity-directed separation led to the isolation of rhodomyrtoxin B and ursolic acid-3-*p*-coumarate as the biologically active materials. The structures of these compounds were elucidated on the basis of spectral analysis. The intercalation interaction of rhodomyrtoxin B with DNA was investigated using molecular mechanics and ab initio molecular-orbital techniques. A favorable π–π interaction between rhodomyrtoxin B and the cytosine–guanine base pair is predicted, but the orientation of the interaction cannot be predicted based on frontier molecular orbitals.

Keywords Cytotoxicity · Antibacterial · *Pilidiostigma tropicum* · Rhodomyrtoxin B · Ursolic acid coumarate · Intercalation

Dedicated to Professor Dr. Paul von Ragué Schleyer on the occasion of his 75th birthday

W. N. Setzer (✉) · G. F. Rozmus · M. C. Setzer ·
J. M. Schmidt · B. Vogler
Department of Chemistry, University of Alabama in Huntsville,
Huntsville, AL 35899, USA
e-mail: wsetzer@chemistry.uah.edu
Tel.: +1-256-8246519
Fax: +1-256-8246349

S. Reeb
Institut für Chemie, Universität Hohenheim,
70599 Stuttgart, Germany

B. R. Jackes
School of Tropical Biology, James Cook University,
Townsville Queensland 4811, Australia

A. K. Irvine
C. S. I. R. O., Tropical Forest Research Centre,
Atherton Queensland 4883, Australia

Introduction

The Myrtaceae is a significant family in herbal medicine throughout the world. Important Australian members of the family include *Eucalyptus globulus* ("blue gum") and *Melaleuca alternifolia* ("tea tree") [1]. We have been interested in Australian tropical rainforest members of the family, [2] and in this work we have examined *Pilidiostigma tropicum* L.S. Smith (common name: Apricot Myrtle) from north Queensland, Australia. To our knowledge, there are no references to any ethnobotanical uses of this tree, nor have there been any previously reported phytochemical or pharmacological investigations.

P. tropicum is endemic to the Wet Tropics region of north Queensland, Australia, between the Paluma Range and just south of Cooktown [3]. Within this region it is a widespread understory tree, up to 18 m. tall, with opposite, simple leaves and flaky, often fissured bark. Its altitudinal range is from near sea level to 1,000 m. and it occurs on a wide range of soil types.

In this paper, we present the isolation and biological activities of the cytotoxic and antibacterial principles from *P. tropicum*. We also present a theoretical investigation of the π–π interactions between rhodomyrtoxin B and the guanine–cytosine base pair in order to determine if the lowest-energy orientations of intercalation can be predicted based upon frontier molecular orbital interactions.

Materials and methods

Collection and extraction of plant material

Bark samples from several specimens of *P. tropicum* were collected 6th August, 1997, from a complex notophyll vine forest [4] at ca. 720 m a.s.l. at the Curtain Fig State Forest, Australia (17°17' S, 145°34' E) under permit issued to A. K. Irvine (C.S.I.R.O.). Voucher specimens have been

Scheme 1 Separation of *Pilidiostigma tropicum* crude bark extract by flash liquid chromatography

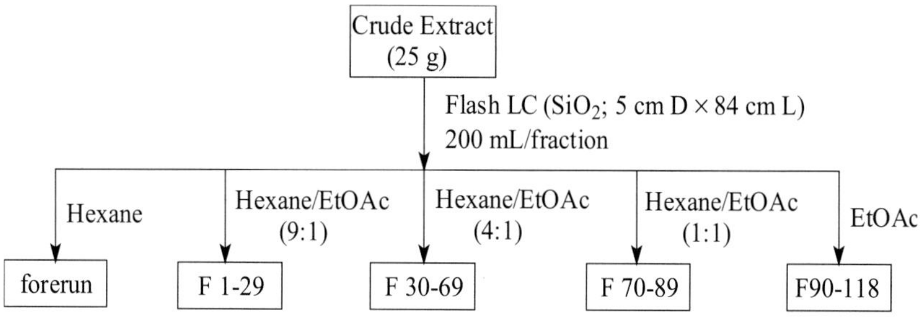

deposited with the C.S.I.R.O. (A.K. Irvine 1707) and the James Cook University Herbaria. The stem bark (988 g) was chopped and extracted with a 1:1 mixture of chloroform/ethanol (24 h) to give 45.8 g extract.

Biological screening

The crude bark extract, chromatographic fractions, and purified compounds were screened for antimicrobial activity against Gram-positive bacteria, *Bacillus cereus* (ATCC No. 14579) and *Staphylococcus aureus* (ATCC No. 29213), and Gram-negative bacteria, *Pseudomonas aeruginosa* (ATCC No. 27853) and *Escherichia coli* (ATCC No. 25922) as previously described [2]. The materials were screened for cytotoxic activity against Human Hep G2 hepatocellular carcinoma cells (ATCC No. HB-8065) [5] and Human MDA-MB-231 mammary adenocarcinoma cells (ATCC No. HTB-26) [6] as described previously [2].

Bioactivity-directed isolation

The crude bark extract (25.0 g) was subjected to bioactivity-directed flash chromatography (Scheme 1); on a silica gel (230–400 mesh) column (84 cm L×5 cm D). Elution was carried out using a hexane/ethyl acetate step gradient (hexane, 9:1 hexane/EtOAc, 4:1 hexane/EtOAc, 1:1 hexane/EtOAc, EtOAc) with detection of eluates by TLC. Bioactivity screening of fractions led to two biologically active compounds: rhodomyrtoxin B (recrystallized from F32-33 to give 867 mg) [7] and ursolic acid-3-*p*-coumarate (recrystallized from F77 to give 86.7 mg). [8] Structures were determined by comparison of ^{1}H and ^{13}C NMR and MS with those reported in the literature. The biological activities of these materials are presented in Table 1.

Computational studies

Hartree-Fock *ab initio* calculations at the 3-21G(*) level were carried out using both GAUSSIAN 98W and SPARTAN '04 computer programs. The π–π interaction of rhodomyrtoxin B with the cytosine–guanine (C–G) base pair was investigated at the post-Hartree-Fock level, using the MP2/6-31G* method, as implemented by the GAUSSIAN 98W computer program. The structures of rhodomyrtoxin B and the base pair (assuming planarity) were optimized at the HF 3-21G(*) level. Single-point MP2/6-31G* calculations were carried out using different coplanar orientations and positions of rhodomyrtoxin B, held at 3.4 Å above the plane of the base pair, to map the energy surface. Molecular mechanics calculations using MM2 were carried out to investigate the intercalation of rhodomyrtoxin B with a C-G dimer, 3.4 Å separation, using the preferred orientation of rhodomyrtoxin B with C-G.

Results

The crude bark extract of *P. tropicum* showed *in vitro* cytotoxic activity against Hep-G2 (human hepatocellular carcinoma) and MDA-MB-231 (human mammary adenocarcinoma) cell lines, as well as antibacterial activity against the Gram-positive bacteria *Bacillus cereus* and *Staphylococcus aureus* (Table 1). Bioactivity-directed flash chromatographic separation (Scheme 1) led to two biologically active compounds: rhodomyrtoxin B and

Table 1 Cytotoxic and antibacterial activities of *Pilidiostigma tropicum* bark extract and constituents

Material	Cytotoxicity (LC_{50})		Antibacterial activity (MIC)	
	Hep-G2	MDA-MB-231	*Bacilluscereus*	*Staphylococcus aureus*
P. tropicum crude bark extract	95.6±0.7% kill at 250 μg/ml	100% kill at 250 μg/ml	<19.5 μg/ml	<19.5 μg/ml
Rhodomyrtoxin B	19.0(±9.0) μM	2.50(±0.27) μM	0.14 μM	0.28 μM
Ursolic acid-3-*p*-coumarate	1.43(±0.36) μM	15.3(±3.1) μM	1.63 μM	6.5 μM
Positive control	1.19(±0.67) μM[a]	9.23(±7.89) μM[a]	2.14 μM[b]	1.07 μM[b]

[a]Vinblastine sulfate
[b]Gentamicin sulfate

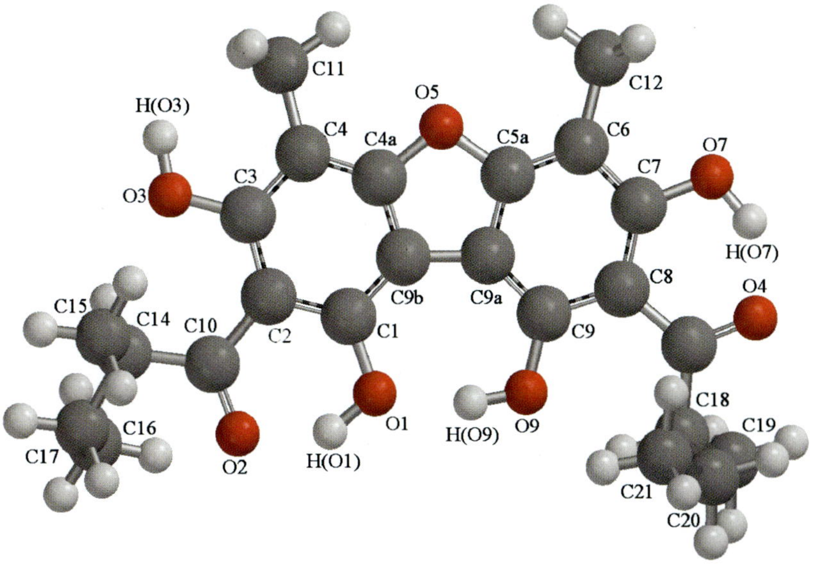

Fig. 1 HF 3-21G(*) optimized structure and numbering scheme of rhodomyrtoxin B

ursolic acid-3-*p*-coumarate. The biological activities of these materials are presented in Table 1.

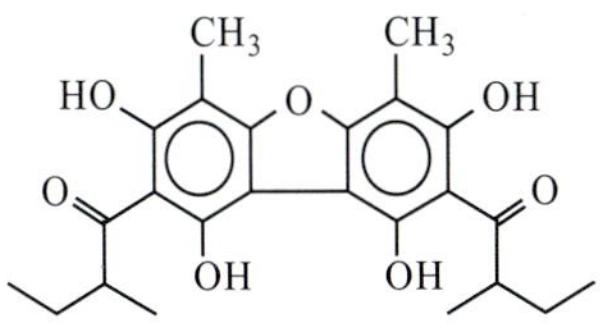

Rhodomyrtoxin B

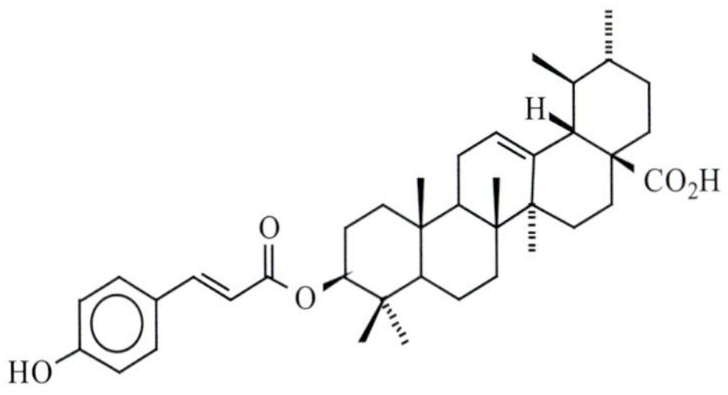

Ursolic acid-3-*p*-coumarate

The 3-21G(*) optimized structures for rhodomyrtoxin B, and the cytosine–guanine base pair are shown in Figs. 1 and 2, respectively. Mulliken atomic charges, HF 3-21G (*), for rhodomyrtoxin B, cytosine, and guanine are listed in Tables 2 and 3. Frontier–molecular–orbital (FMO) energies for C–G are: HOMO=−7.35 eV, LUMO=2.39 eV; and for rhodomyrtoxin B are: HOMO=−7.69 eV, LUMO=1.89 eV. Thus, the important frontier–molecular–orbital interactions are expected to be the HOMO of cytosine–guanine and the LUMO of rhodomyrtoxin B. These FMO's are shown in Fig. 3. Mapping the π–π interaction energy between rhodomyrtoxin B and the C-G base pair at the MP2/6-31G* level revealed six low-energy orientations (Figs. 4, 5, 6, 7, 8 and 9). Molecular mechanics (MM2) calculations indicate that the lowest-energy orientation from the ab initio study cannot be accommodated in an intercalative model. Thus, the MM2 strain energy increase in intercalation of rhodomyrtoxin B into the C–G complementary dimer is 82.7 kcal mol^{-1} (see Fig. 10).

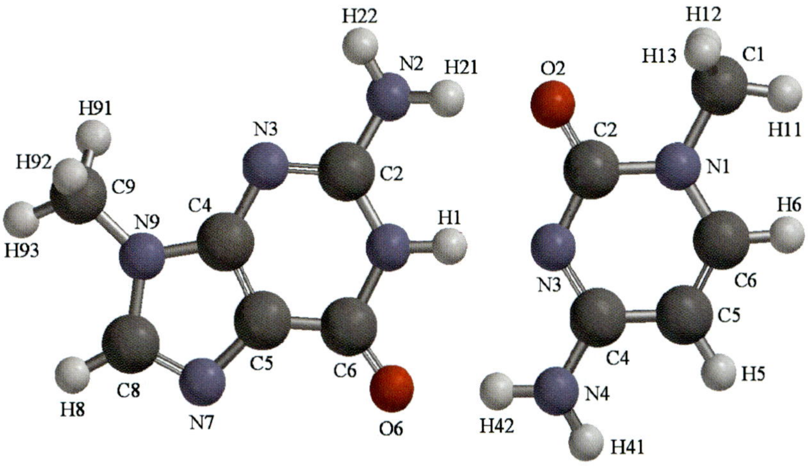

Fig. 2 HF 3-21G(*) optimized structure (constrained to be planar) and numbering scheme of guanine–cytosine

Table 2 Mulliken atomic charges, HF 3-21G(*), of rhodomyrtoxin B

Atom	Charge	Atom	Charge
C1	0.498	O3	−0.760
C2	−0.267	H(O3)	0.426
C3	0.431	C11	−0.616
C4	−0.135	Ha(C11)	0.275
C4a	0.445	Hb(C11)	0.221
O5	−0.768	Hc(C11)	0.220
C5a	0.425	C12	−0.544
C6	−0.153	Ha(C12)	0.233
C7	0.517	Hb(C12)	0.232
C8	−0.296	Hc(C12)	0.215
C9	0.460	O7	−0.790
C9a	−0.077	H(O7)	0.458
C9b	−0.064	C13	0.732
O1	−0.847	O4	−0.681
H(O1)	0.481	C18	−0.433
C10	0.731	H(C18)	0.297
O2	−0.680	C19	−0.548
C14	−0.415	Ha(C19)	0.228
H(C14)	0.278	Hb(C19)	0.200
C15	−0.556	Hc(C19)	0.199
Ha(C15)	0.221	C20	−0.380
Hb(C15)	0.214	Ha(C20)	0.226
Hc(C15	0.212	Hb(C20)	0.208
C16	−0.379	C21	−0.600
Ha(C16)	0.236	Ha(C21)	0.218
Hb(C16)	0.206	Hb(C21)	0.202
C17	−0.593	Hc(C21)	0.197
Ha(C17)	0.213		

Table 3 Mulliken atomic charges, HF 3-21G(*), of cytosine and guanine

Guanine		Cytosine	
Atom	Charge	Atom	Charge
N1	−1.048	N1	−0.943
H1	0.477	C1	−0.327
C2	1.020	H11	0.209
N2	−0.915	H12	0.248
H21	0.399	H13	0.249
H22	0.346	C2	1.141
N3	−0.813	O2	−0.674
C4	0.772	N3	−0.885
C5	0.028	C4	0.814
C6	0.946	N4	−0.940
O6	−0.696	H41	0.342
N7	−0.629	H42	0.440
C8	0.344	C5	−0.423
H8	0.266	H5	0.259
N9	−0.933	C6	0.266
C9	−0.321	H6	0.291
H91	0.249		
H92	0.225		
H93	0.216		

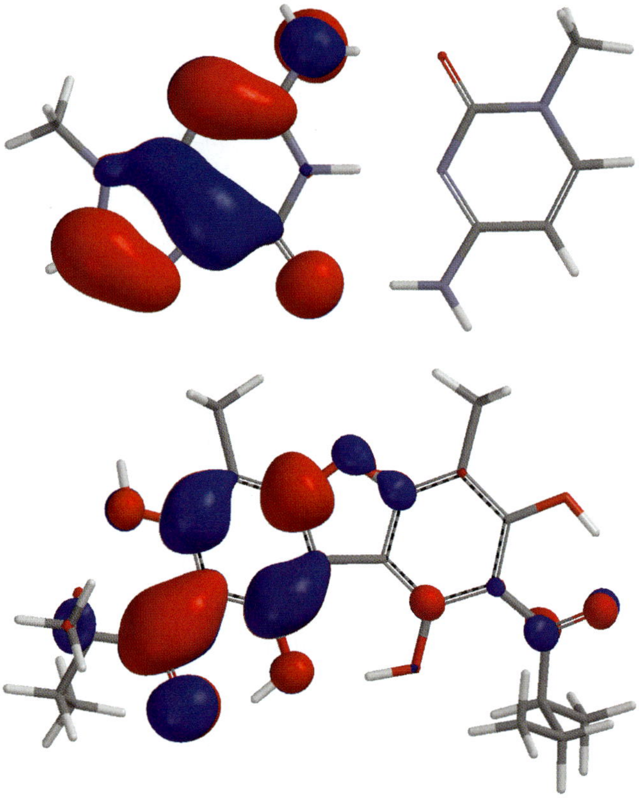

Fig. 3 HOMO of C–G and LUMO of rhodomyrtoxin B

Discussion

Rhodomyrtoxin B, isolated previously from *Rhodomyrtus macrocarpa*, [7] is a dibenzofuran derivative. This class of compounds has exhibited remarkable bioactivity, most notably the polychlorinated dibenzofurans, [9] but natural dibenzofurans are active as well [10–14]. Dibenzofurans are known to bind to the aryl hydrocarbon receptor [15] as well as to intercalate into DNA [16–20].

Ursolic acid [21] as well as coumarate esters of terpenoids have also shown cytotoxic activity [22–24]. Ursolic acid itself has been shown to inhibit the enzymes

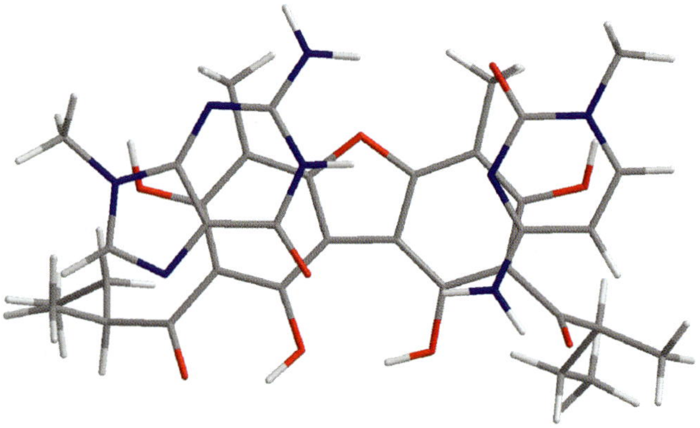

Fig. 4 Lowest-energy orientation of rhodomyrtoxin B with cytosine–guanine

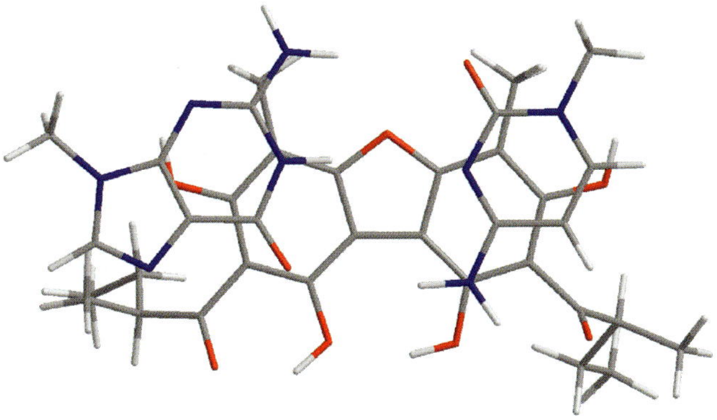

Fig. 5 Low-energy (E_{rel}=1.47 kcal mol^{-1}) orientation of rhodomyrtoxin B with cytosine–guanine

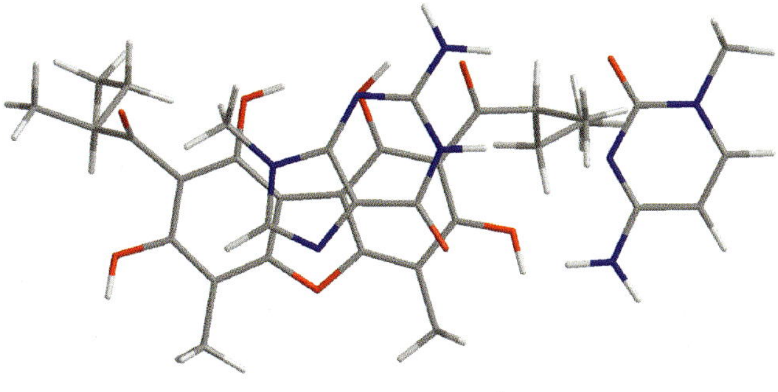

Fig. 7 Low-energy (E_{rel}=2.20 kcal mol^{-1}) orientation of rhodomyrtoxin B with cytosine–guanine

topoisomerase I and II, DNA polymerase, lipoxygenase, and tyrosine kinase, and induces apoptosis possibly involving activation of caspases [25]. Hydroxycinnamic esters probably exhibit their bioactivity by inhibition of respiratory chain-linked electron transport. The lipophilic triterpenoid allows penetration into membranes and the phenolic moiety produces stable free radicals, thus interfering with redox reactions [26, 27].

Interactions between π-systems such as benzene dimer [28, 29], naphthalene dimer [30], substituted benzenes with benzene [31, 32], substituted benzenes with aromatic nitrogen bases, [33] and guanine assemblies, [34] have been studied computationally as model systems for π–π interactions [35]. Dispersion interactions, which have their origins in electron correlation, are important facets in π–π interactions, and therefore, electron correlation must be included [36, 37]. In this study, the second order Møller–Plesset perturbation (MP2) [38] was used to account for correlation effects.

Intercalating antitumor drugs [39] function by forming a ternary cleavable complex with DNA and various DNA polymerases. These drugs intercalate or otherwise insert themselves into DNA and form a complex with proteins such as topoisomerase I and II that replicate and repair DNA such that they cannot continue to move along the DNA strand [40–43]. These agents have the greatest effect on cells that replicate rapidly and/or frequently, as tumor cells are known to do [40, 44]. DNA-intercalating ligands are a diverse group of compounds containing an aromatic chromophore that bind reversibly to double-stranded DNA by insertion between base pairs [45, 46]. Drugs that are currently used to treat cancers and leukemias such as doxorubicin and amsacrine [47] have been shown to intercalate into the DNA of these rapidly proliferating cells. Intercalating agents like these are typically aromatic heterocycles comprised of three fused six-membered rings. It is not known whether these agents exhibit sequence selectivity beyond two or three base pairs.

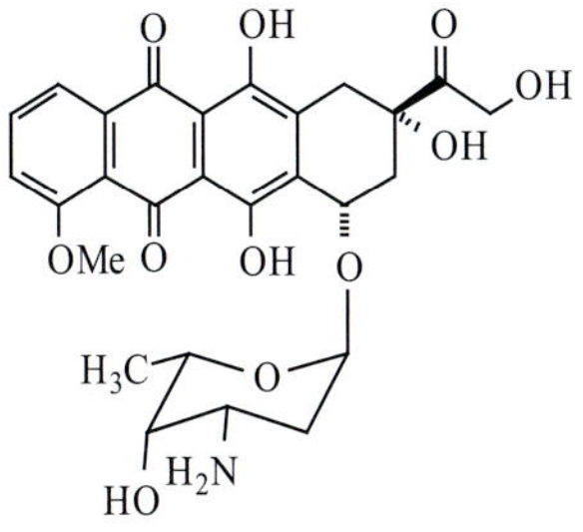

Doxorubicin

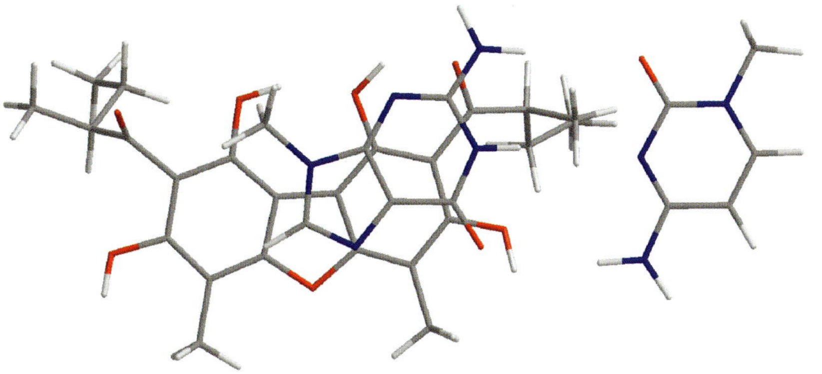

Fig. 6 Low-energy (E_{rel}= 1.95 kcal mol^{-1}) orientation of rhodomyrtoxin B with cytosine–guanine

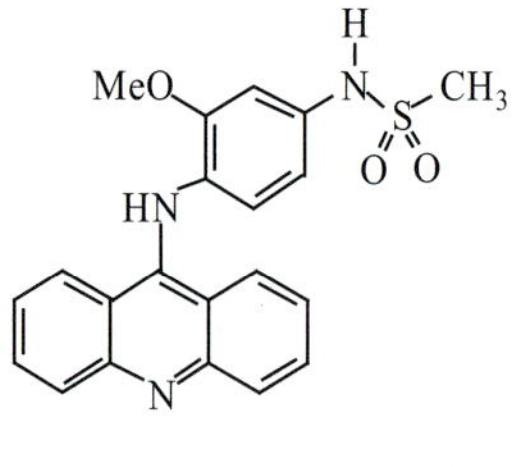

Amsacrine

Experimental structural studies of intercalating ligands and oligonucleotides have been carried out in the solid state using X-ray crystallography [48, 49] and in solution using high-resolution NMR techniques [50–56]. Interaction of DNA intercalators has been modeled using molecular mechanics/molecular dynamics [47, 57–61]. These modeling studies have concentrated on electrostatic interactions and hydrogen bonding. Although it is commonly accepted [47] that intercalating ligands align themselves within the DNA helix to "effect optimal π overlap with the two base pairs that constitute the binding site, there is no theory available that adequately explains the nature of the hydrophobic forces in nucleic acids" [62]. Nucleotide base stacking has been examined using high-level ab initio methods [63–65]. These workers have concluded that electron correlated ab initio calculations are strictly required for obtaining reliable energies for π–π interactions, and that density functional (DFT) methods fail to adequately describe nucleotide stacked base pairs. Recently, Reha and co-workers [66] have examined the nature of intercalative interactions of ethidium, daunomycin, ellipticine, and 4',6-diaminide-2-phenylindole with DNA base pairs using ab initio, density functional theory (DFT), and empirical potential methods. These workers have concluded that only theoretical procedures properly covering dispersion, polarization, and charge-transfer effects can be used to study intercalation. Thus, simple Hartree-Fock (HF) methods or DFT/B3LYP methods are not recommended; whereas inclusion of Møller-Plesset second order electron correlation (MP2) or DFTB-D methods provide the dispersion energy necessary for π–π stacking.

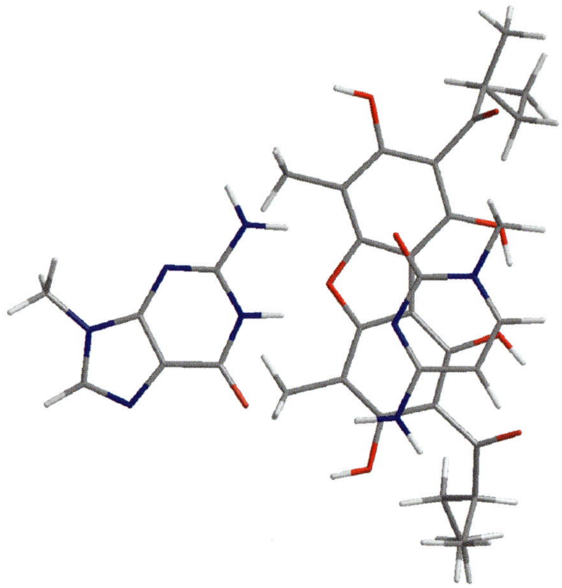

Fig. 9 Low-energy (E_{rel}=2.59 kcal mol^{-1}) orientation of rhodomyrtoxin B with cytosine–guanine

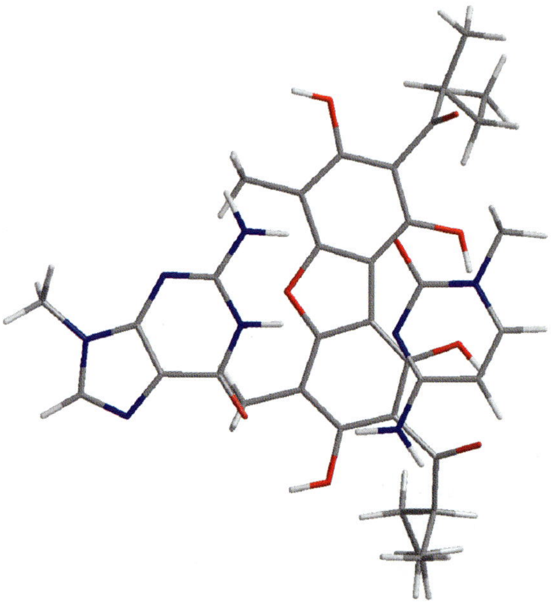

Fig. 8 Low-energy (E_{rel}=2.53 kcal mol^{-1}) orientation of rhodomyrtoxin B with cytosine–guanine

Frontier–molecular–orbital interactions have been considered in π-stacking (self organization) of naphthalene with pyromellitimide [67]. The self-organized π-stacked naphthalene-pyromellitimide-naphthalene sandwich shows favorable orbital overlap between the highest occupied molecular orbital (HOMO) of naphthalene and the lowest unoccupied molecular orbital (LUMO) of pyromellitimide. In the case of rhodomyrtoxin B and the C–G base pair, the HOMO of the base pair is located on guanine while the LUMO of rhodomyrtoxin B is located predominantly on one of the benzene rings (see Fig. 3). However, the lowest energy orientation of rhodomyrtoxin B with the C–G base pair (Fig. 4) does not allow maximum complementary overlap of the HOMO of C–G with the LUMO of rhodomyrtoxin B. Indeed, the orientation that gives the best frontier molecular orbital overlap pictorially is 13.5 kcal mol^{-1} higher in energy than the lowest-energy orientation. Thus, in this case, favorable frontier–molecular–orbital overlap cannot be used to predict the preferred orientation of the intercalator.

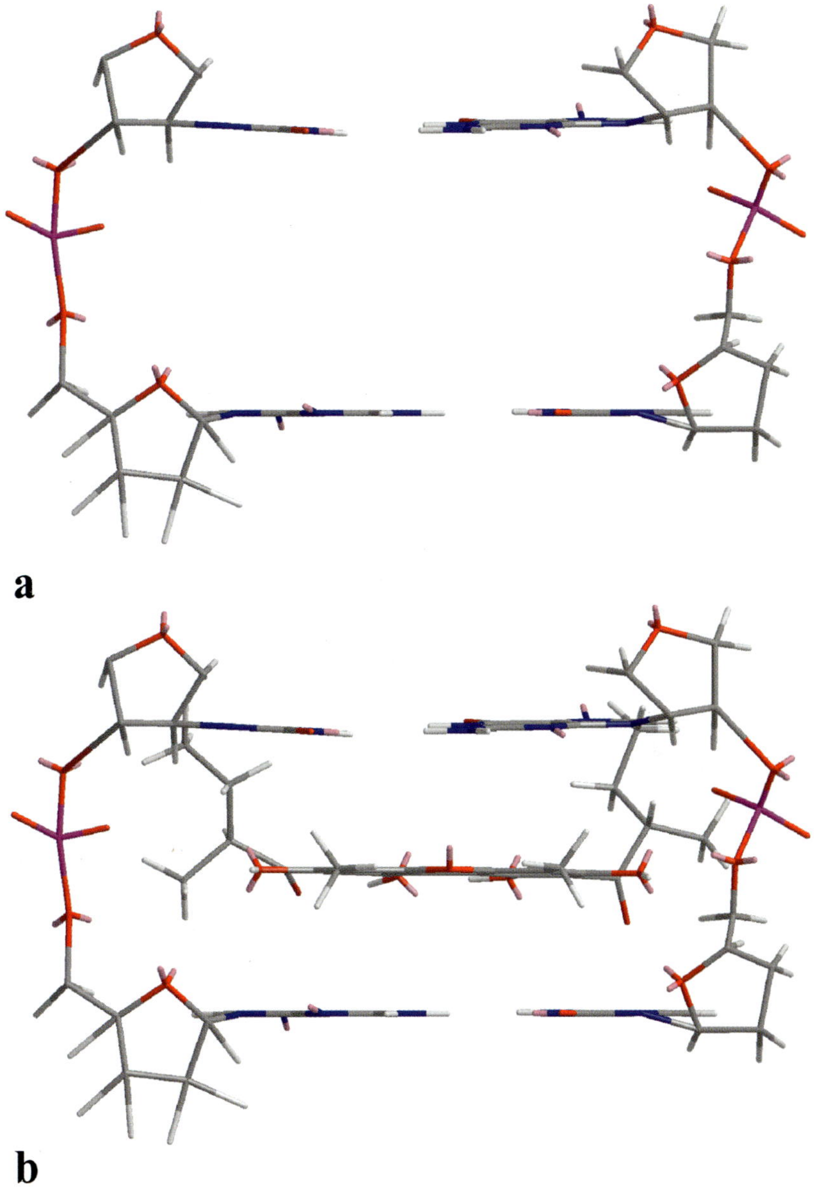

Fig. 10 Intercalation model of rhodomyrtoxin B with a C–G duplex (**a** base-pair separation of 6.8 Å. **b** intercalation of rhodomyrtoxin B)

Electrostatic interactions are generally considered to be important for DNA intercalation. Thus, for example, electrostatic interactions play a major role in the positioning of 9-aminoacridine as a DNA intercalator and a significant stabilizing role in the binding of ethidium in DNA [68]. Examination of the atomic charges on rhodomyrtoxin B and the C–G base pair does not reveal compelling evidence of favorable electrostatic interactions in this case, however. Dipole–dipole interactions may also be important in DNA intercalation. It has been suggested that intercalation of methylene blue between DNA base pairs is stabilized by favorable dipole-dipole interactions as well as π–π interactions [69]. The molecular dipoles of rhodomyrtoxin B and the C–G base pair in the lowest-energy orientation do, in fact, align in nearly opposite directions (Fig. 11). Thus, dipole–dipole interactions may be the important interaction governing the orientation of rhodomyrtoxin B in DNA.

Conclusions

The bioactive agents responsible for the cytotoxic and antimicrobial activity of the crude bark extract of *Pilidiostigma tropicum* are rhodomyrtoxin B and ursolic acid-3-*p*-coumarate. These compounds exhibit bioactivities comparable to the antitumor drug vinblastine or the antibiotic gentamicin. The computational results from this study are consistent with an intercalative mechanism as the possible source of the cytotoxic activity of rhodomyrtoxin B. Although lowest-energy positions and orientations of rhodomyrtoxin B with the C–G base pair have been

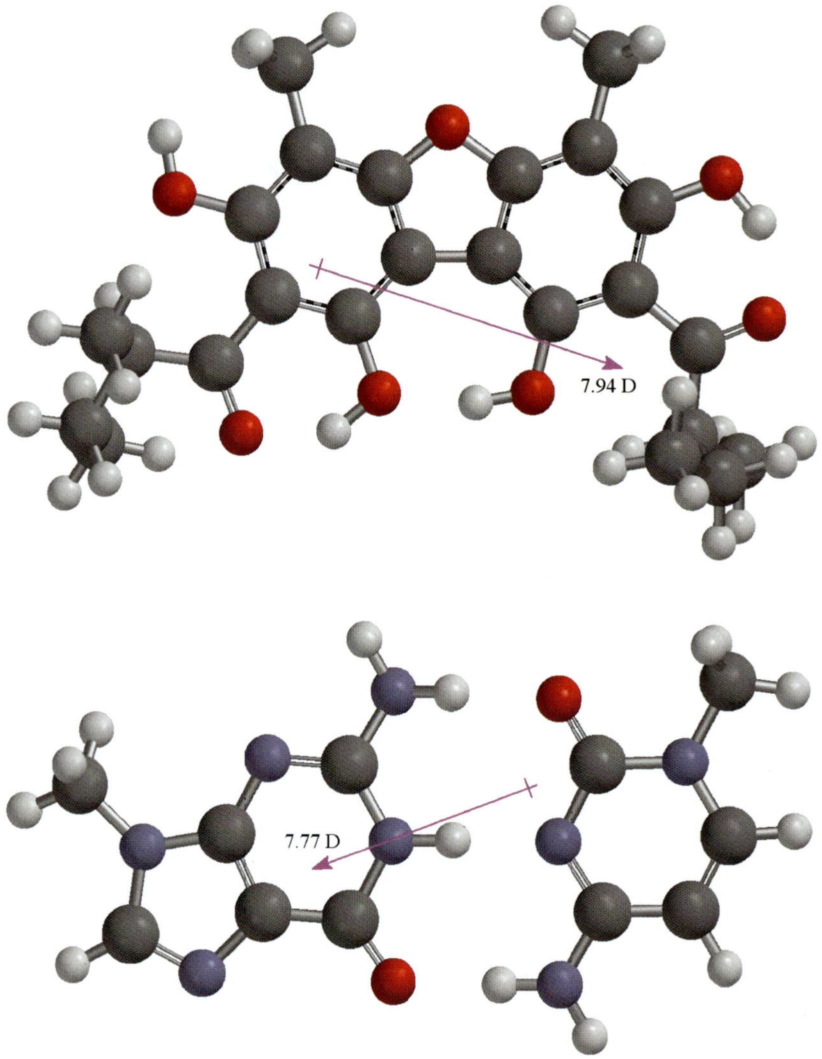

Fig. 11 Calculated molecular dipoles of rhodomyrtoxin B and C–G base pair

calculated using the MP2 method, these positions cannot be predicted based upon simple frontier-molecular–orbital interactions, electrostatic interactions, or dipole-dipole interactions. Steric interactions of the alkyl side chains of rhodomyrtoxin B may also preclude these orientations in DNA itself.

Acknowledgements Support for this research was provided in part by a grant from the National Institutes of Health (Grant No. 1 R15 AIOD39740-01). We are grateful to the Queensland Forest Service for allowing us access to the Curtain Fig State Forest. We thank Ms. Susan M. Fields for assistance with the cytotoxicity screening.

References

1. Lassak EV, McCarthy T (1990) Australian medicinal plants. Mandarin Press, Melbourne, Australia
2. Setzer MC, Setzer WN, Jackes BR, Gentry GA, Moriarity DM (2001) Pharmaceut Biol 39:67–78
3. Hyland BPM, Whiffen T (1993) Australian tropical rain forest trees, vol 2. C.S.I.R.O., Melbourne, Australia
4. Tracey JG (1982) The vegetation of the humid tropical region of North Queensland. C.S.I.R.O., Melbourne, Australia
5. Knowles BB, Howe CC, Aden DP (1980) Science 209: 497–499
6. Cailleau R, Young R, Olive M, Reeves WJ (1974) J Natl Cancer Inst 53:661–674
7. Igboechi CA, Parfitt RT, Rowan MG (1984) Phytochemistry 23:1139–1141
8. Yasue M, Sakakibara J, Ina H (1973) Yakugaku Zasshi 93: 687–691
9. Safe SH (1994) Crit Rev Toxicol 24:87–149
10. Carney JR, Scheuer PJ (1993) Tetrahedron Lett 34:3727–3730
11. Kokubun T, Harborne JB, Eagles J, Waterman PG (1995) Phytochemistry 38:57–60
12. Kokubun T, Harborne JB, Eagles J, Waterman PG (1995) Phytochemistry 39:1039–1042
13. Kumar KC, Muller K (1999) J Nat Prod 62:821–823
14. Ingolfsdottir K (2002) Phytochemistry 61:729–736
15. Hahn ME (1998) Comp Biochem Physiol C 121:23–53
16. Wang S, Hall JE, Tanious FA, Wilson WD, Patrick DA, McCurdy DR, Bender BC, Tidwell RR (1999) Eur J Med Chem 34:215–224
17. Yamada M, Kato K, Shindo K, Nomizu M, Haruki M, Sakairi N, Ohkawa K, Yamamoto H, Nishi N (2001) Biomaterials 22:3121–3126

18. Yamada M, Kato K, Nomizu M, Ohkawa K, Yamamoto H, Nishi N (2002) Environ Sci Technol 36:949–954
19. Zhao C, Liu X, Nomizu M, Nishi N (2004) J Coll Interface Sci 275:470–476
20. Liu XD, Murayama Y, Matsunaga M, Nomizu M, Nishi N (2005) Int J Biol Macromol 35:193–199
21. Lee KH, Lin YM, Wu TS, Zhang DC, Yamagishi T, Hayashi T, Hall IH, Chang JJ, Wu RY, Yang TH (1988) Planta Med 54:308–311
22. Numata A, Yang P, Takahashi C, Fujiki R, Nabae M, Fujita E (1989) Chem Pharm Bull 37:648–651
23. Kashiwada Y, Zhang DC, Chen YP, Cheng CM, Chen HT, Chang HC, Chang JJ, Lee KH (1993) J Nat Prod 56:2077–2082
24. Setzer WN, Setzer MC, Bates RB, Nakkiew P, Jackes BR, Chen L, McFerrin MB, Meehan EJ (1999) Planta Med 65:747–749
25. Setzer WN, Setzer MC (2003) Mini Rev Med Chem 3:540–556
26. Kubo I, Fujita KI, Nihei KI, Masuoka N (2003) Bioorg Med Chem 11:573–580
27. Sakihama Y, Cohen MF, Grace SC, Yamasaki H (2002) Toxicology 177:67–80
28. Tsuzuki S, Honda K, Uchimaru T, Mikami M, Tanabe K (2002) J Am Chem Soc 124:104–112
29. Sinnokrot MO, Valeev EF, Sherrill CD (2002) J Am Chem Soc 124:10887–10893
30. Tsuzuki S, Uchimaru T, Matsumura K, Mikami M, Tanabe K (2000) Chem Phys Lett 319:547–554
31. Sinnokrot MO, Sherrill CD (2004) J Am Chem Soc 126: 2690–2697
32. Mignon P, Loverix S, Geerlings P (2005) Chem Phys Lett 401:40–46
33. Chandra PP, Jain A, Sapse AM (2004) J Mol Model 10:1–5
34. Di Felice R, Calzolari A, Molinari E, Garbesi A (2002) Phys Rev B 65:045104
35. Waters ML (2002) Curr Opin Chem Biol 6:736–741
36. Pickholz M, Stafström S (2001) Chem Phys 270:245–251
37. Pickholz M, dos Santos MC (2005) J Mol Struct Theochem 717:99–106
38. Møller C, Plesset MS (1934) Phys Rev 46:618–622
39. Berman HM, Young PR (1981) Annu Rev Biophys Bioeng 10:87–114
40. Vivas-Mejia PE, Cox O, Gonzalez FA (1998) Mol Cell Biochem 178:203–212
41. Baez A, Riou JF, Le Pecq JB, Riou G (1989) Mol Pharmacol 37:377–382
42. Chen H, Patel DJ (1995) J Mol Biol 246:164–179
43. Tabarrini O, Cecchetti V, Fravolini A, Nocentini G, Barzi A, Sabatini S, Miao H, Sissi C (1999) J Med Chem 42:2136–2144
44. Dassonneville L, Lansiaux A, Wattelet A (2000) Eur J Pharmacol 409:9–18
45. Denny WA (1989) Anti-Cancer Drug Design 4:241–263
46. Baguley BC (1991) Anti-Cancer Drug Design 6:1–35
47. Neidle S, Jenkins TC (1991) Meth Enzymol 203:433–458
48. Todd AK, Adams A, Thorpe JH, Denny WA, Wakelin LPG, Cardin CJ (1999) J Med Chem 42:536–540
49. Adams A, Guss JM, Collyer CA, Denny WA, Wakelin LPG (1999) Biochemistry 38:9221–9233
50. Bunkenborg J, Stidsen MM, Jacobsen JP (1999) Bioconjugate Chem 10:824–831
51. Bostock-Smith CE, Gimenez-Arnau E, Missailidis S, Laughton CA, Stevens MFG, Searle MS (1999) Biochemistry 38: 6723–6731
52. Bible KC, Bible RH, Kottke TJ, Svingen PA, Xu K, Pang YP, Hajdu E, Kaufmann SH (2000) Cancer Res 60:2419–2428
53. Li Z, Tamura PJ, Wilkinson AS, Harris CM, Harris TM, Stone MP (2001) Biochemistry 40:6743–6755
54. Favier A, Blackledge M, Simorre JP, Crouzy S, Dabouis V, Gueiffier A, Marion D, Debouzy JC (2001) Biochemistry 40:8717–8726
55. Brown K, Hingerty BE, Guenther EA, Krishnan VV, Broyde S, Turteltaub KW, Cosman M (2001) Proc Natl Acad Sci USA 98:8507–8512
56. Han X, Gao X (2001) Curr Med Chem 8:551–581
57. Hill GC, Wunz TP, MacKenzie NE, Gooley PR, Remers WA (1991) J Med Chem 34:2079–2088
58. Cardozo MG, Hopfinger AJ (1991) Mol Pharmacol 40: 1023–1028
59. van der Klein-de Gunst FJM, van Boom JH, Liskamp RMJ (1992) J Comput Aided Mol Des 6:33–46
60. de Pascual-Teresa B, Gallego J, Ortiz AR, Gago F (1996) J Med Chem 39:4810–4824
61. Fisher G, Pindur U (1999) Pharmazie 54:83–93
62. Voet D, Voet JG (1995) Biochemistry, 2nd edn. Wiley, New York, pp 848–872
63. Sponer J, Leszczynski J, Hobza P (1996) J Phys Chem 100:5590–5596
64. Hobza P, Sponer J (1999) Chem Rev 99:3247–3276
65. Elstner M, Hobza P, Frauenheim T, Suhai S, Kaxiras E (2001) J Chem Phys 114:5149–5155
66. Reha D, Kabelac M, Ryjacek F, Sponer J, Sponer JE, Elstner M, Suhai S, Hobza P (2002) J Am Chem Soc 124:3366–3376
67. Cooray AS, de Silva KMN (2004) J Mol Struct Theochem 678:223–231
68. Medhi C, Mitchell JBO, Price SL, Tabor AB (1999) Biopolymers 52:84–93
69. Zhang LZ, Tang GQ (2004) J Photochem Photobiol B 74: 119–125

J Mol Model (2006) 12: 713–721
DOI 10.1007/s00894-005-0040-8

ORIGINAL PAPER

Oliver Weingart · Igor Schapiro · Volker Buss

Bond torsion affects the product distribution in the photoreaction of retinal model chromophores

Received: 23 May 2005 / Accepted: 26 July 2005 / Published online: 10 November 2005

Abstract Ab initio molecular dynamics (MD) calculations have been performed to study the photoisomerization of a 3-double-bond retinal model chromophore, the all-*trans*-4, 6-dimethylpenta-3, 5-dieniminium cation, and the possible influence of non-planar distortions on the product distribution. In total, 171 trajectories have been generated for four different conformations of the structure, a planar one and three in which the C4–C5 and the C5 = C6 bonds were increasingly twisted out of plane. Starting geometries randomly distributed about the equilibrium geometry were generated by zero-point energy sampling; trajectories were calculated using CASSCF-BOMD methodology and were followed until the photoproduct and its configuration could be assigned. For the latter, two different approaches were applied, one involving the CASSCF configuration vectors, the other an analysis of the MD at the first possible hopping event. Isomerization was found to occur almost exclusively about the central C3 = C4 double bond in the case of the planar model compound. Twisting the conjugated π-system shifts the isomerization site from the central double bond to the terminal C5 = C6 double bond. With both the C4–C5 and the C5 = C6 bonds twisted by 20°, about 35% of the trajectories lead to the configurationally inverted 5-*cis* product. The results are discussed with reference to the highly selective and efficient photo-induced isomerization of the retinal chromophore in rhodopsin.

Keywords CASSCF · Molecular dynamics · Rhodopsin · Retinal · Isomerization

Dedicated to Professor Dr. Paul von Ragué Schleyer on the occasion of his 75th birthday.

O. Weingart · I. Schapiro · V. Buss (✉)
Department of Chemistry, University of Duisburg-Essen, Campus Duisburg, 47048 Duisburg, Germany
E-mail: theobuss@uni-duisburg.de
Tel.: +49-203-3793315
Fax: +49-203-3792772

Abbreviations MD: Molecular dynamics · BOMD: Born–Oppenheimer molecular dynamics · ZPES: Zero-point energy sampling · Rho: Rhodopsin · bR: Bacteriorhodopsin · Batho: Bathorhodopsin · pSb: Protonated Schiff base · SA: State-averaged · CASSCF: Complete active space self consistent field · MCSCF: Multiconfigurational SCF

Introduction

One of the most efficient photoconversions devised by nature is the reaction which initiates the process of vision in the rod cells of the vertebrate eye. Isomerization of 11-*cis*-retinal, the chromophore of the visual protein rhodopsin (rho), occurs ultra fast and is highly selective and effective: the first spectral evidence of the photoproduct is found after 200 fs; the single product is all-*trans*-retinal, and the quantum yield of the reaction is 0.67 [1]. After isomerization of the chromophore, the protein passes through a sequence of intermediates in which the configurational change of the chromophore is transferred stepwise into structural changes of the protein, until the active state is reached and the visual cascade is started [2]. A reaction of comparable efficiency initiates the photocycle of bacteriorhodopsin (bR), like rho, a retinal binding 7-helical membrane protein with, however, a different function: bR pumps protons from the cytoplasmatic to the extracellular side of the membrane. The chromophore of bR is all-*trans*-retinal, and the photoreaction causes isomerization from all-*trans*-to 13-*cis* [3]. In rho and in bR, the chromophores are covalently bound via protonated Schiff base (pSb) linkages to the ε-amino side chain of lysine residues, Lys296 and Lys216, respectively (Fig. 1) and are buried deep within the binding pockets of their respective proteins. Both configuration and conformation of the rho [4, 5] and of the bR chromophore [6, 7] have been verified with high-resolution X-ray crystallography.

The unsurpassed efficiency of the protein-bound retinal photoreaction has been studied in several theoretical

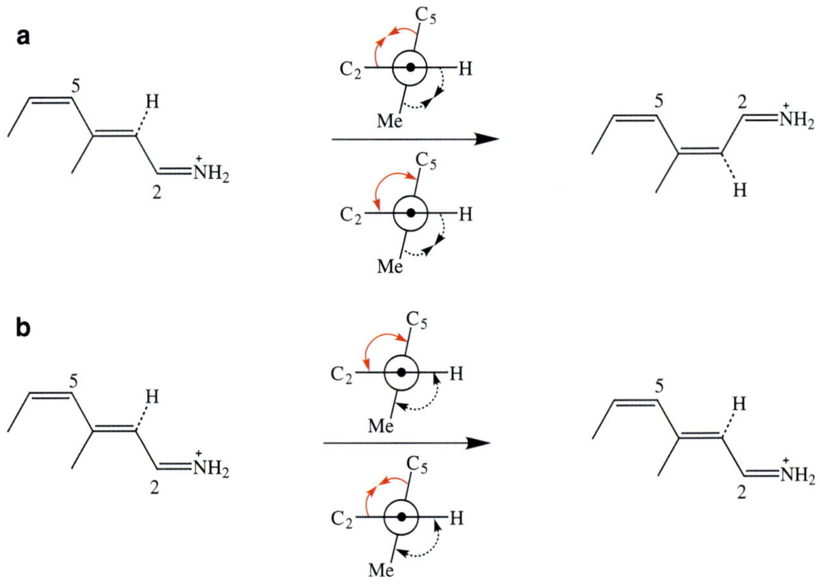

Fig. 2 Relationship between torsional modes and outcome of the photoisomerization. The reaction is productive only when at the decay point the H–C3 = C4–Me angle decreases, regardless of the C2–C3 = C4–C5 direction of motion (**a**). When this angle increases the starting configuration is regenerated (**b**)

cis-pentadieniminium cation, we were able to reproduce the product distribution obtained by time-dependent surface hopping exactly. It has also been used to determine the product distribution when starting from the *trans*-pentadieniminium cation [29]. It is the time-saving effect that has allowed the product analysis, i.e., the determination of the Z/E ratio, of the 171 MD runs of this study.

Results and discussion

In this section, we will first describe and discuss the features and general aspects of the calculated molecular trajectories. We will then present and compare the product distribution for chromophore **1** based on the two different methods for determining the hopping event. Finally, we will present the statistical analysis for all four chromophores.

"Productive" and "unproductive" photo-events

Figure 3 shows how several key molecular parameters evolve in time during a typical MD calculation. The left column of panels refers to a productive photoreaction in which one of the double bonds, C3 = C4 in this particular case, isomerizes from *trans* to *cis*. The right column

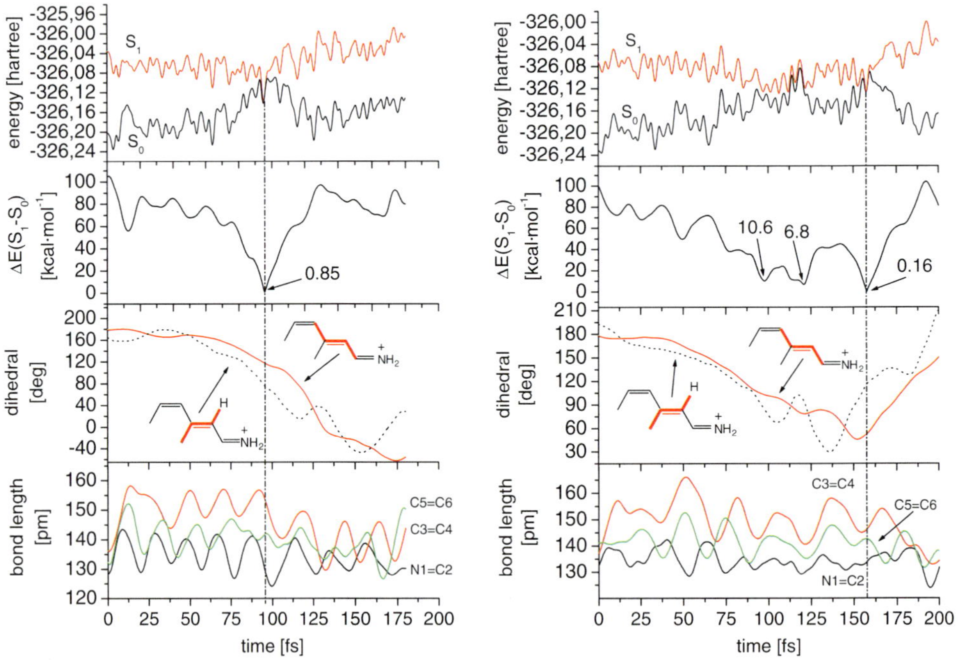

Fig. 3 Two trajectories of model **1**; *left*, trajectory that hops at the first close approach of S_1 and S_0 yielding the C3 = C4 *trans* product; *right*, trajectory that regenerates the starting configuration at the third approach of S_1 and S_0. The panels show (from *top*) the time development of the S_1 and S_0 energies; the energy difference, the C2–C3 = C4–C5 (*red full line*) and H–C3 = C4–Me (*dotted line*) torsion angles, and the double-bond lengths

shows how the parameters change in the case of an unproductive reaction, i.e., a bond starts to rotate, but at some point in time the rotation reverts and starting product is obtained.

The plots start with the Franck–Condon excitation from S_0 to S_1. The potential energy curves of both states show the high-frequency modulation caused by the fast C–H stretching vibrations. This is a result of the zero-energy sampling process in which all nuclei including the hydrogens have starting momenta and start to vibrate around their equilibrium positions. In the energy difference graph shown below, the C–H vibrational energy changes cancel out, in contrast to the energy changes due to the stretching modes of the double bonds, which are activated as a consequence of the electronic excitation. There is a conspicuous drop in the energy difference during the first 10 fs after excitation when all three double bonds start to adjust in-phase to the inverted electron density alternation of the S_1 state. This energy drop is singular since the vibrations start to dephase, which is obvious from an inspection of the bottom panel where the double-bond lengths are plotted against time. The strongest vibration is induced in the central double bond, which reaches a maximum length of close to 160 pm.

The third panel from top shows how the dihedral angles along the rotating bond changes along the trajectory. It reveals that the molecule is still essentially planar 60 fs after the excitation, when it slowly starts to rotate about the central bond. Concomittant with the rotation, the energy difference between S_1 and S_0 begins to drop sharply since the excited state is stabilized, and the ground state destabilized, by this torsional motion. After 95 fs, the energy difference has reached a value of less than 1 kcal mol^{-1} and the molecule hops back to the ground state. The continuing rotation carries the system quickly out of the hopping region, and the planar inverted geometry is reached after a total of 130 fs.

In the right trajectory, the molecule has already passed two close approaches—after 100 and 120 fs with energy differences of 10.6 and 6.8 kcal mol^{-1}, respectively, until it finally hops with an energy difference of 0.16 kcal mol^{-1} after 160 fs. The dihedral angle of the rotating bond is only 55° at the hopping event, i.e., the molecule has advanced significantly further toward the *cis* geometry than in the former trajectory, but this is not sufficient to realize a stable *cis*-configurated product. Instead, immediately after the hop the rotation of the bond reverts and the starting product is formed back. Note the rather smooth variation of the C2–C3 = C4–C5 dihedral angle in contrast to the erratic behaviour of the H–C3 = C4–Me torsion angle, especially in the neighborhood of close approaches of the energy curves. Also note that at the hopping event both angles are increasing again, allowing the prediction—as it turns out to be the case—that the product of this reaction will be the original all-*trans* configurated chromophore.

Product distribution: the vector rotation method

Forty-three trajectories were calculated and analyzed starting with the planar geometry of the chromophore model **1**. Despite the strain of the C5 = C6 bond due to the methyl groups at C4 and C6, there was only one trajectory in which this strain is released by torsion of this bond; 42 trajectories gave products resulting from C3 = C4 bond torsion. There is a broad distribution of hopping geometries, as shown in Fig. 4 (left). The Z/E ratio (i.e., productive vs. unproductive events) is large when the torsion angle is still large and hopping occurs at the first close approach. With decreasing torsion angle, the probability of unsuccessful approaches of the energy surfaces increases, and so does the fraction of unproductive events. In most of these cases, two or more approaches are necessary before the molecule actually decays. The calculated quantum yield is 0.6. Like the hopping geometries, the lifetimes of the excited states (Fig. 4, right) are spread out, and again there is a preference for high Z/E ratios when the lifetime is short.

Product distribution: hopping at the first close approach

Figure 5 shows what happens when the molecules are assumed to hop at the first close approach of the surfaces. Not unexpectedly the results differ substantially from those obtained with the vector rotation method,

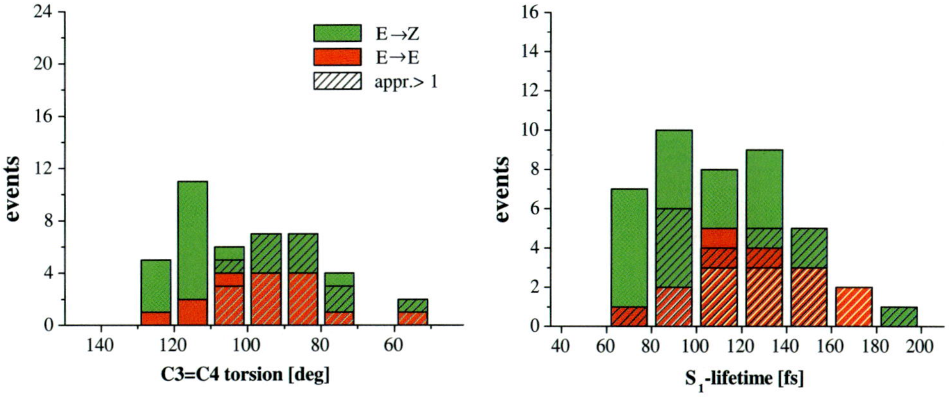

Fig. 4 Distribution of torsion angles at the hop point (*left*) and excited state lifetimes (*right*) for model **1** determined with the vector rotation method. *Lined boxes* denote trajectories not decaying at the first approach of S_1 and S_0

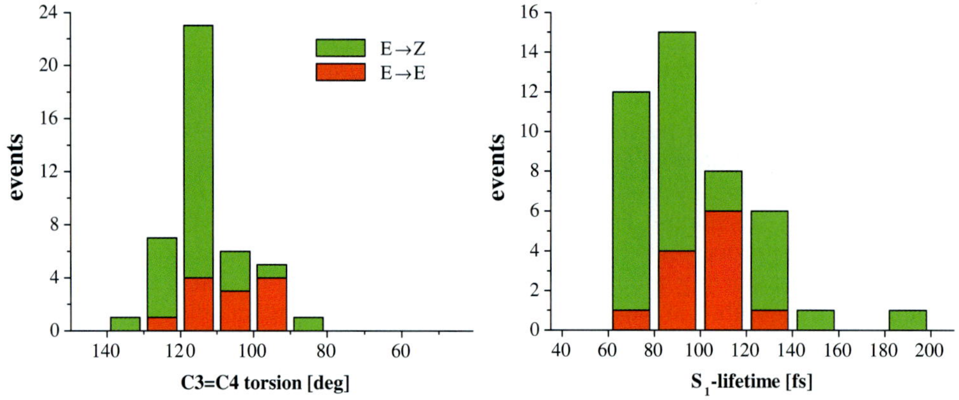

Fig. 5 *Left*, distribution of torsion angles and expected product distribution at the first close approach of S_1 and S_0 for model **1**; *right*, excited state lifetimes until the first close approach is reached

since the time that the molecules spend on the excited state surface is significantly reduced. One indication for this is the narrow distribution of torsion angles at the possible hopping event: since the molecules have less time for equilibration in the excited state, the hopping geometries are more similar. Fast decay is obviously connected with productive isomerization; more than 50% of the *cis* product is formed in less than 100 fs, while the majority of trajectories yielding starting material have lifetimes between 100 and 120 fs. In total, 30 out of 42 trajectories, i.e., 71% yield the *cis* product, compared to 60% when the vector rotation method is applied.

The two methods yield upper and lower bounds to the correct solution based on time-dependent quantum theory: the first possible event for surface hopping occurs when the surfaces approach sufficiently, whereas at a surface crossing the molecules will necessarily decay to the ground state.

The distribution of energy gaps (Fig. 6) is similar to that found in a non-adiabatic surface hopping MD of the *cis*-pentadieniminium cation [28, 29]. In nearly 60% of the trajectories we observe very small energy gaps of less than 4 kcal mol^{-1}, where surface hops are in fact highly probable.

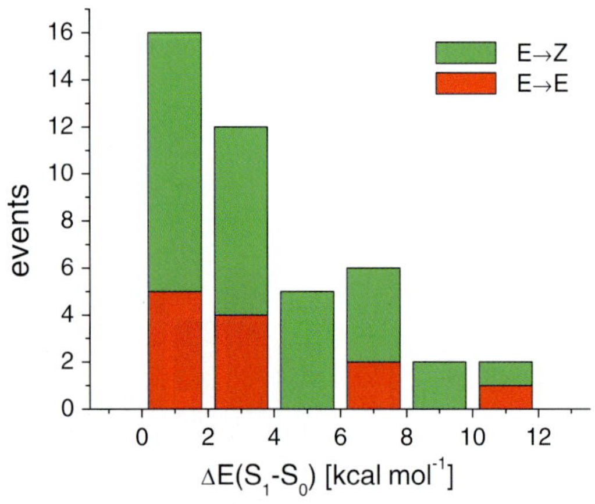

Fig. 6 Distribution of energy gaps at the first close approach of S_1 and S_0 for model system **1**

Effect of bond torsion on the product distribution

In total, 171 trajectories were calculated for the model chromophores **1**–**4** and analyzed with respect to the product distribution. The question which bond is going to isomerize can be answered very early in the trajectory as the rotatory motion is funneled into a particular site of the molecule. Whether the trajectory will yield the *Z*-product after hopping, or will reverse itself and return to the starting *E*-configuration, has been decided by an analysis of the chromophore dynamics at the first close approach of S_1 and S_0 as described in a preceding section. The results are summarized in Table 1, which also lists averaged values for the dihedral angles, the energy differences at the first close approach, and the time needed to arrive at this point.

Turning first to the question whether the non-planar distortion of the chromophore determines the position that is going to be affected by the photoreaction, we note that the fraction of molecules in which the central C3 = C4 bond isomerizes decreases significantly

Table 1 Summary of calculations for chromophore models **1**–**4** showing the total number of trajectories calculated and the number of C3 = C4 and C5 = C6 isomerization events

	1	2	3	4
Number of trajectories	43	44	43	41
C3 = C4 rotations	42	39	27	34
Productive	30	31	18	28
Clockwise	24	23	19	21
ϕ_{av}	114.3	112.2	109.6	114.0
ΔE_{av}	3.6	4.2	4.9	4.4
τ_{av}	98.4	96.3	99.7	93.6
C5 = C6 rotations	1	5	16	6
Productive	1	4	15	5
Clockwise	1	5	16	6
ϕ_{av}	80.2	82.5	80.3	92.2
ΔE_{av}	0.8	6.9	4.5	3.45
τ_{av}	114.7	83.0	81.8	87.7

For both types of isomerization, the table lists the number of productive runs yielding the *Z*-product; the number of trajectories following clockwise rotation; the average torsion angle ϕ (in degrees) of the rotating bond at the first close approach, the S_1-S_0 energy difference ΔE (in kcal mol^{-1}) at the first close approach and the excited state life time τ (in fs)

Fig. 7 Product distribution in the MD simulations of models **1**–**4**. The percentage of trajectories that lead to either C3 = C4 or C5 = C6 rotation is given beside the bar graphs. The *green* and the *red* portions of the bars represent, respectively, the productive and the unproductive events with respect to that particular rotation estimated from the torsion angles at the first close approach of the energy surfaces

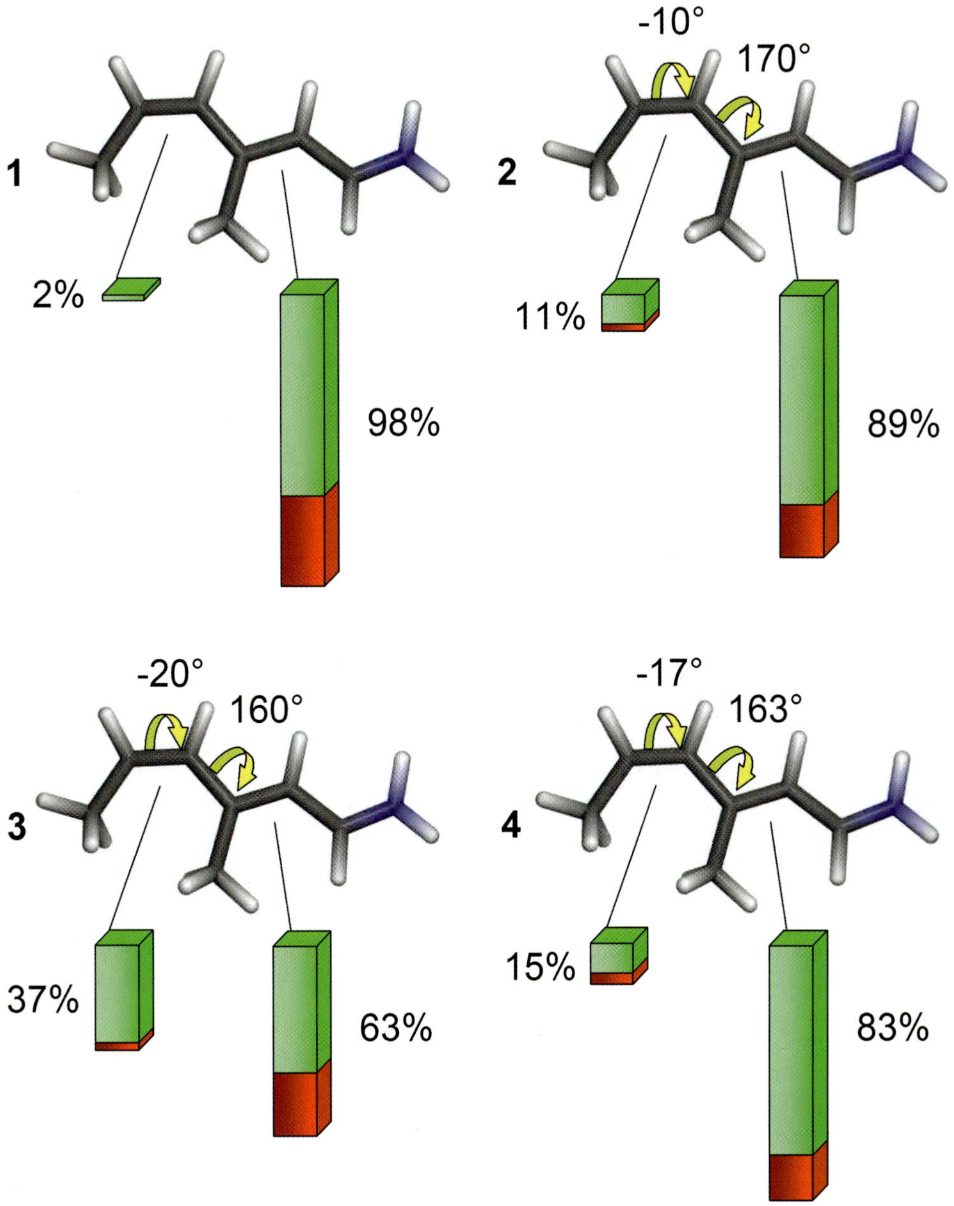

as the twist of the two terminal bonds increases. The number of C3 = C4 isomerizations drops from 42 in the planar model **1** to 39 in **2** and finally to 27 in **3**, the model with the largest twist. In the same order the number of trajectories involving C5 = C6 rotations increases first from 1 to 5 and then to 16. The shift toward isomerization of the terminal double bond is not linear with the distortion, but picks up momentum as the pre-twist angles increase. This is also evident from the data of model **4** in which the out-of-plane torsion is much closer to **3**, but the distribution is more similar to **2**. There is no clear pattern in the *Z*/*E* product ratio of C3 = C4 bond isomerization; in contrast the *E* product yield is very high throughout the model chromophores when the pre-twisted C5 = C6 bond isomerizes. These results are also summarized graphically in Fig. 7.

The sense of rotation of the isomerizing double bond is not relevant in the planar model **1**: both the clockwise and the counterclockwise rotation are equally probable. Within the limits set by the number of trajectories studied, this is borne out by the statistics of this chromophore. With the twist angles preset in a clockwise manner in the chromophores **2** to **4**, it is not surprising that the trajectories of C5 = C6 isomerization follow in the same direction without exception. Whether the preference for clockwise rotation about the (planar) C3 = C4 bond in these cases is significant or accidental cannot be decided on the basis of the available data.

With respect to the averaged data concerning the hopping event, there is a significant difference in the hopping geometries: for the C3 = C4 double bond most events take place between 120° and 110° of torsion, as observed already in the statistical analysis of **1**. Most of the E/Z isomerizations of the terminal double bond happen between 75° and 85° of torsion, i.e., much closer to the product geometry. Also, in the latter case the average lifetime of the excited state is definitely shorter than for the central bond, possibly on account of the initial twist imparted on this bond. A report about how the reaction velocity is influenced by this effect has appeared recently [10].

Conclusions

The ultra fast, highly selective, and efficient photoisomerization of the retinal chromophore in the retinal binding proteins rho and bR has been probed on a small all-*E*-configurated 3-double-bond model chromophore using ab initio excited state MD methodology and employing a technique for analyzing the product distribution by following the motion of high-amplitude hydrogen-out-of-plane modes at possible decay events.

The following are our key findings:

- In the planar chromophore isomerization occurred almost exclusively at the central double bond with an overall yield of 60% of the *Z*-product; 40% of the trajectories returned to the initial *E*-configuration of the starting product. The average lifetime of the excited state was between 98 and 114 fs, depending on the method used for determining the surface-hopping event. For a trajectory to yield the *Z*-product, the S_1 to S_0 surface hopping should occur at the first close approach when the internal vibrational and rotational modes are still largely coupled. The probability for a return to the *E*-configuration increases with the number of unsuccessful close approaches, even though the continuous rotation of the double bond shifts the geometry of the molecule at the hopping event further toward the *Z*-product.
- Bond torsion significantly affects the product distribution of the model chromophores. With an initial torsion of the terminal bonds by 10° each, isomerization of the central double bond is still favored; 89% of the trajectories involve rotation of the C3 = C4 bond. However, 11% of the photoproduct show that isomerization has occurred at the C5 = C6 bond. Doubling the dihedral angles shifts the distribution further, to 63 and 37%, respectively. Whether the observed changes of the E/Z ratio are significant and can be correlated with observed quantum yields remains to be seen considering the limited number of trajectories available. The results show, however, convincingly that the initial torsion about certain bonds has a directing effect on the photochemistry of the chromophore.

The calculated isomer ratios, C3 = C4 versus C5 = C6, are far from the selectivity observed in real proteins. There are several factors that can contribute to this discrepancy. Our model chromophores are very short and cannot possibly reflect the whole complexity of a 6-double-bond chromophore. The preference for isomerizing the central double bond shown in model **1** would work very much in favor of the C11 = C12 bond in rho, in addition to it being *cis*-configurated and having a significant pre-twist. Larger models will have to be studied to come to a conclusive answer. Also, more work is needed to decide whether bond torsion can also be held accountable for the selective C13 = C14 bond isomerization in bR.

Acknowledgments This work was funded by the DFG Forschergruppe "Retinal Protein Action." We thank Dr. A.-N. Bondar and Dr. M. Sugihara for helpful discussion.

References

1. Mathies RA, Lugtenburg J (2000) The primary photoreaction of rhodopsin. In: Stavenga DG, DeGrip WJ, Pugh EN Jr (eds) Handbook of biological physics, vol 3. Elsevier, Amsterdam, pp 55–90
2. Okada T, Ernst OP, Palczewski K, Hofmann KP (2001) Trends Biochem Sci 26:318–324
3. Haupts U, Tittor J, Oesterhelt D (1999) Annu Rev Biophys Biomol Struct 28:367–399
4. Palczewski K, Kumasaka T, Hori T, Behnke CA, Motoshima H, Fox BA, Le Trong I, Teller DC, Okada T, Stenkamp RE, Yamamoto M, Miyano M (2000) Science 289:739–745
5. Okada T, Sugihara M, Bondar AN, Elstner M, Entel P, Buss V (2004) J Mol Biol 342:571–583
6. Luecke H, Schobert B, Richter HT, Cartailler JP, Lanyi JK (1999) J Mol Biol 291:899–911
7. Belrhali H, Nollert P, Royant A, Menzel C, Rosenbusch JP, Landau EM, Pebay-Peyroula E (1999) Structure 7:909–917
8. Vreven T, Bernardi F, Garavelli M, Olivucci M, Robb MA, Schlegel HB (1997) J Am Chem Soc 119:12687–12688
9. Garavelli M, Bernardi F, Olivucci M, Vreven T, Klein S, Celani P, Robb MA (1998) Faraday Discuss 110:51–70
10. Buss V, Weingart O, Sugihara M (2000) Angew Chem 39:2784–2786
11. Warshel A, Chu ZT (2001) J Phys Chem B 105:9857–9871
12. Ferré N, Olivucci M (2001) J Am Chem Soc 125:6868–6869
13. Andruniow T, Ferré N, Olivucci M (2004) Proc Natl Acad Sci USA 101:17908–17913
14. Freedman K, Becker RS (1986) J Am Chem Soc 108:1245–1251
15. Sugihara M, Buss V, Entel P, Hafner J, Bondar AN, Elstner M, Frauenheim T (2004) Phase Trans 77:31–45
16. Hayashi S, Tajkhorshid E, Schulten K (2002) Biophys J 83:1281–1297
17. Cembran A, Bernardi F, Olivucci M, Garavelli M (2004) J Am Chem Soc 126:16018–16037
18. Aharoni A, Hou B, Friedman N, Ottolenghi M, Rousso I, Ruhman S, Sheves M, Ye T, Zhong Q (2001) Biochemistry (Moscow) 66:1210–1219
19. Sugihara M, Buss V, Entel P, Elstner M, Frauenheim T (2002) Biochemistry 41:15259–15266
20. Frisch MJ, Trucks GW, Schlegel HB, Scuseria GE, Robb MA, Cheeseman JR, Zakrzewski VG, Montgomery JA, Stratmann RE, Burant JC, Dapprich S, Millam JM, Daniels AD, Kudin KN, Strain MC, Farkas O, Tomasi J, Barone V, Cossi M, Cammi R, Mennucci B, Pomelli C, Adamo C, Clifford S, Ochterski J, Petersson GA, Ayala PY, Cui Q, Morokuma K, Malick DK, Rabuck AD, Raghavachari K, Foresman JB, Cioslowski J, Ortiz JV, Baboul AG, Stefanov BB, Liu G, Liashenko A, Piskorz P, Komaromi I, Gomperts R, Martin RL, Fox DJ, Keith T, Al-Laham MA, Peng CY, Nanayakkara A, Challacombe M, Gill PMW, Johnson B, Chen W, Wong MW, Andres JL, Gonzalez C, Head-Gordon M, Replogle ES, Pople JA (1998) Gaussian 98, Revision A.9. Gaussian Inc., Pittsburgh

21. Terstegen F, Carter EA, Buss V (1999) Int J Quantum Chem 75:141–145
22. Andersson K, Barysz M, Bernhardsson A, Blomberg MRA, Cooper DL, Fülscher MP, de Graaf C, Hess BA, Karlström G, Lindh R, Malmqvist PÅ, Nakajima T, Neogrády P, Olsen J, Roos BO, Schimmelpfennig B, Schütz M, Seijo L, Serrano-Andrés L, Siegbahn PEM, Stålring J, Thorsteinsson T, Veryazov V, Widmark PO (2002) MOLCAS Version 5.4. Lund University, Sweden
23. Fox DJ, Osamura Y, Hoffmann MR, Gaw JF, Fitzgerald G, Yamaguchi Y, Schaefer HF (1983) Chem Phys Lett 102:17–19
24. Sloane CS, Hase WL (1977) J Chem Phys 66:1523–1533
25. Tully JC, Preston RK (1971) J Chem Phys 55:562–572
26. Paterson MJ, Hunt PA, Robb MA, Takahashi O (2003) J Phys Chem A 106:10494–10504
27. Groenhof G, Bouxin-Cademartory M, Hess B, De Visser SP, Berendsen HJC, Olivucci M, Mark AE, Robb MA (2004) J Am Chem Soc 126:4228–4233
28. Weingart O, Migani A, Olivucci M, Robb MA, Buss V, Hunt P (2004) J Phys Chem A 108:4685–4693
29. Weingart O, Buss V, Robb MA (2005) Phase Trans 78:17–24

J Mol Model (2006) 12: 723–729
DOI 10.1007/s00894-005-0081-z

ORIGINAL PAPER

Rainer Koch · Torsten Bruhn

Theoretical ^{49}Ti NMR chemical shifts

Received: 24 May 2005 / Accepted: 7 November 2005 / Published online: 29 March 2006

Abstract ^{49}Ti chemical shifts for a total of 20 titanium complexes are reported, and several levels of theory are evaluated in order to identify a reliable approach for the calculation of titanium NMR data. The popular B3LYP/6–31G(d)//B3LYP/6–31G(d) proves to give very good agreement with experimental data over a range from 1,400 to −1,300 ppm. The MP2/6–31G(d)//MP2/6–31G(d) level computes even smaller average deviations but fails for TiI_4. This behavior together with its huge demand for computational resources requires careful handling of this theoretical level. In addition, NMR data for five titanium fulvene (or related) complexes are given.

Keywords NMR chemical shifts · Density functional calculations · ^{49}Ti NMR

Introduction

Nuclear magnetic resonance spectroscopy has always been an invaluable tool for obtaining information on geometries and electronic structures of mainly “organic" molecules [1]. In recent years, NMR spectroscopy of transition metal complexes has developed to a fully validated method that also allows examination of transition metal nuclei [2, 3]. Paralleling this development, many theoretical calculations have been carried out to assist the spectroscopist in analyzing and understanding the spectra obtained, and as the ultimate goal, to predict them [4–6]. Considerable methodological effort has been made in order to improve the treatment of transition metal elements, in particular the heavier third-row nuclei, where relativistic effects are a requirement for correct description of NMR chemical shifts and spin-spin coupling constants [7–9]. First-row and most of the second-row elements can be treated at a non-relativistic level, and there are numerous computational studies demonstrating the applicability of density functional theory, [10–15] mainly by Michael Bühl. This is also true for ^{49}Ti, where two theoretical studies have been published: the first shows good agreement between theory (IGLO, individual gauge for localized orbitals) and experiment for δ (^{49}Ti) for the $TiCl_nMe_{4-n}$ set of molecules [16]. However, deficiencies in the derivation of chemical shifts for later transition metals with the Hartree–Fock level of theory, [15, 17] which was used in this work, require a more robust testing of additional levels of theory. The second thorough study deals mainly with DFT-based approaches and gives reasonably good agreement for a somewhat larger but still limited test set [18]. It also suggests that line widths, which can become an issue for the NMR-active ^{47}Ti and ^{49}Ti, cannot be reproduced using B3LYP-based electric field gradients.

Due to the still small number of molecules included in the evaluation of theoretical approaches and the neglect of perturbation theory in the calculation of chemical shifts, we set out to extend the existing theoretical knowledge to a broader foundation, using both a larger test set and more levels of theory.

Dedicated to Professor Dr. Paul von Ragué Schleyer on the occasion of his 75th birthday

R. Koch (✉) · T. Bruhn
Institut für Reine und Angewandte Chemie,
Carl von Ossietzky Universität Oldenburg,
P.O. Box 2503, 26111 Oldenburg, Germany
e-mail: rainer.koch@uni-oldenburg.de
Tel.: +49-441-7983653
Fax: +49-441-7983329

Materials and methods

All geometries were fully optimized at different levels of theory with the Gaussian 03 program package [19]. The first employs the hybrid functional by Becke and Lee, Yang and Parr (denoted B3LYP) [20, 21] together with the standard 6–31G(d) basis set [22, 23] for all atoms with the exception of iodine and additional calculations for bromine, where the Stuttgart–Dresden effective-core potentials (SDD) were used. The second level differs only in the use of second order perturbation theory (MP2) [24, 25] instead of B3LYP, while the third again uses B3LYP in connection with a larger 6–311G(3df,2p) basis set. The nature of all local minimum structures found was confirmed by the lack of an imaginary harmonic vibrational frequency. Absolute magnetic shieldings were determined with the GIAO (gauge—including atomic orbital) method [26–28] at the HF, [29] B3LYP and MP2 levels of theory as implemented in Gaussian 03. The shielding tensors obtained were referenced against titanium tetrachloride to yield relative chemical shifts. Basis sets used herein range from the relatively small 6–31G(d) to the more demanding 6–311G(3df,2pd). It also includes the IGLO II [30] basis set, which was used for all atoms except Ti where 6–31G(d) and 6–311G(2df) were employed. For details on method/basis set combinations, see Table 1. Calculations including solvent effects used the polarized continuum model (PCM) [31–33] and dichloromethane as solvent (ε=8.93).

Results and discussion

In order to obtain reliable relative chemical shifts, it is important to verify that the geometry of a molecule under investigation is in agreement with experimental results. Table 2 therefore lists important geometrical parameters of the optimized molecules studied herein, together with experimental data if available.

All three levels of theory present reliable results for the molecules studied. The structures derived from the DFT

Table 1 Selected structural data for titanium complexes (bond lengths in Å, angles in °)

Compound	Variable	Experiment	B3LYP/ 6–31G(d)	B3LYP/ 6–311G(3df,2p)	MP2/ 6–31G(d)
$TiCl_4$ [b]	Ti–Cl	2.170	2.178	2.177	2.172
$MeTiCl_3$ [c]	Ti–Cl	2.185	2.193	2.193	2.178
–	Ti–C	2.047	2.029	2.039	2.054
–	Cl–Ti–C	105.6	104.8	105.1	104.4
Me_2TiCl_2 [d]	Ti–Cl	2.196	2.212	2.211	2.197
–	Ti–C	2.058	2.043	2.051	2.054
–	Cl–Ti–Cl	117.3	118.7	118.1	120.5
–	C–Ti–C	102.8	103.8	104.4	101.9
Me_3TiCl	Ti–Cl	–	2.231	2.229	2.220
–	Ti–C	–	2.058	2.066	2.065
Me_4Ti	Ti–C	–	2.076	2.082	2.080
$CpTiCl_3$ [e]	Ti–Cl	2.224	2.233	2.232	2.219
–	Ti–x [a]	2.010	2.051	2.052	2.043
Cp_2TiCl_2 [f]	Ti–Cl	2.364	2.349	2.344	2.334
–	Ti–x [a]	2.058	2.099	2.103	2.069
–	x–Ti–x [a]	130.9	131.6	131.5	130.6
–	Cl–Ti–Cl	94.4	96.3	96.6	98.8
$TiBr_4$ [g]	Ti–Br	2.339	2.319	–	2.326
$TiBr_4$ (Br=SDD)	Ti–Br	2.339	2.346	–	–
TiI_4 (I=SDD) [g]	Ti–I	2.546	2.560	–	2.571
$MeTiBr_3$	Ti–Br	–	2.332	–	2.327
–	Ti–C	–	2.032	–	2.058
Cp_2TiBr_2 [h]	Ti–Br	2.490	2.495	–	2.499
–	Ti–x [a]	2.050	2.101	–	2.071
–	x–Ti–x [a]	131.2	131.7	–	131.0
–	Br–Ti–Br	94.7	96.4	–	98.3
Cp_2TiBr_2 (Br=SDD)	Ti–Br	2.490	2.565	–	–
–	Ti–x [a]	2.050	2.085	–	–
–	x–Ti–x [a]	131.2	132.1	–	–
–	Br–Ti–Br	94.7	95.4	–	–
Cp_2TiF_2	Ti–F	–	1.802	–	1.825
–	Ti–x [a]	–	2.125	–	2.081
–	x–Ti–x [a]	–	132.1	–	131.5
–	F–Ti–F	–	98.5	–	99.8
Cp^*TiCl_3 [i]	Ti–Cl	2.245	2.247	–	2.231
–	Ti–x^* [a]	2.021	2.053	–	2.046
$Ti(CO)_6^{2-}$ [j]	Ti–C	2.038	2.051	–	2.038
$Ti(NMe_2)_4$	Ti–N	–	1.918	–	1.916

[a] x is the centre of the cyclopentadienyl ring
[b] [40]
[c] [41]
[d] [42]
[e] [43]
[f] [44] and [45]
[g] [46]
[h] [47] and [48]
[i] [49]
[j] [50]

optimization are largely unaffected by the basis set: the much larger triple-ζ basis does not give better results compared to the standard 6–31G(d) basis set. The calculations using an SDD ECP for bromine also do not lead to significant improvement. The only minor difficulties in reproducing the experimental data are encountered for titanium cyclopentadienyl complexes: some of the DFT-calculated Ti-centroid distances tend to be too long by about 3–5 ppm. MP2 ge-

Table 2 Experimental and calculated relative chemical shifts δ (in ppm), errors and unsigned averaged deviations for a test set of titanium compounds

	Experimental	B3LYP/6–31G(d)// B3LYP/6–31G(d)	Deviation	B3LYP/6–311G(3df,2pd)// B3LYP/6–31G(d)	Deviation	HF/6–31+G(d,p)// B3LYP/6–31G(d)	Deviation
$TiCl_4$[d]	0	0.0	0.0	0.0	0.0	0.0	0.0
$MeTiCl_3$[d]	613	456.0	157.0	452.1	160.9	488.0	125.0
Me_2TiCl_2[d]	907	815.6	91.4	815.3	91.7	718.8	188.3
Me_3TiCl[d]	1,188	1,098.8	89.2	1,105.8	82.2	884.9	303.1
Me_4Ti[d]	1,325	1,320.5	4.5	1,334.7	−9.7	1,029.5	295.5
$CpTiCl_3$[e]	−390	−385.8	−4.2	−457.8	67.8	−536.1	146.1
Cp_2TiCl_2[f]	−773	−785.8	12.8	−898.5	125.5	−983.6	210.6
Average deviation	–	–	59.9	–	89.6	–	211.4
–		MP2/6–31G(d)// B3LYP/6–31G(d)	Deviation	B3LYP/6–311G(3df,2p)// B3LYP/6–311G(3df,2p)	Deviation	MP2/6–31G(d)// MP2/6–31G(d)	Deviation
$TiCl_4$	0	0.0	0.0	0.0	0.0	0.0	0.0
$MeTiCl_3$	613	430.2	182.8	465.4	147.6	438.8	174.2
Me_2TiCl_2	907	885.8	21.2	836.8	70.2	917.8	−10.8
Me_3TiCl	1,188	1,192.8	−4.8	1,133.9	54.1	1,237.1	−49.1
Me_4Ti	1,325	−1,411.2	−86.2	1,371.1	−46.1	1,450.7	−125.7
$CpTiCl_3$	−390	−328.7	−61.3	−461.6	71.6	−363.0	−27.0
Cp_2TiCl_2	−773	−722.6	−50.4	−904.6	131.6	−766.9	−6.1
Average deviation	–	–	67.8	–	86.9	–	65.5
–	Experimental	B3LYP/IGLOII[a]// MP2/6–31G(d)	Deviation	B3LYP/IGLOII[b]// MP2/6–31G(d)	Deviation	B3LYP/6–31+G(d,p)// MP2/6–31G(d)	Deviation
$TiCl_4$	0	0.0	0.0	0.0	0.0	0.0	0.0
$MeTiCl_3$	613	431.3	181.7	423.9	189.1	435.0	178.0
Me_2TiCl_2	907	794.8	112.2	790.2	116.8	799.8	107.2
Me_3TiCl	1,188	1,076.2	111.8	1,088.6	99.4	1,077.9	110.1
Me_4Ti	1,325	1,286.6	38.4	1,326.8	−1.8	1,287.6	37.4
$CpTiCl_3$	−390	−448.8	58.8	−505.7	115.7	−435.2	45.2
Cp_2TiCl_2	−773	−874.8	101.8	−967.6	194.6	−833.8	60.8
Average deviation	–	–	100.8	–	119.6	–	89.8
–		HF/II′[c]//BP86/AE1	Deviation	BPW91/II′[c]//BP86/AE1	Deviation	B3LYP/II′[c]//BP86/ AE1	Deviation
$TiCl_4$	0	0	0	0	0	0	0
$MeTiCl_3$	613	512	101	431	182	443	170
Me_2TiCl_2	907	776	131	800	107	796	111
Me_3TiCl	1,188	978	210	1,103	85	1,077	111
Me_4Ti	1,325	1,163	162	1,369	−44	1,317	8
$CpTiCl_3$	−390	−512	122	−424	34	−454	64
Cp_2TiCl_2	−773	−951	178	−864	91	−892	119
Average deviation	–	–	150.7	–	90.5	–	97.2

[a] 6–31G(d) basis set for Ti
[b] 6–311G(2df) basis set for Ti
[c] II′ denotes an augmented Wachters basis for Ti, IGLO II for all other atoms except H (double-zeta basis). For details see [18]
[d] [16]
[e] [38]
[f] [39]

by about 1,800 ppm, although the corresponding DFT data are very reasonable. Interestingly, HF-based calculations predict the Ti signal 1,000 ppm at a lower field compared to the experimental data! From our results one can clearly see that perturbation theory cannot handle the NMR determination of iodine-containing systems, in contrast to DFT approaches. Without inclusion of these deficiencies, both MP2 NMR calculations are at least as good as the DFT results, in the case of underlying MP2 geometries, there is a reduction of the total average error to 60 ppm! Although the deviation is somewhat smaller, due to its very demanding need for computational resources, the general use of MP2/6–31G(d)//MP2/6–31G(d) is only recommended for smaller systems, thus allowing more economic studies with B3LYP/6–31G(d)//B3LYP/6–31G(d). Figure 1 shows a plot of the experimental versus calculated values for the large dataset at B3LYP/6–31G(d)//B3LYP/6–31G(d) with remarkable agreement (R^2=0.9967) and an almost ideal slope of the regression line of 1.002.

As can be seen from Table 2, additional optimizations of several compounds with halogen atoms have been performed employing the SDD basis set/ECP for the halogen atoms. Subsequent NMR calculations with SDD on halogen do not, however, lead to better results. As one might expect, the data for $TiBr_4$, Cp_2TiBr_2, and Cp_2TiF_2 are of equal quality compared to the B3LYP/6–31G(d).

Due to the quadrupole moment of the NMR-active titanium nuclei, line widths are considered as important. Although the use of electric field gradients for estimating line widths has proven successful for many other nuclei, it had been shown that this concept does not work in the case of titanium [18].

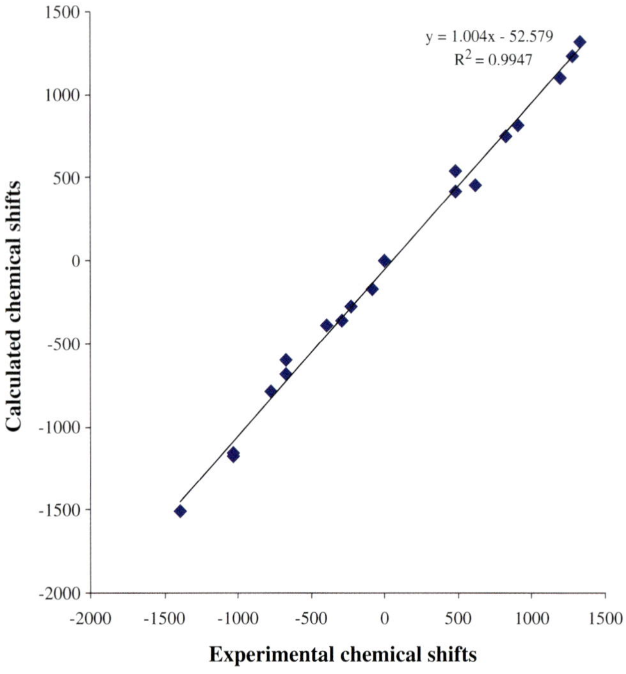

Fig. 1 Experimental and calculated ^{49}Ti chemical shifts δ obtained from Tables 1 and 4 at B3LYP/6–31G(d)//B3LYP/6–31G(d)

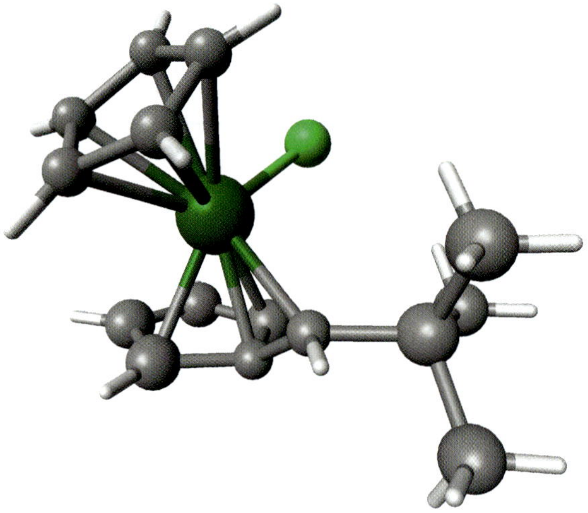

Fig. 2 Calculated structure of CpTi(6-*tert*.butyl-fulvene)Cl at B3LYP/6–31G(d)//B3LYP/6–31G(d)

We extended the investigation to a series of fulvene systems and related complexes (an example is shown in Fig. 2) which have been synthesized [37] in the Beckhaus group in our institute. The DFT results, included in Table 4, indicate a variation of the ^{49}Ti signal from −950 to 100 ppm. Unfortunately, the experimentalists have not been able to obtain a clear ^{49}Ti signal for those complexes, most likely due to very wide signals, so that a comparison with experimental data is not possible.

Conclusions

The use of an extended reference set of molecules and the extension of tested levels of theory for the evaluation of ^{49}Ti chemical shifts results in significant improvements over the present "best" choice [18] (B3LYP/II′//BP86/AE1). The average error is reduced from 107 to 68 ppm with the relatively inexpensive B3LYP/6–31G(d)//B3LYP/6–31G(d). Even better results can be obtained using the MP2/6–31G(d)//MP2/6–31G(d) level of theory. However, it fails dramatically when investigating TiI_4, so the approach has to be handled with care. Furthermore, its huge demand for computational resources limits its general applicability. The excellent performance of the relatively small basis sets used herein and in other studies has to be treated with caution and cannot be generalized. Apparently due to error cancellation, these theoretical approaches give better results than larger basis sets, which are systematically better. The demonstrated influence of a simulated solvent environment on the chemical shifts leaves the question open for a general solvation inclusion for these calculations.

Acknowledgements We would like to thank Dr. M. Bühl and K. Xaver for helpful discussions. This work was supported by generous allocation of computer time at the Norddeutscher Verbund für Hoch- und Höchstleistungsrechnen (HLRN), Hannover, Berlin, and at the Hochschulrechenzentrum, Universität Oldenburg. Financial support by the Deutsche Forschungsgemeinschaft is gratefully acknowledged.

References

1. Claridge TDW (1999) High-resolution NMR techniques in organic chemistry. In: Baldwin JE, Williams M (eds) Tetrahedron organic chemistry series, vol 19. Pergamon, Amsterdam
2. Benn R, Rufinska A (1986) Angew Chem 98:851–871
3. von Philipsborn W (1999) Chem Soc Rev 28:95–105
4. Bühl M (2004) In: Kaupp M, Bühl M, Malkin VG (eds) Calculation of NMR and EPR parameters. Theory and applications. Wiley, Weinheim
5. Bühl M, Kaupp M, Malkin VG, Malkina OL (1999) J Comput Chem 20:91–105
6. Schreckenbach G, Ziegler T (1998) Theor Chem Acc 99:71–82
7. Autschbach J (2004) In: Kaupp M, Bühl M, Malkin VG (eds) Calculation of NMR and EPR parameters. Theory and applications. Wiley, Weinheim
8. Autschbach J, Ziegler T (2000) J Chem Phys 113:936–947
9. Wolff SK, Ziegler T, van Lenthe E, Baerends EJ (1999) J Chem Phys 110:7689–7698
10. Bühl M (1997) Chem Phys Lett 267:251–257
11. Bühl M, Hamprecht FA (1998) J Comput Chem 119:113–122
12. Bühl M, Gaemers S, Elsevier CJ (2000) Chem Eur J 6:3272–3280
13. Bühl M (2002) Theor Chem Acc 107:336–342
14. Godbout N, Oldfield E (1997) J Am Chem Soc 119:8065–8069
15. Bühl M (1999) Chem Eur J 5:3514–3522
16. Berger S, Bock W, Frenking G, Jonas V, Müller V (1995) J Am Chem Soc 117:3820–3829
17. Bühl M (1997) Organometallics 16:261–267
18. Bühl M, Mauschick FT (2004) Magn Res Chem 42:737–744
19. Frisch MJ, Trucks GW, Schlegel HB, Scuseria GE, Robb MA, Cheeseman JR, Montgomery JA Jr, Vreven T, Kudin KN, Burant JC, Millam JM, Iyengar SS, Tomasi J, Barone V, Mennucci B, Cossi M, Scalmani G, Rega N, Petersson GA, Nakatsuji H, Hada M, Ehara M, Toyota K, Fukuda R, Hasegawa J, Ishida M, Nakajima T, Honda Y, Kitao O, Nakai H, Klene M, Li X, Knox JE, Hratchian HP, Cross JB, Bakken V, Adamo C, Jaramillo J, Gomperts R, Stratmann RE, Yazyev O, Austin AJ, Cammi R, Pomelli C, Ochterski JW, Ayala PY, Morokuma K, Voth GA, Salvador P, Dannenberg JJ, Zakrzewski VG, Dapprich S, Daniels AD, Strain MC, Farkas O, Malick DK, Rabuck AD, Raghavachari K, Foresman JB, Ortiz JV, Cui Q, Baboul AG, Clifford S, Cioslowski J, Stefanov BB, Liu G, Liashenko A, Piskorz P, Komaromi I, Martin RL, Fox DJ, Keith T, Al-Laham MA, Peng CY, Nanayakkara A, Challacombe M, Gill PMW, Johnson B, Chen W, Wong MW, Gonzalez C, Pople JA (2004) Gaussian 03, revision C02. Gaussian Inc, Wallingford, Connecticut
20. Becke AD (1988) Phys Rev A 38:3098–3100
21. Lee C, Yang W, Parr RG (1988) Phys Rev B 37:785–789
22. Hehre WJ, Ditchfield R, Pople JA (1972) J Chem Phys 56:2257–2261
23. Hariharan PC, Pople JA (1973) Theor Chim Acta 28:213–222
24. Møller C, Plesset MS (1934) Phys Rev 46:618–622
25. Head-Gordon M, Pople JA, Frisch MJ (1988) Chem Phys Lett 153:503–509
26. Ditchfield R (1974) Mol Phys 27:789–807
27. Wolinski K, Hinton JF, Pulay P (1990) J Am Chem Soc 112:8251–8260
28. Cheeseman JR, Frisch MJ, Devlin FJ, Stephens PJ (2000) J Phys Chem A 104:1039–1046
29. For a text book see: Jensen F (1999) Introduction to computational chemistry. Wiley, Chichester
30. Fleischer U (1992) Ph D Thesis. Bochum
31. Cancès MT, Mennucci B, Tomasi J (1997) J Chem Phys 107:3032–3041
32. Mennucci B, Tomasi J (1997) J Chem Phys 106:5151–5158
33. Cossi M, Barone V, Mennucci B, Tomasi J (1998) Chem Phys Lett 286:253–260
34. Koch R, Wiedel B, Wentrup C (1997) J Chem Soc, Perkin II 1851–1859
35. Koch R, Bruhn T, Veedu RN, Wentrup C (2004) J Mol Struct (THEOCHEM) 686:31–36
36. van Eikema Hommes NJR, Clark T (2005) J Mol Mod 11:175–185
37. Beckhaus R, Lützen A, Haase D, Saak W, Stroot J, Becke S, Heinrichs J (2001) Angew Chem 113:2112–2115 and unpublished results
38. Hafner A, Okuda J (1993) Organometallics 12:949–950
39. Gassmann PG, Campbell WH, Macomber DW (1984) Organometallics 3:385–387
40. Morino Y, Uehara U (1966) J Chem Phys 45:4543–4550
41. Briant P, Green J, Haaland A, Mollendal A, Rypdal K, Tremmel J (1989) J Am Chem Soc 111:3434–3436
42. McGrady GS, Downs AJ, McKean DC, Haaland A, Schere W, Verne HP, Volden MV (1996) Inorg Chem 35:4713–4718
43. Engelhardt LM, Papasergio RI, Raston CL, White AH (1984) Organometallics 3:18–20
44. Tkachev VV, Atovmyan L (1972) Zh Strukt Khim 134:287–290
45. Clearfield A, Warner DK, Saldarriaga-Molina CH, Ropal R, Bernal I (1975) Can J Chem 53:1622–1629
46. Callomon JH, Hirota E, Iijima K, Kuchitsu K, Lafferty WJ (1987) In: Hellwege KH, Hellwege AM (eds) Landolt–Börnstein new series supplement II/15. Springer, Berlin
47. Klouras N, Nastopoulos V, Demakopoulos I, Leban I (1993) Z Anorg Allg Chem 619:1927–1930
48. Jones P, Kienitz C, Thone CZ (1994) Z Kristallogr 209:85–86
49. Pevec A (2003) Acta Chim Slov 50:199–206
50. Ellis J, Chi KM (1990) J Am Chem Soc 112:6022–6025
51. Hao N, Sayer BG, Denes G, Bickley DG, Detellier C, McGlinchey MJ (1982) J Magn Reson 50:50–63
52. Pregosin PS (ed) Transition metal nuclear magnetic resonance. Elsevier, Amsterdam
53. Berger S, Bock W, Marth CF, Raguse B, Reetz MT (1990) Magn Res Chem 28:559–560

J Mol Model (2006) 12: 731–737
DOI 10.1007/s00894-005-0048-0

ORIGINAL PAPER

Rainer Glaser · Hong Wu · Francisca von Saint Paul

Chemical carcinogens in non-enzymatic cytosine deamination: 3-isocyanatoacrylonitrile

Received: 25 May 2005 / Accepted: 29 September 2005 / Published online: 13 January 2006

Abstract Uracil has long been known as the main product of nitrosative cytosine deamination in aqueous solution. Recent mechanistic studies of cytosinediazonium ion suggest that the cation formed by its dediazoniation can ring-open to N-protonated (*Z,s-cis*)-3-isocyanatoacrylonitrile **7**. Stereochemical preferences are discussed of the 3-isocyanatoacrylonitriles (*Z,s-cis*)-**10**, (*E,s-cis*)-**11**, (*Z,s-trans*)-**12**, and (*E,s-trans*)-**13**. The electronic structures of **7** and **10**–**13** have been analyzed and a rationale is provided for the thermodynamic preference for (*Z,s-cis*)-**10**. It is shown that *s-cis*/*s-trans*-interconversion occurs *via* C–N rotation–inversion paths with barriers below 3 kcal mol^{-1}. The proton affinities of 3-isocyanatoacrylonitrile **10** and water are nearly identical and, thus, 3-isocyanatoacrylonitriles can and should be formed in aqueous media from **7** along with 3-aminoacrylonitriles **9**. The results highlight the relevance of the chemistry of 3-isocyanatoacrylonitriles for the understanding of the chemical toxicology of nitrosation of the nucleobase cytosine.

Keywords Isocyanate · Acrylonitrile · Rotational barrier · Inversion · Ab initio · Non-covalent bonding · Electronic structure · Chemical toxicology · Carcinogenesis

Introduction

DNA cytosine methyltransferases [1] deaminate cytosine (C, **1**) to uracil (U, **3**), and cytosine deamination also can be effected by nitrosation [2, 3]. C-to-U damage can be repaired *via* enzymatic base excision [4] and, if left unrepaired, causes the G:C→A:T mutation [5, 6]. The nitrosative C-to-U process had been thought to occur *via* the transient cytosinediazonium ion **2** and its hydrolysis by direct nucleophilic aromatic substitution (Fig. 1), which is very much S_N1Ar-like [7]. However, theoretical study of the unimolecular dediazonation of cytosinediazonium ion [8] showed that the classical diazonium ion **2** (**5**←N≡N) is merely a shallow minimum and *less stable* than free **5** and N_2 and that an electrostatic complex **4** (**5**···N≡N) is bound by only 5.4 kcal mol^{-1}. The HO-tautomer of cytosinediazonium ion **2′** (not shown in Fig. 1) is 3.7 kcal mol^{-1} more stable than **2** itself. **2′** has a classical diazonium-ion structure and it is bound by 10.8 kcal mol^{-1} relative to **5′** and N_2. Hence, cytosine deamination essentially produces a free cation, **5** or **5′**. More recently, it has been shown that the ions **5** or **5′** are best described as cyclic nitrilium ions **6** and **6′** and that the dative bond easily breaks to form the more stable ring-opened cations **7** and **7′**. These results seriously put in question whether uracil is the only product [9] of nitrosative cytosine deamination.

If the *R*-group is ribose or 2′-deoxyribose, **6** and **7** are the appropriate models for the deamination of cytidine, its nucleotides CMP, CDP, and CTP, and their 2′-deoxy derivatives. Ubiquitous water may react with **7** by addition, substitution (*R*≠H), or deprotonation (*R*=H). Water addition forms of carbamic acid [10] and subsequent decarboxylation [11] leads to (*Z*)-3-aminoacrylonitrile **9**. (*Z*)-3-aminoacrylonitrile is susceptible to base- [12, 13] and acid-catalyzed [14] nucleophilic addition to the C=C and C≡N bonds and this chemistry also might lead to DNA adducts. Alternatively, deglycation by substitution (*R*≠H) or deprotonation (*R*=H) would lead to (*Z*)-3-isocyanatoacrylonitrile **10**. Unsaturated isocyanates are toxic [15] and 3-isocyanatoacrylonitrile can form adducts and cross-links [16, 17].

Here, we report the results of a study of the formation of **10** by deprotonation of **7** and **7′**, of the *E*/*Z*-preferences of 3-isocyanatoacrylonitrile **10** and **11** and of their conformers **12** and **13**, and of C–N rotation–inversion in both *E*/*Z*-isomers *via* **TS1** and **TS3** (Fig. 2).

Dedicated to professor Dr. Paul von Ragué Schleyer on the occasion of his 75th birthday

Electronic Supplementary Material Supplementary Material is available for this article at http://dx.doi.org/10.1007/s00894-005-0048-0

R. Glaser (✉) · H. Wu · F. von Saint Paul
Department of Chemistry, University of Missouri-Columbia, Columbia, MO 65211, USA
e-mail: glaserr@missouri.edu
Tel.: +1-573-8820331
Fax: +1-573-8822754

Fig. 1 Nitrosative deamination of cytosine and cytidine

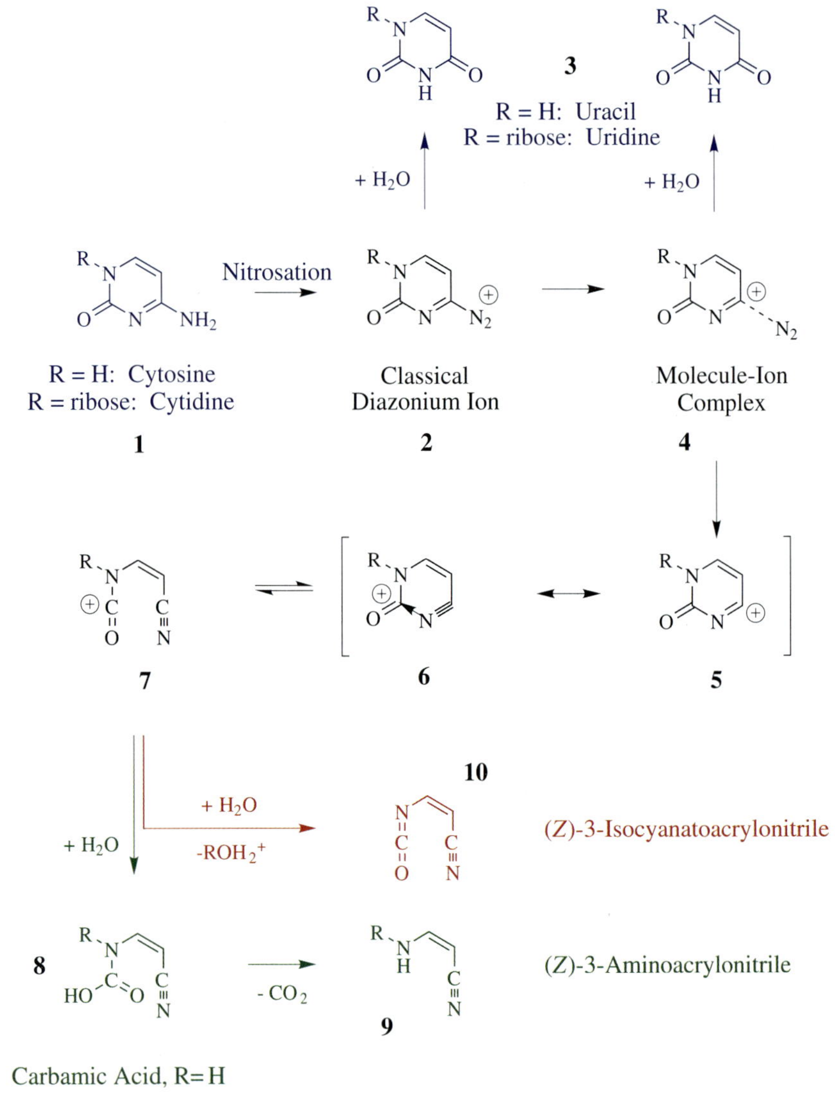

Fig. 2 Scope of the present study of 3-isocyanatoacrylonitrile

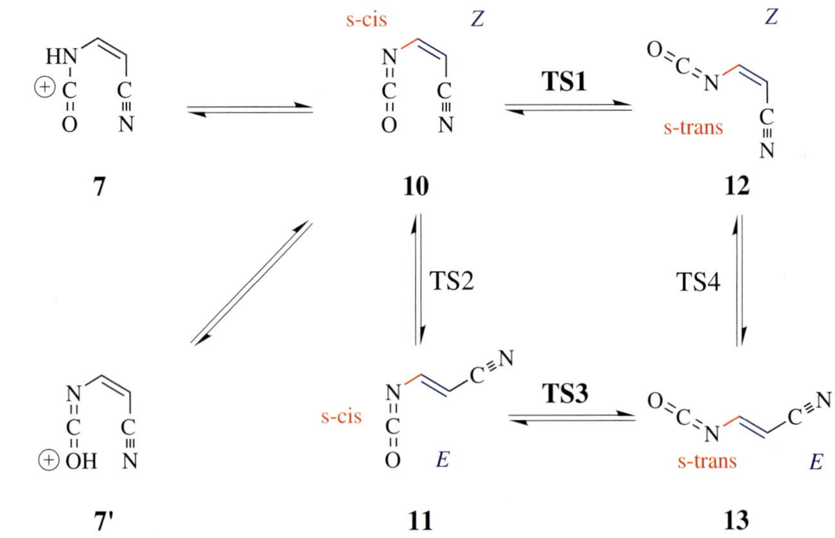

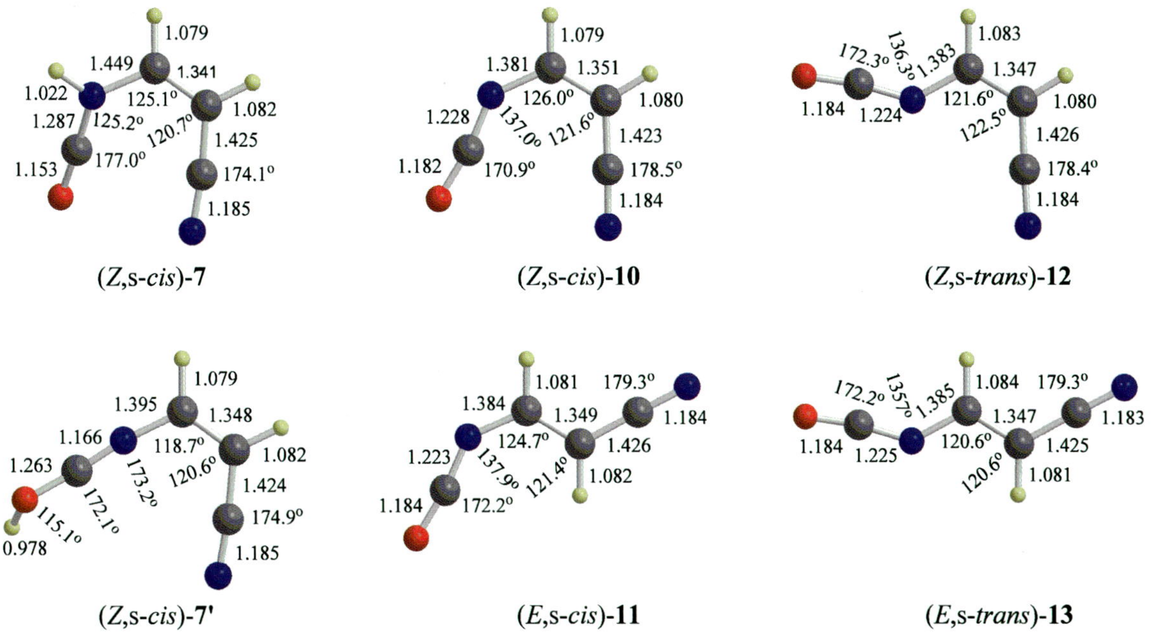

Fig. 3 MP2(full)/6-31G** structures of **7**–**13**

Computational methods

Geometry optimizations and vibrational analyses were carried out at the MP2(full)/6-31G** level [18] with the program *Gaussian03* [19]. We employed perturbation theory deliberately because it is preferable to (hybrid) density functional theory for the present purpose. Structure (*Z*,*s-cis*)-**10** benefits from through-space interaction and such dispersion interactions are not properly accounted for by the standard DFT models [20]. Atomic charges were calculated with the natural bond orbital (NBO) method [21, 22] at the same level. Molecular models of the minima and major structural data and atom and fragment charges are reported in Figs. 3 and 4, respectively. Molecular models of the transition-state structures are shown in Fig. 5. Coordinates of all stationary structures are available as Supporting Information.

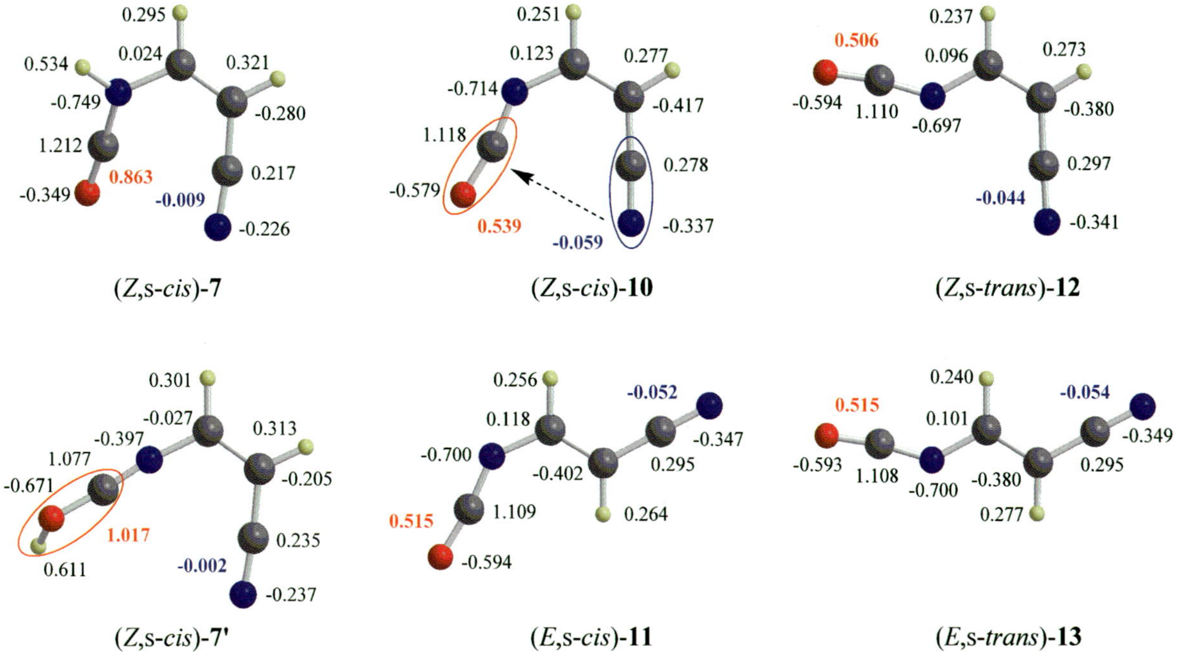

Fig. 4 NBO analysis of **7**–**13**. Numbers printed in bold are fragment charges

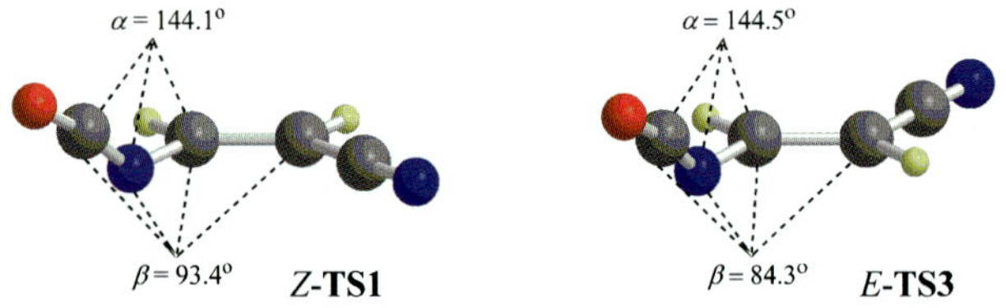

Fig. 5 Molecular models of the MP2(full)/6-31G** structures of TS1 and TS3

Total energies E (in Hartrees), vibrational zero-point energies $VZPE$ (kcal mol^{-1}), thermal energies TE (kcal mol^{-1}, 298.15 K), and entropies S (cal mol^{-1} K^{-1}) are reported in Table 1. In Tables 2 and 3 are reported the values ΔE, $\Delta E_0 = \Delta E + \Delta VZPE$, $\Delta H_{298} = \Delta E + \Delta TE + \Delta RT$, and $\Delta G_{298} = \Delta H_{298} - 0.29815{\cdot}\Delta S$. The discussion refers to Gibbs free energies unless otherwise noted.

Results and discussion

Stereochemical preferences Studies of α,β-unsaturated isocyanates are scarce [23, 24] and the stereochemistry of vinyl isocyanates with additional functionality remains unexplored. Structures **10-13** (Fig. 3) are almost isoenergetic (Table 2). (*Z,s-cis*)-**10** is the most stable isocyanatoacrylonitrile and the relative energies of isomers and conformers only are 0.1–0.9 kcal mol^{-1}. *Z*-preferences occur for the *s-cis-* and *s-trans*-conformers and they are higher for the *s-cis*-structures. If steric interactions dominated, one would expect preferences for the *E*-configuration and the *s-trans* conformation; (*E,s-trans*)-**13**. The calculated relative energies of 0.87 kcal mol^{-1} for (*E,s-cis*)-**11** *vs.* (*E,s-trans*)-**13** and of 0.28 kcal mol^{-1} for (*Z,s-trans*)-**11** *vs.* (*E,s-trans*)-**13** are in line with expectation based on steric demand. Assuming additivity, one would thus estimate an energy of about 1.15 kcal mol^{-1} for (*Z,s-cis*)-**10** relative to (*E,s-trans*)-**13** for steric reasons. Yet, (*Z, s-cis*)-**10** is 0.05 kcal mol^{-1} *more stable* than (*E,s-trans*)-**13** and, hence, **10** benefits from through-space neighboring group attraction of about 1.2 kcal mol^{-1}.

Electronic structures and through-space interactions The NBO analysis shows highly polar electronic structures and the main effect of the isocyanato group on the acrylonitrile

Table 2 Relative energies, *trans*-preferences, *Z*-preferences and proton affinities (kcal mol^{-1})

Parameter	ΔE	ΔE_0	ΔH_{298}	ΔG_{298}
(*Z,s-trans*)-**12** *vs.* (*Z,s-cis*)-**10**	0.59	0.46	0.51	0.33
(*E,s-cis*)-**11** *vs.* (*Z,s-cis*)-**10**	1.11	1.00	1.10	0.92
(*E,s-trans*)-**13** *vs.* (*Z,s-cis*)-**10**	0.24	0.01	0.14	0.05
s-Trans-preference of *Z*-isomers	−0.59	−0.46	−0.51	−0.33
s-Trans-preference of *E*-isomers	0.87	0.99	0.96	0.87
Z-preference of *s-cis*-conformers	1.11	1.00	1.10	0.92
Z-preference of *s-trans*-conformers	0.35	0.45	0.37	0.28
PA(**10** → **7**)	179.84	172.59	173.27	166.41
PA(**10** → **7′**)	168.28	161.88	162.18	156.70

[14] is an increase of the electron-deficiency of the CH group to which it is attached (Fig. 4). The CN groups in **10**–**13** are highly dipolar but their overall charge is very small. On the other hand, the NCO groups in **10**–**13** are charged significantly, they carry charges of about −0.2, and they are extremely quadrupolar [25]. These findings suggest that the attractive neighboring-group interaction in **10** is due in part to polarization of the cyano group in the field of the overall charge of the isocyanato group and in part to dipole–dipole attraction. The dipole-dipole attraction between the CN bond dipoles in the NCO and CN groups appears to dominate the repulsion between CO bond dipole of the NCO group and the CN group dipole. The NBO charge analysis thus provides a consistent rationale based on simple and widely accepted concepts of interaction and reactivity. The proposed interactions can be quantified *via* fragment-interaction analysis [26, 27] and recent progress in the development of orbital deletion analysis [22, 28] also might provide for interesting tests of the proposed rationale and any possible hyperconjugation.

The major effect of the protonation of the NCO group is a withdrawal of electron density from the alkene fragment; in the cations the HNCO and alkene fragments carry charges of about 0.6 and 0.4, respectively. Note that the protonation of either one of the heteroatoms of the NCO group does very little to increase the electrophilicity of the heterocumulene's center. Instead, there is a rearrangement of electron density from one heteroatom to the protonated one.

Table 1 Total energies and thermodynamical data

Molecule	Sym.	*E*	*VZPE*	TE	*S*	N
(*Z,s-cis*)-**7**	C_1	−337.692206	41.98	46.29	82.39	0
(*Z,s-cis*)-**7′**	C_1	−337.673798	41.14	45.82	85.89	0
(*Z,s-cis*)-**10**	C_s	−337.405620	34.73	38.83	81.51	0
(*Z,s-trans*)-**12**	C_s	−337.404675	34.60	38.74	82.10	0
(*E,s-cis*)-**11**	C_s	−337.403858	34.63	38.83	82.13	0
(*E,s-trans*)-**13**	C_s	−337.405230	34.49	38.73	81.82	0
(*Z*)-TS1	C_1	−337.401737	34.47	38.15	78.29	1
	C_s	−337.400516	34.31	37.61	75.62	2
(*E*)-TS3	C_1	−337.401528	34.35	38.15	78.86	1
	C_s	−337.400358	34.20	37.60	75.15	2

Table 3 Activation barriers for C–N rotation–inversion (kcal mol^{-1})

Process	ΔE	ΔE_0	ΔH_{298}	ΔG_{298}
(*Z*,*s-cis*)-**10** → (*Z*,*s-trans*)-**12**	2.44	2.17	1.72	2.72
	1.84	1.71	1.25	2.39
(*E*,*s-cis*)-**11** → (*E*,*s-trans*)-**13**	1.46	1.18	0.78	1.75
	2.32	2.18	1.74	2.62

Rotation-Inversion A priori the *s-cis*/*s-trans* interconversion might involve in-plane N-inversion or rotation about the OCN–CH bond. The N-inversion would involve an N-hybridization change from sp^2 to sp and shorten both bonds to N. Hybridization changes would be small along the rotation path and there would be some loss of π-conjugation.

The ideal inversion process would involve a planar transition-state structure. We searched for these transition-state structures under the constraint to C_s-symmetry, and found the resulting structures C_s-**TS1** (*i*118, *i*101 cm^{-1}) and C_s-**TS3** (*i*119, *i*81 cm^{-1}) to be second-order saddle points on the potential-energy surface. The true transition-state structures **TS1** and **TS3** were located (Fig. 5) and the activation barriers are less than 3 kcal mol^{-1} (Table 3).

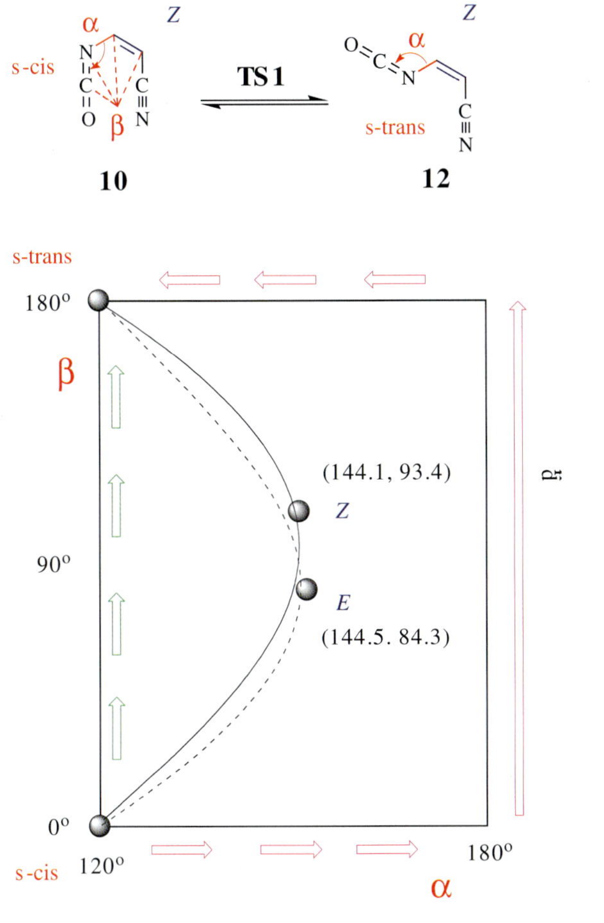

Fig. 6 Illustration of the rotation–inversion paths and the position of the transition state structures in the (α, β)-map. The "pure" rotation and inversion paths follow the *green* and *purple arrows*, respectively. (*Horizontal axis* is not drawn to linear scale)

The ∠(C–C–N–C) dihedral angles β in **TS1** and **TS3** are 93.4° and 84.3°, respectively, and close to 90°. However, the ∠(C–N–C) angles α in **TS1** and **TS3** are 144.1° and 144.5°, respectively, and it is clear that these are not the structures of "pure" rotational transition states.

The rotation–inversion paths are illustrated schematically in Fig. 6. The "pure" rotation would leave $\alpha\approx120°$ relatively unchanged as the rotation proceeds from β=0° to β=180°. The "pure" inversion would change α from $\alpha\approx120°$ to $\alpha\approx180°$ while β=0° and then from $\alpha\approx180°$ to $\alpha\approx120°$ while β=180°. The change in β at α=180° causes the discontinuity on the right in Fig. 6. In the present case, neither of these ideal paths exists and rotation–inversion paths occur instead.

Proton affinities Isocyanate **10** shows a pronounced preference of about 10 kcal mol^{-1} for protonation at N rather than at O (Table 2). The proton affinity for N-protonation is 166.4 kcal mol^{-1} and very close to the proton affinity of water, which is about 167 kcal mol^{-1} [29]. Given that **7** is formed in aqueous solution, this result provides strong evidence that the reaction H_2O + **7** ⇄ **10** + H_3O^+ occurs under typical conditions of nitrosative deamination.

Conclusions

Pyrimidine ring-opening in nitrosative deamination of guanine, guanosine, and its nucleotide derivatives has been well established by theoretical study [20, 30, 31] and experimentation [32, 33]. The unimolecular dediazoniations of adeninediazonium and cytosinediazonium ions can proceed without ring-opening and it was realized only more recently that the cations which are produced by dediazoniation of adeninediazonium ion [34] and cytosinediazonium ion [9] can ring-open with very little kinetic hindrance.

The proton affinity of **10** suggests that **10** and its isomers and conformers are formed in nitrosative cytosine deamination in aqueous solution. The possibility for recyclization of **10** to uracils has been explored experimentally and it is not likely [35]. Thus, toxicological studies of nitrosative cytosine deamination need to direct attention at 3-aminoacrylonitrile **9** and 3-isocyanatoacrylonitrile **10**. It remains to be seen whether **10** can be formed by hydrolytic deglycation from the respective cytidine derivatives.

Once **10** is formed, it can react with water to carbamic acid **8**, with alcohols to carbamates **14**, with amines to ureas **15**, and with thiols to carbamothioates **16**, etc. as shown in Fig. 7 for conformer **12** [36]. Note that this is a *second* path to **8** and on to **9**, and this path differs from the sequence **7** → **8** → **9** in one very important way: **10** has some lifetime and **8** and **9** would be formed at some distance from the site of the deamination event while **7** survives merely until a nucleophile diffuses to it (usually water). The 3-isocyanatoacrylonitrile thus has the potential to cause biological damage in a larger region around the site of nitrosation and this neutral carcinogen can be much

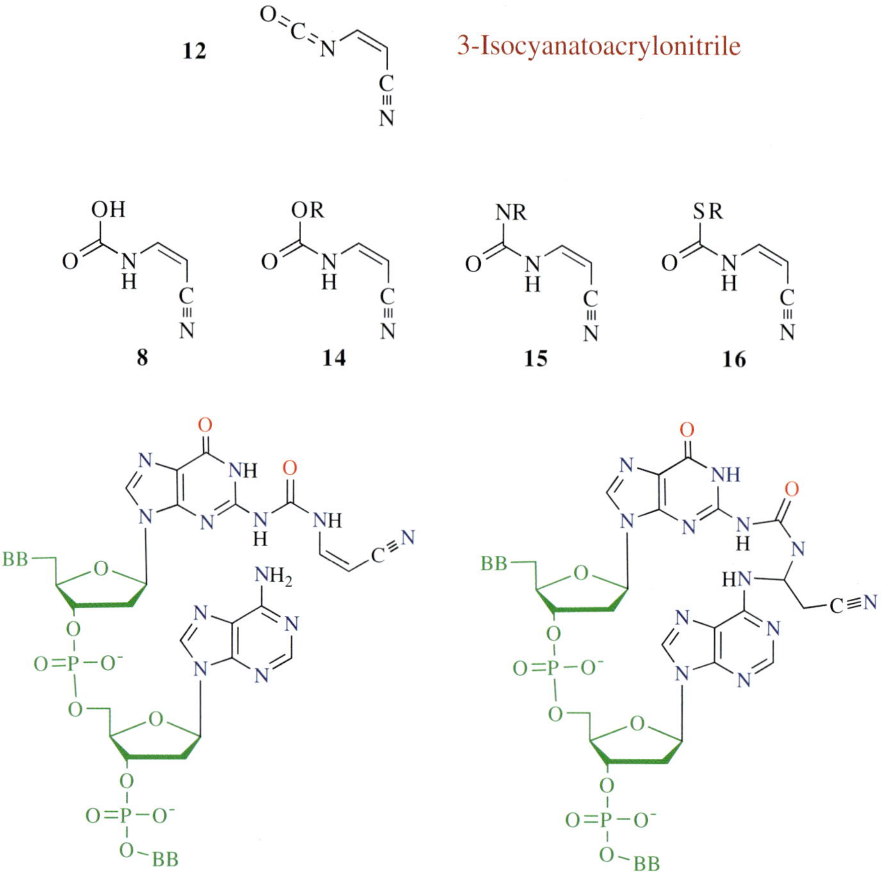

Fig. 7 Possible products of nucleophilic addition to **12** and examples for adduct and cross-link formation

more selective in its reactions. In addition to the types of isocyanate adducts shown in Fig. 7 (and their isomers and conformers), there also exist further possibilities for additions to the acrylonitrile moiety [14], either alternatively or subsequently. Fig. 7 provides an example for a urea adduct formation with dG and subsequent intrastrand dG-to-dA cross-link formation. The exploration of the chemistry of 3-isocyanatoacrylonitrile promises to be interesting and complicated.

Acknowledgements This work was funded by the U.S. National Institutes of Health (GM61027) and the U.S. National Science Foundation (DUE-0231354). FvSP was a fellow of the MU Life Sciences Undergraduate Research Opportunity Program (LS-UROP).

References

1. (a) Wyszynski M, Gabbara S, Bhagwat AS (1994) Proc Natl Acad Sci USA 91:1574–1578 (b) Yebra MJ, Bhagwat AS (1995) Biochem 34:14752-14757 (c) Sharath AN, Weinhold E, Bhagwat A S (2000) Biochem 39:14611–14616
2. Marnett LJ (1996) Chem Res Toxicol 9:807–808
3. (a) Caulfield JL, Wishnok JS, Tannenbaum SR (1998) J Biol Chem 273:12689–12695 (b) Wink DA, Kasprzak KS, Maragos CM, Elespuru RK, Misra M, Dunams TM, Cebula TA, Koch WH, Andrews AW, Allen JS, Keefer LK (1991) Science 254:1001–1003
4. (a) Lindahl T (1974) Proc Natl Acad Sci USA 71:3649-3653 (b) Tye BK, Chien J, Lehman IR, Duncan BK, Warner HR (1978) Proc Natl Acad Sci USA 75:233–237 (c) Wallace SS (1988) Environ Mol Mutagen 12:431–477
5. (a) Bridges BA (1999) Science 284:62–63 (b) Sagher D, Strauss B (1983) Biochem 22:4518–4526 (c) Loeb LA, Cheng KC (1990) Mutat Res 238:297–304 (d) Duncan BK, Miller JH (1980) Nature 287:560–561
6. Suzuki T, Nakamura T, Yamada M, Ide H, Kanaori K, Tajima K, Morii T, Makino K (1999) Biochem 38:7151–7158
7. Wu Z, Glaser R (2004) J Am Chem Soc 126:10632–10639
8. Glaser R, Rayat S, Lewis M, Son M-S, Meyer S (1999) J Am Chem Soc 121:6108–6119
9. Rayat S (2003) Dissertation, University of Missouri-Columbia
10. Raspoet G, Nguyen MT, McGarraghy M, Hegarty AF (1998) J Org Chem 63:6867–6877
11. (a) Satchell DPN, Satchell RS (1975) Chem Soc Rev 4:231–250 (b) Saunders JH, Frisch KC (1962) Polyurethanes, chemistry and technology. Interscience, New York
12. Yamamoto T, Hirasawa S, Muraoka M (1985) Bull Chem Soc Jpn 58:771–772
13. (a) Alberola A, Antolin LF, Gonzalez AM, Laguna MA, Pulido F (1986) J Heterocyclic Chem 23:1035–1038 (b) Cocco MT, Congiu C, Onnis V (1995) J Heterocyclic Chem 32:1679–1682
14. Wu H, Glaser R (2005) Chem Res Toxicol 18:111–114
15. (a) Nakashima K, Takeshita T, Morimoto K (2002) Env Health and Prevent Med 7:1–6 (b) Karol MH, Jin R (1991) Chem Res Tox 4:503–509
16. (a) Trembley P, Lesagem J, Ostiguy C, Van Tra H (2003) Analyst 128:142–149 (b) Schwetlick K, Noack R, Stebner F (1994) J Chem Soc Perkin Trans 2 3:599–608
17. Tiger RP, Levina MA, Entelis SG, Andreev MA (2002) Kinetics & Catalysis 43:662–666

18. Hehre WJ, Radom L, Schleyer PvR, Pople JA (1986) Ab initio molecular orbital theory. Wiley, New York
19. Frisch MJ, Trucks GW, Schlegel HB, Scuseria GE, Robb MA, Cheeseman JR, Montgomery JA, Vreven Jr T, Kudin KN, Burant JC, Millam JM, Iyengar SS, Tomasi J, Barone V, Mennucci B, Cossi M, Scalmani G, Rega N, Petersson GA, Nakatsuji H, Hada M, Ehara M, Toyata K, Fukuya R, Hasegawa J, Ishida M, Nakajima T, Honda Y, Kitao O, Nakai H, Klene M, Li X, Knox JE, Hratchian HP, Cross JB, Adamo C, Jaramillo J, Gomperts R, Stratmann RE, Yazyev O, Austin AJ, Cammi R, Pomelli C, Ochterski JW, Ayala PY, Morokuma K, Voth GA, Salvador P, Dannenberg JJ, Zakrzewski VG, Dapprich S, Daniels AD, Stain MC, Farkas O, Malick DK, Rabuck AD, Raghavachari K, Foresman JB, Ortiz JV, Cui Q, Baboul AG, Clifford S, Cioslowski J, Stefanov BB, Liu G, Liashenko A, Piskorz P, Komaromi I, Martin RL, Fox DJ, Keith T, Al-Laham MA, Peng CY, Nanayakkara A, Challacombe M, Gill PMW, Johnson B, Chen W, Wong MW, Gonzalez C, Pople JA (2003) Gaussian 03. Gaussian Inc, Pittsburgh, PA
20. Rayat S, Glaser R (2003) J Org Chem 68:9882–9892
21. (a) Foster JP, Weinhold F (1980) J Am Chem Soc 102:7211–7218 (b) Reed AE, Weinhold F (1983) J Chem Phys 78:4066–4073 (c) Reed AE, Curtiss LA, Weinhold F (1988) Chem Rev 88:899–926
22. Weinhold F, Landis CR (2005) Valency and bonding—a natural bond orbital donor–acceptor perspective. Cambridge University Press, Cambridge, UK
23. Badawi HM, Förner W, Al-Saadi AJ (2001) Mol Struct (Theochem) 535:183–197
24. Badawi HM, Förner W, Abu-Sharkh BF, Olorigebe YS (2002) J Mol Model 8:44–49
25. (a) Glaser R, Lewis M, Wu Z (2002) J Phys Chem A 106:7950–7957 (b) Glaser R, Wu Z, Lewis M (2000) J Mol Struct 556:131–141
26. Adamovic I, Gordon MS (2005) J Phys Chem A 109:1629–1636
27. (a) Glaser R, Horan CJ (1996) Can J Chem 74:1200–1214 (b) Glaser R, Horan C, Nelson E, Hall MK (1992) J Org Chem 57:215–228
28. (a) Mo Y, Wu W, Song L, Lin M, Zhang Q, Gao J (2004) Angew Chem Int Ed 43:1986–1990 (b) Mo Y, Jiao H, Schleyer PvR (2004) J Org Chem 69:3493–3499 (c) Mo Y, Schleyer PvR, Wu W, Lin M, Zhang Q, Gao J (2003) J Phys Chem A 107:10011–10018
29. Collyer SM, McMahon TB (1983) J Phys Chem 87:909–911
30. Rayat S, Wu Z, Glaser R (2004) Chem Res Toxicol 17:1157–1169
31. (a) Glaser R, Wu H, Lewis M (2005) J Am Chem Soc 127:7346–7358 (b) Glaser R, Lewis M (1999) Org Lett 1:273–276
32. Rayat S, Majumdar P, Tipton P, Glaser R (2004) J Am Chem Soc 126:9960–9969
33. (a) Qian M, Glaser R (2005) J Am Chem Soc 127:880–887 (b) Qian M, Glaser R (2004) J Am Chem Soc 126:2274–2275
34. Hodgen B, Rayat S, Glaser R (2003) Org Lett 5:4077–4080
35. Rayat S, Qian M, Glaser R (2005) Chem Res Toxicol 18:1211–1218
36. (a) Lewis M, Glaser R (2002) Chem Eur J 8:1934–1944 (b) Lewis M, Glaser R (1998) J Am Chem Soc 120:8541–8542

J Mol Model (2006) 12: 739–747
DOI 10.1007/s00894-005-0079-6

ORIGINAL PAPER

Ralph Puchta · Timothy Clark · Walter Bauer

The formation of *endo*-complexes between calixarenes and amines—a reinvestigation

Received: 9 November 2005 / Accepted: 21 November 2005 / Published online: 1 April 2006

Abstract The formation of an *endo*-complex between *p*-allylcalix[4]arene and *t*-butylamine was described by Gutsche in 1985. However, for a comparable system, it has been shown using NOE methods that the amine does *not* reside inside the calix. Instead, an *exo*-calix complex is formed. A reevaluation suggests that the previous conclusion is an artifact due to improper NMR-data processing. DFT (RB3LYP/6–31G(d)) calculations confirm the higher stability of the *exo*-complex over its *endo*-counterpart.

Keywords DPFGSE-NOE · Nuclear Overhauser effect · Field gradient · Calixarene · Amine · *Endo*-complex · *Exo*-complex

Introduction

In 1872, Adolf von Baeyer first synthesized calixarenes by condensation of phenol derivatives and formaldehyde, [1] but it took another 70 years until the nature and structure of the *p-tert*-butylcalix[4]arene was suggested [2] and the great potential of this class of compounds was realized. The outstanding synthetic work by, among others, Gutsche, [3] Hayes and Hunter, [4] Kämmerer, [5] Böhmer, [6] and Shinkai [7] led to an enormous number of experimental [8] and theoretical [9] studies on calixarenes. Since the beginning, the main interest has been focused on the possibility of endohedral complexation of molecules in the calixarene cavity in order to mimic, e.g., protein receptors by a simple model [10]. The hydrophobic substrate pocket of calixarenes is formed by the aromatic phenol derivatives, which constitute the "walls". Thus, calixarenes may complex neutral [11] and charged [12] guests through CH–π-, [13] π–π- [14] and ion–π interactions [15]. These types of interaction contrast with those of classical cryptands, [16, 17] coronands, [18] and their metalla-analogues, [19] which almost exclusively complex metal ions or NH_4^+ [20]. Today, there is a wide range of possible tasks for calixarenes and their derivatives: ligand systems for potential catalytically interesting transition metal complexes, [21] chemosensors, [22] separation of lanthanoids and actinoids in radioactive waste, [23] selective complexation of ions, [24] the encapsulation of small compounds, [25] as platforms for dendrimers, [26] MRI-contrast agents [27] and chromatography [28]. Moreover, aesthetically pleasing molecules like the molecular equivalent of the Soccer World Cup by Soi and Hirsch [29] have been realized.

Dedicated to Professor Dr. Paul von Ragué Schleyer on the occasion of his 75th birthday.

R. Puchta · T. Clark
Computer-Chemie-Centrum,
Friedrich-Alexander-Universität Erlangen-Nürnberg,
Nägelsbachstrasse,
25, 91052 Erlangen, Germany

W. Bauer (✉)
Institut für Organische Chemie,
Friedrich-Alexander-Universität Erlangen-Nürnberg,
Henkestrasse,
42, 91054 Erlangen, Germany
e-mail: bauer@chemie.uni-erlangen.de

Materials and methods

Experimental

NMR spectra were recorded on a JEOL Alpha 500 spectrometer (11.7 T; 500 MHz). For the spectra of Figs. 3, 4 and 5, ca. 15 mg of *tert*-butylcalix[4]arene **1b** (Aldrich) was suspended in an NMR tube in CD_3CN and *tert*-butylamine was slowly added under stirring until the calixarene had completely dissolved. The clear solution was degassed and the tube was sealed. Integration of the ^1H NMR spectrum revealed a stoichiometric ratio of **1b** and *tert*-butylamine of 1:23. The NOESY spectrum (Fig. 4) was recorded in magnitude mode on an inverse probe-head with actively shielded gradient coils; 512 complex points in t_2, zero-filled to 1,024, four dummy scans, 16 scans per t_1-increment, acquisition time in t_2=0.1105 s, spectral width in f_1 and f_2=4,634 Hz, 64 t_1-increments, zero-filled to 256, relaxation delay 4.0 s, mixing time 200 ms, 90° ^1H-pulse width 6.6 μs, sine bell window in t_1 and t_2. For the symmetrized spectrum (Fig. 5), the built-in JEOL routine was used. The individual 1D-NOE spectra (Fig. 3b–f) were recorded under the following conditions: pulse sequence DPFGSE-NOE, [30] inverse probe-head with actively shielded gradient coils, spectral width 4,554 Hz, 16,384 data points, acquisition time 3.6 s, relaxation delay 4 s, 512 scans with four dummy scans, hard 90° pulse width 6.2 μs, mixing time 600 ms, gradient duration 1 ms, gradient strength ratio 19.6:8.4:5.6:−5.6 G/cm. For the selective irradiations, an attenuated 40-ms 180° Gaussian pulse was applied.

Quantum chemical methods

All structures were fully optimized at the B3LYP/6–31G(d) density-functional theory (DFT) level [31] using the Gaussian 03 program [32]. Frequencies were computed at the same level to characterize stationary points and obtain zero-point vibrational energies (ZPEs). All frequencies are unscaled. DFT, in particular B3LYP, has been shown to provide accurate geometries and good harmonic vibrational frequencies for a broad range of molecules and ions [33]. As shown recently, the level of theory selected is well suited for NMR [34, 35] and NICS calculations [36].

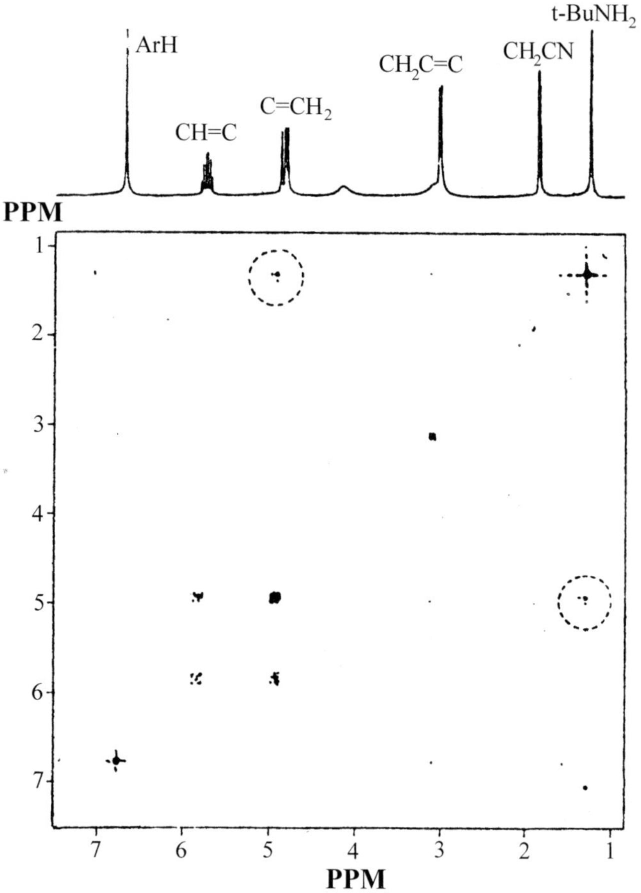

Fig. 2 Facsimile of the NOESY spectrum (300 MHz) of a 1:1 mixture of *tert*-butylamine and *p*-allylcalix[4]arene in CD_3CN at 17.8 °C. The *circled cross peaks* are claimed to be indicative of an *endo*-calix complex. Reproduced with permission from [8e]

Results and discussion

A recently established course on supramolecular chemistry at the Institut für Organische Chemie of the University of Erlangen–Nürnberg prompted us to set up an easy NMR experiment for students. In this experiment, the *endo*-complexation of an amine within a calixarene was to be demonstrated using homonuclear (^{1}H) Overhauser effect (NOE) methods. We chose to focus on an experiment described earlier by Gutsche, [8d] in which the reaction of various calix[4]arenes with *tert*-butylamine was thought to proceed via a proton transfer from one phenol-OH to the amine as a first step to form an ion pair. In a second step, the formation of an *endo*-complex between the ammonium ion and the calixarene-phenolate was postulated. These findings were corroborated in a follow-up paper [8e] by a 2D-NOESY experiment. The relevant reaction scheme and the relevant NOESY spectrum are reproduced from the original paper [8e] in Figs. 1 and 2.

In [8e], a pair of cross peaks in the original NOESY spectrum (Fig. 2) (circled) was claimed to indicate an *endo*-calix complex between *p*-allylcalix[4]arene and *tert*-butylamine. These cross peaks involve the protons of the amine *tert*-butyl group and the terminal CH_2-protons of the allyl group. Surprisingly, a cross peak

Fig. 1 Facsimile of the postulated reaction scheme of calix[4] arenes with *tert*-butylamine to form a postulated *endo*-calix complex. Reproduced with permission from [8e]

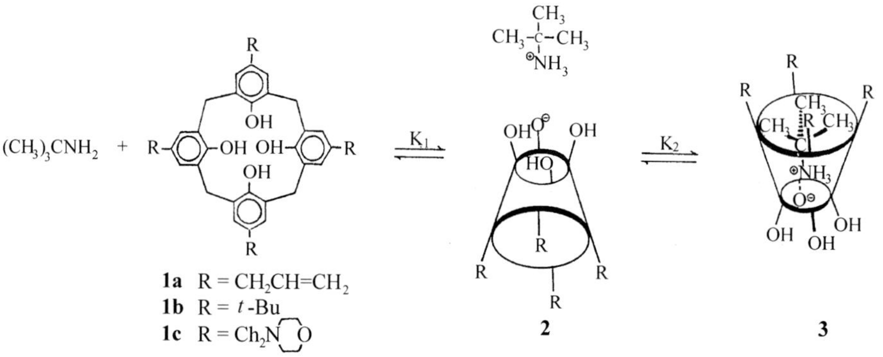

Formation of *exo*-calix and *endo*-calix complexes from amines and calix[4]arenes.

between the *tert*-butyl group and the protons at the aromatic rings of the calix is missing. In the text of [8e], the authors state (p. 4317): *Although the ^{1}H NMR data in Tables I–V are commensurate with either the exo-calix or endo-calix postulate, the NOE data provide strong evidence for the latter.* On the other hand, the authors themselves doubt this result a few lines above: *As an additional check, however, one-dimensional NOE difference spectral determinations were also made as well as a simple 2D HOMCOR NMR determination which showed no cross resonances arising from the tert-butylamine protons and the calixarene protons.*

In order to prove the presence or the absence of an *endo*-calix complex we repeated the NOE measurements of [8e] under comparable conditions. Instead of *p*-allylcalix[4] arene we used *tert*-butylcalix[4]arene **1b**.

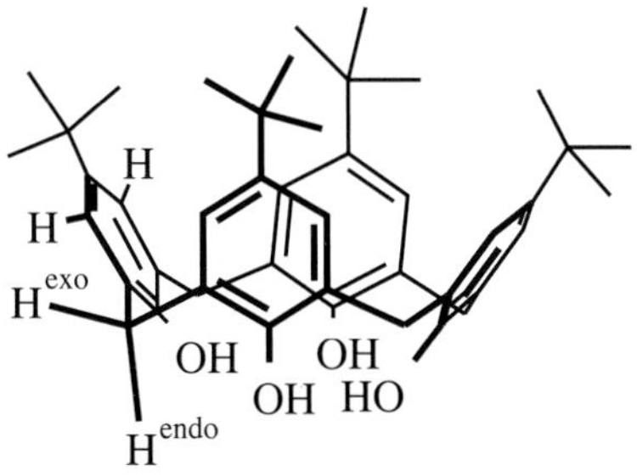

1b

In contrast to the conditions employed by Gutsche, [8e] we chose to add *tert*-butylamine in large excess in order to shift the equilibria shown in Fig. 1 to the right hand side. A degassed sample of **1b** admixed with *tert*-butylamine in a 1:23 molar ratio (solvent: CD_3CN) was prepared and a series of one-dimensional NOE was recorded. The pulse sequence used was the DPFGSE (double pulsed field-gradient spin echo) method, [30] which has become a routine measurement in many NMR labs. In contrast to the conventional difference NOE method, DPFGSE yields NOE spectra that are essentially free from subtraction artifacts, thus enabling the detection of even very weak NOEs (down to 0.01%). Hence, spatial relationships of protons up to ca. 5.5 Å can be detected.

Figure 3 shows a series of DPFGSE-NOE spectra in which all individual resonances in the ^{1}H-spectrum were irradiated sequentially. Selective excitation of the calix-aryl protons (Fig. 3b) yields strong NOEs for the *endo*- and *exo*-protons of the calix CH_2-groups. While the NOE at the *exo*-proton is expected, the *endo*-proton NOE seems surprising at first. However, this NOE comes from rapid exchange of the *exo*- and *endo*-protons on the NMR time scale. This exchange has long been known for calixarenes and involves the well-established cone-partial cone-1,2-alternate-1,3-alternate exchange. A detailed study of these dynamics can be found in [37]. Thus, the observed NOE at the *endo*-CH_2-proton arises from a genuine *exo*-CH_2-proton NOE, which is subsequently transferred to the *endo*-position due to chemical exchange. In addition, Fig. 3b shows an expected strong NOE that involves the protons of the calix *tert*-butyl group. However, no or a vanishingly weak NOE is observed for the protons of the amine *tert*-butyl group. This is strong evidence against an *endo*-calix complex. Were the amine inside the calix in the way described by Gutsche, [8e] a strong NOE must be expected for the amine *tert*-butyl protons as well. It may be concluded from Fig. 3b that a proton transfer between the calixarene and the amine has taken place. However, the calixarene aryl-protons and the protons of the amine *tert*-butyl group are very remote.

Irradiating the resonance line of the *endo*-CH_2 proton (Fig. 3c) leads to the expected complimentary NOE at the calix aryl protons. In addition, an NOE to the OH protons is found; the signal at δ=2.0 ppm is the averaged resonance of the calix-OH protons, of the amine NH_2 protons, and of the residual water protons (from the solvent). Figure 3c shows an additional downward signal at the position of the *exo*-CH_2 proton. This comes from the chemical *exo–endo* exchange described above. No or vanishingly small NOEs are found in Fig. 3c for the calix- and the amine *tert*-butyl groups. The observed weak signals originate from residual subtraction artifacts.

Irradiation of the *exo*-CH_2 proton resonance (Fig. 3d) leads to a spectrum that is identical to Fig. 3c. Again, the reason is rapid *exo–endo* exchange. Irradiating the OH/NH_2-line (Fig. 3e) gives weak NOEs at the *exo*- and *endo*-CH_2 position, complimentary to Fig. 3c and e. When the calix *tert*-butyl resonance is irradiated (Fig. 3f), a strong NOE is found for the calix aryl-H resonance, in agreement with the results from Fig. 3b. Finally, irradiation of the amine *tert*-butyl resonance (Fig. 3g) leads to an NOE at the OH/NH_2-proton resonance line, an expected NOE. A very weak NOE in Fig. 3g is also observed for the calix aryl-proton line. However, this NOE does not indicate close proximity of the amine and the calix. Rather, the bandwidth of the selective 180° Gaussian pulse means that partial irradiation of the close calix *tert*-butyl resonance cannot be avoided, thus giving the expected NOE.

In summary, all individual spectra of Fig. 3 are consistent with *remote* calixarene and amine units. From the chemical behavior during the sample preparation, it is evident that a proton transfer between the calixarene-OH protons and the amine NH_2 group must have taken place: calixarene **1b** is virtually insoluble in CD_3CN. However, it dissolves upon addition of the amine. This proton transfer is well documented in the literature, and we will discuss it in more detail in the Calculational section with respect to ^{13}C-chemical shifts. On the other hand, no formation of an *endo*-calix complex is observed, in contrast to Gutsche's postulate. These findings are summarized in Scheme 1. The nature of the complex between the calixarene-phenolate and the ammonium can be an ion pair or an *exo*-calix complex. However, *the amine has no close contacts with the upper rim of the calixarene.*

A recent paper [38] strongly supports the above conclusions that an *endo*-calix complex is not formed. In this work, conductometric and spectrophotometric mea-

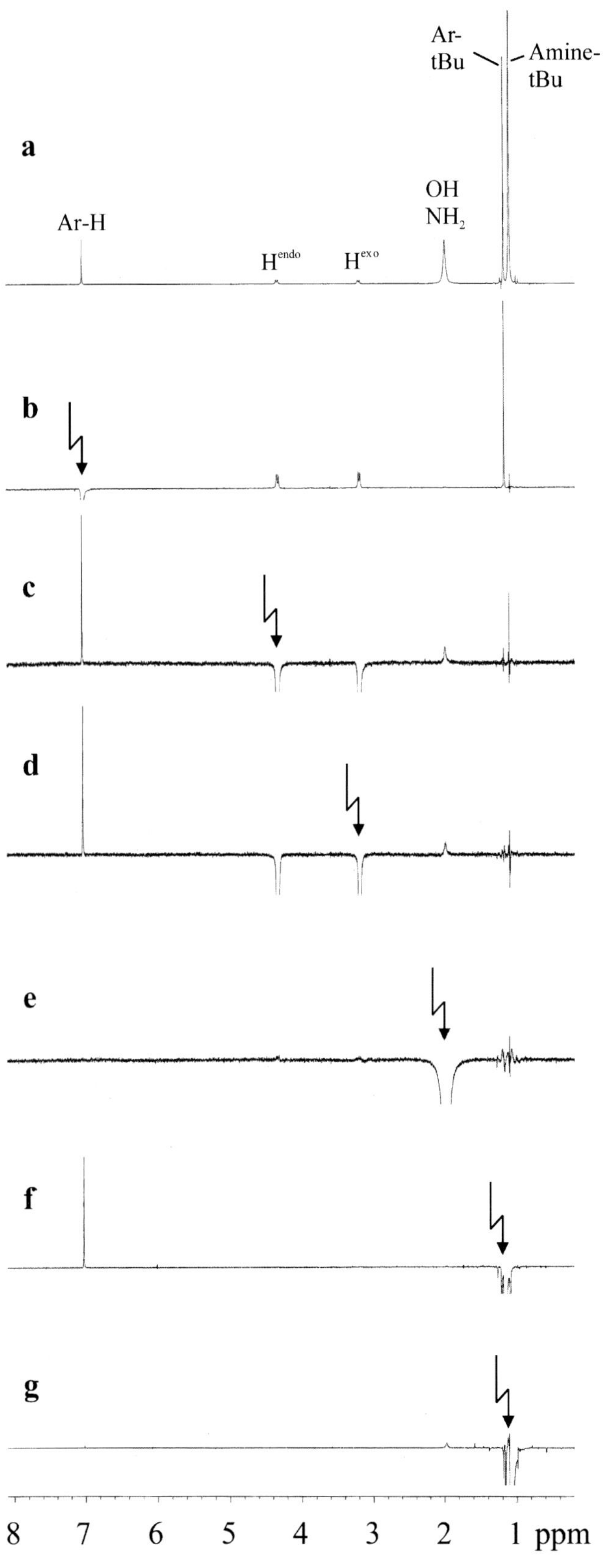

Fig. 3 Series of DPFGSE-NOE spectra on a 1:23 molar-ratio mixture of *tert*-butylcalix[4]arene **1b** and *tert*-butylamine in CD_3CN at +25 °C. The residual solvent signal resides *beneath* the OH/NH_2-signal under these conditions. **a** Normal ^1H NMR spectrum with assignments. **b–g** NOE spectra with irradiation at the positions indicated by *arrows*. For conditions, see the Methods section

surements at various concentrations led to similar conclusions of the formation of an *exo*-calix complex with Coulombic attraction between the calixarene-phenolate and the ammonium ion.

What is the origin of the postulate of an *endo*-calix complex by Gutsche which, in our view, must be wrong? This postulate is based, as outlined above, on the presence of cross peaks in the NOESY spectrum shown in Fig. 2. Virtually all NMR software-processing packages offer the possibility of symmetrizing COSY and NOESY spectra. The idea behind this procedure is to improve spectra cosmetically. COSY and NOESY spectra are principally symmetrical with respect to the diagonal; in principle data points appear with pair-wise equal intensities. In practice, however, 2D-spectra show more or less intense "t_1-ridges" that run vertically through the spectrum at the resonance-line positions. These artificial signals arise from spectrometer and temperature instabilities during a long-term (several hours) 2D-measurement. The symmetrization procedure consists of intensity comparison of the data point pairs above and below the diagonal. The lower of the two intensity numbers is then stored back at both positions in the 2D-matrix.

Gutsche et al. carried out the symmetrization procedure on their 2D-NMR data (see Experimental Section in [8e]). Authors of NMR textbooks usually warn strongly against this cosmetic "spectrum polishing" since artifacts may be introduced. A spectacular example can be found in the well-known book of Derome [39] on p. 225 (Fig. 8.32; *The symmetrisation trap; beware of falling into it*). There, a sample consisting of $CHCl_3$ and CH_2Cl_2 was used to record a COSY spectrum. Of course, no cross peaks between the two singlets observed in the proton NMR spectrum can be expected. Instead, in the non-symmetrized 2D-plot, strong t_1-ridges are observed; the spectrum looks quite awful to a non-expert. Symmetrization of these data leads to a spectrum which, from a "cosmetic" point of view, looks much better. However, spurious cross peaks are now present between the resonances of the two components that may lead to the erroneous conclusion of scalar coupling between chloroform and dichloromethane.

Analogous to [8e] we recorded a 2D-NOESY spectrum on the 1:23 molar ratio mixture of *tert*-butylcalix[4]arene **1b** and *tert*-butylamine in CD_3CN. The recording conditions and spectral parameters were comparable to those described by Gutsche. The processed but not symmetrized 2D-plot is shown in Fig. 4.

The spectrum (in magnitude mode) shows cross peaks between the calix aryl resonances and the resonances of the *exo*- and *endo*-CH_2-protons. In addition, intense exchange cross peaks are observed among the *exo*- and the *endo*-CH_2-protons. These findings are in complete agreement with the results from Fig. 3. However, the spectrum also exhibits strong t_1-noise artifacts reaching from top to

Scheme 1 Proton transfer between calixarene **1b** and tert-butylamine

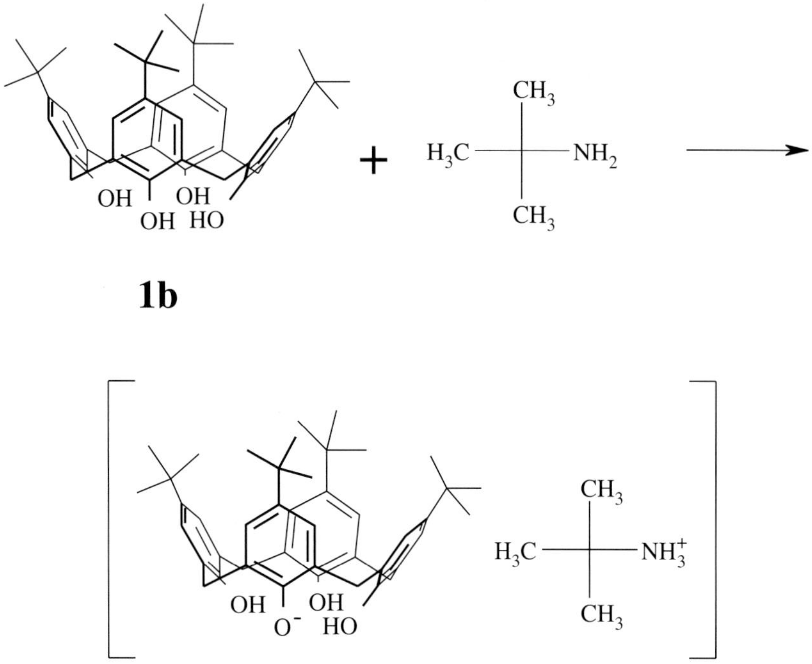

4

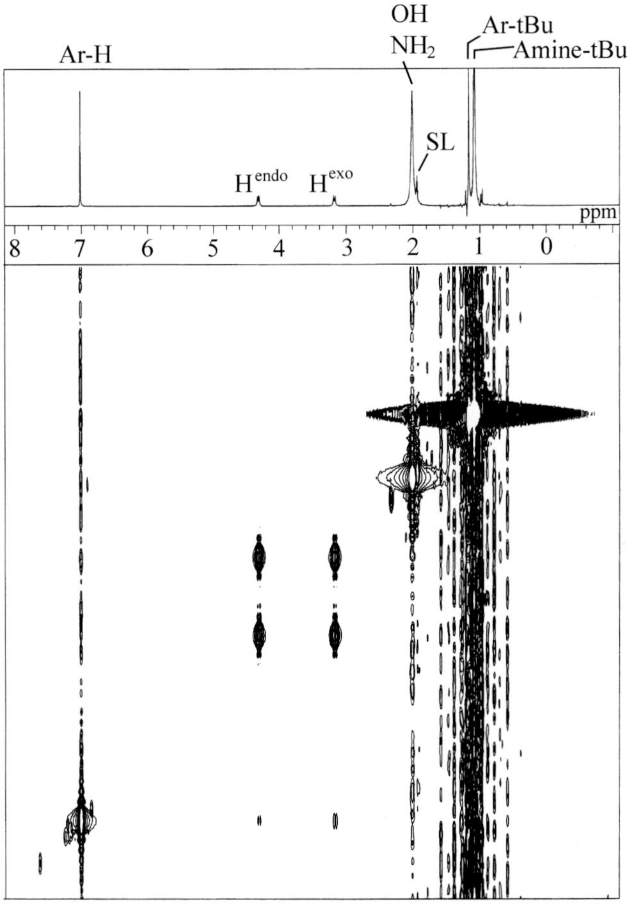

Fig. 4 2D-NOESY spectrum (magnitude mode) of a 1:23 molar-ratio mixture of *tert*-butylcalix[4]arene **1b** and *tert*-butylamine in CD_3CN at +25 °C; SL=solvent. No symmetrization processing has been applied. For spectral parameters, see the Methods section

bottom at all resonance lines. The origins of these artificial signals have been outlined above.

When the symmetrization procedure is applied to the data of Fig. 4, the spectrum shown in Fig. 5 is obtained. To the untrained observer, this spectrum looks much "better" and "cleaner". Note, however, that the genuine cross peaks between the calix aryl-proton resonances and the signals of the CH_2-group have now disappeared, thus forming a potential source of misinterpretation. The most crucial artifact comes from the circled cross peaks, of which one is magnified in Fig. 5b. Here, a genuine cross peak manifests itself between the calix aryl-proton resonance and the calix *tert*-butyl group. However, a second cross peak (highlighted with a dashed box) arises between the calix aryl-protons and the amine *tert*-butyl group. Based on this cross peak (which is analogous to the circled cross peaks in Fig. 2), one might conclude close proximity (and, hence, an *endo*-calix complex) between the groups involved. Clearly, this particular cross peak is an artifact, introduced by the weird symmetrization procedure.

From the findings of Figs. 4 and 5, we again conclude that Gutsche's postulate of an *endo*-calix complex between the calixarene and the amine must be based on erroneous and artificial cross peaks introduced by a cosmetic NMR-data processing routine.

Calculations

^{13}C NMR chemical shifts

The experimental ^{13}C NMR spectra of *tert*-butylcalix[4]arene **1b** and its deprotonated form **4** (in presence of excess *tert*-butylamine and with *tert*-butylammonium as counter-

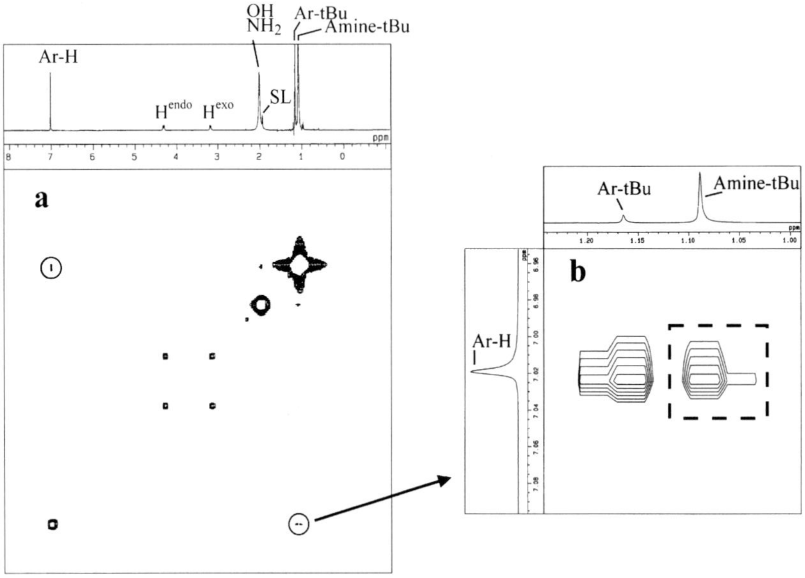

Fig. 5 2D-NOESY spectrum (magnitude mode) of a 1:23 mixture of *tert*-butylcalix[4]arene **1b** and *tert*-butylamine in CD_3CN at +25 °C; SL=solvent. The data shown in Fig. 4 were used with an additional symmetrization procedure. **a** Full data matrix. The *circled cross peaks* are the crucial ones that indicate the presence or the absence of an *endo*-calix complex. One of the *circled cross peaks* is magnified in (**b**); the *left cross peak* is genuine, whereas the *right cross peak circled with a dashed box* is artificial and has been introduced by the symmetrization step. Based on this *cross peak*, an erroneous conclusion might be drawn on the existence of an *endo*-calix complex. For spectral parameters, see the Methods section

ion) differ significantly. We take this as strong evidence for the established proton transfer from the phenol to the amine. The relevant experimental and calculated data are shown in Table 1.

The largest difference, as expected, is observed for the phenolic carbon. The experimental number is averaged between three intact phenols and one phenolate.

For the calculations, we omitted the *tert*-butyl groups on the calixarene. Instead, the unsubstituted calix[4]arene **5** and its conjugate phenolate systems **6** and **7**, were used. Figure 6 shows the calculated structures of **5**, **6** and **7**. Hence, substituent effects of the *tert*-butyl residue on the ^{13}C-chemical shifts are not included, and the shift data of the C-position para to the OH group have not been included in Table 1. Nonetheless, the calculated data of Table 1 show good agreement with the experimental chemical shifts and again confirm the proton transfer from the calixarene to the amine. Clearly, the experimental data for **4** show better agreement with those for the calculated *exo*-complex **6** than for the *endo*-complex **7**. We take this as further evidence for the existence of the *exo*-complex in CD_3CN solution.

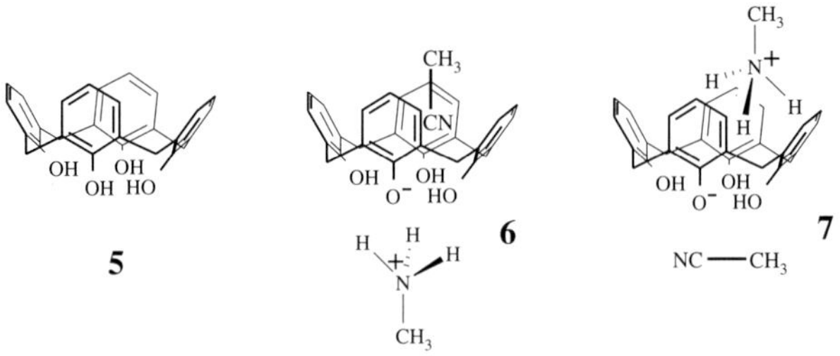

Exo-complexation of the amine is further corroborated by DFT calculations. As a model we employed the unsubstituted calix[4]arene and methylamine. To reach the complex formed by the deprotonated calixarene anion and the methyl ammonium cation, the uncharged moieties must be united first, no matter if an inclusion complex or an

Table 1 Experimental and calculated ^{13}C-chemical shifts of calixarenes and their conjugate phenolates (with *tert*-butylammonium and methylammonium, as counterions)

C-position	**1b** (exp.)[a]	**5** (calc.)	**4** (exp.)[b]	**6** (calc.)[c]	**7** (calc.)[c]
CH_2	32.62	33.82	33.80	35.68	35.31
Ar-CH	125.93	128.77	125.20	128.33	128.45
Ar-C-CH_2	127.69	129.55	131.65	131.81	132.18
Ar-C-O	146.67	149.66	152.74	154.95	156.60

Structure **6** is the *exo*-calix complex with one CH_3CN inside the calix; structure **7** is the *endo*-calix complex with one CH_3CN outside the calix

[a]in $CDCl_3$ at +25°C

[b]in CD_3CN in presence of excess *tert*-butylamine

[c]averaged numbers of three C-OH and one C-O^- moieties

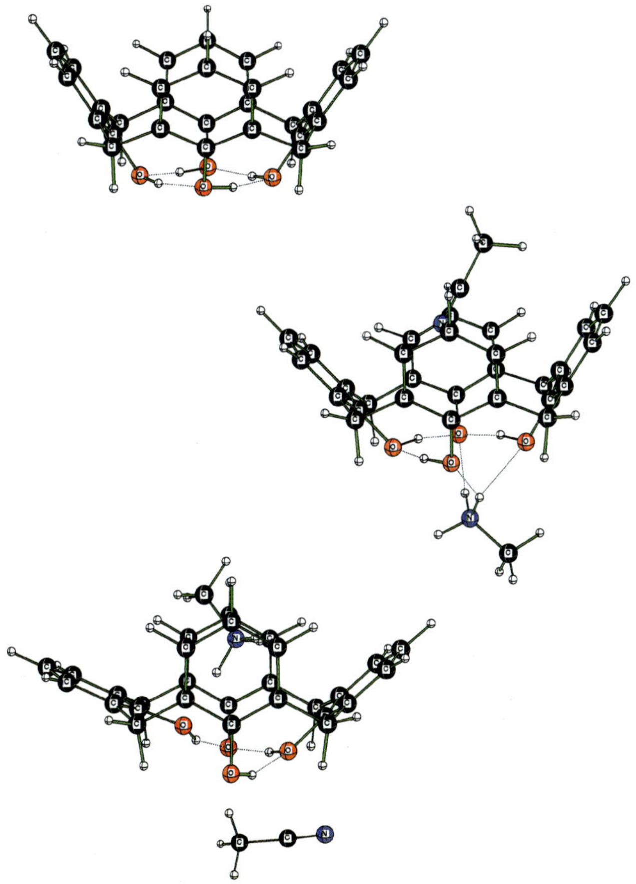

Fig. 6 Calculated (RB3LYP/6-31G(d)) structures of **5** (top), **6** (center) and **7** (bottom). H-bonds are indicated by *dashed lines*

exo-complex are formed. If an inclusion complex is formed in the first step, the NH_2–CH_3 is included in a reaction that is exothermic by only 0.3 kcal mol^{-1}. With −5.3 kcal mol^{-1} the formation of the *exo*-complex between calix[4]arene and methyl amine is clearly exothermic and favors the formation of the *exo*-complex, which is the precursor for the system observed by NMR. The next step is the proton transfer. Transferring the proton from the calix[4]arene to the included amine costs 0.4 kcal mol^{-1}. In the case of the *exo*-complexed amine, 10.1 kcal mol^{-1} are necessary. This energy is certainly overestimated. While in solution charge separation is stabilized by solvent effects, in gas-phase DFT-calculations, charge separation is unfavorable and the energy differences are therefore too high. Implicit solvent methods, such as the polarizable continuum model with an isodensity molecular surface (IPCM) offer a possibility to approximate such effects, but these methods often fail for technical reasons, as in this case, and could therefore not be used [40].

NICS-calculations

The main interaction for complexing guests inside the calixarene nest is based on the cyclically delocalized π-electrons in the benzene moieties. NICS has become one of the most used tools for investigating aromatic systems in recent years. Comparing the total NICS(1)—values inside (−11.0) and outside (−10.7) of the model calix[4]arene funnel shows that the total value inside is slightly more negative than outside. Comparing the NICS values inside and outside with the NICS(1)—total value of 2,6-dimethylphenol (−10.7) the calixarene appears to be a normal aromatic compound substituted by two methyl and one hydroxy groups. This finding is supported by the NICS values of the *endo*- and *exo*-complexes. A real perturbation of the aromatic system is the deprotonation of one hydroxyl group. While deprotonation of 2,6-dimethylphenol moves the NICS(1) value to −7.9, the NICS shift after removing a proton from a calixarene hydroxyl group is only shifted to −9.8 in mono-deprotonated calix[4]arene. We attribute the modest change in the NICS value after mono-deprotonation to the hydrogen-bond network established in the calixarenes investigated. The aromaticity of calix[4]arene is best described as four absolutely independent 2,6-dimethylphenol unities that do not communicate with each other.

Conclusions

A DPFGSE-NOE study of a 1:23 molar ratio mixture of *tert*-butylcalixarene **1b** and *tert*-butyl amine in CD_3CN has demonstrated that *no endo*-calix complex is formed. Rather, an *exo*-calix complex is present after the proton-transfer step. The contradicting results of Gutsche [8e] are almost certainly based on the misinterpretation of a 2D-NOESY spectrum where artificial cross peaks were introduced by the cosmetic symmetrization procedure. This strongly supports all warnings never to apply symmetrization to 2D-NMR data. Other erroneous conclusions that may have been drawn with respect to *endo*-calix complexes and related by other authors should be investigated.

Acknowledgements We thank Dr. Frank Beierlein and Dr. Nico J. R. van Eikema Hommes for helpful discussions.

References

1. (a) v Baeyer A (1872) Ber Dtsch Chem Ges 5:25–26; (b) v Baeyer A (1872) Ber Dtsch Chem Ges 5:280–282; (c) v Baeyer A (1872) Ber Dtsch Chem Ges 5:1094
2. (a) Zinke A, Ziegler E (1944) Wiener Chem Ztg 47:151–161; (b) Zinke A, Ziegler E (1944) Ber Dtsch Chem Ges 77:264–272
3. (a) Gutsche CD, Muthukrishnan RJ (1978) Org Chem 43:4905–4906; (b) Gutsche CD, Iqbal M (1990) Org Synth 68:234–237; (c) Gutsche CD, Dhawan B, Leonis M, Stewart D (1990) Org Synth 68:238–242; (d) Munch JH, Gutsche CD (1990) Org Synth 68:243–246
4. Hayes BT, Hunter RF (1958) J Appl Chem 8:743–748
5. (a) Kämmerer H, Happel G, Caesar F (1972) Makromol Chem 162:179–197; (b) Happel G, Mathiasch B, Kämmerer H (1975) Makromol Chem 176:3317–3334; (c) Kämmerer H, Happel G (1978) Makromol Chem 179:1199–1207; (d) Kämmerer H, Happel G(1980) Makromol Chem 181:2049–2062; (e) Kämmerer H, Happel G, Mathiasch B, (1981) Makromol Chem 182:1685–1694

6. (a) Böhmer V, Chim P, Kämmerer H (1979) Makromol Chem 180:2503–2506; (b) Böhmer V, Marschollek F, Zetta L (1987) J Org Chem 52:3200–3205; (c) Böhmer V, Merkel L, Kunz U (1987) J Chem Soc, Chem Commun 896–897; (d) Böhmer V (1995) Angew Chem 107:785–818; (e) Böhmer V (1995) Angew Chem, Int Ed Engl 34:713–745
7. (a) Shinkai S, Ikeda A (1999) Pure Appl Chem 71:275–280; (b) Lhotak P, Nakamura R, Shinkai S (1997) Supramolecular Chem 8:333–334; (c) Tsudera T, Ikeda A, Shinkai S (1997) Tetrahedron 53:13609–13620; (d) Lhotak P, Shinkai SJ (1997) Phys Org Chem 10:273–285; (e) Ikeda A, Shinkai S (1997) Chem Rev 97:1713–1734; (f) Araki K, Akao K, Ikeda A, Suzuki T, Shinkai S (1996) Tetrahedron Lett 37:73–76; (g) Harada T, Shinkai SJ (1995) Chem Soc, Perkin Trans 2 2231–2242; (h) Harada T, Ohseto F, Shinkai S (1994) Tetrahedron 50:13377–13394; (i) Iwamoto K, Ikeda A, Araki K, Harada T, Shinkai S (1993) Tetrahedron 49:9937–2115; (j) Harada T, Rudzinski JM, Shinkai SJ (1992) Chem Soc Perkin Trans 2a 2109–2115
8. Gutsche CD (1983) Acc Chem Res 16:161–170; (b) Gutsche CD (1984) Topics in Current Chemistry 183:161–170; (c) Gutsche CD, Bauer LJ (1985) J Am Chem Soc 107: 6052–6059; (d) Gutsche CD, Bauer LJ (1985) J Am Chem Soc 107:6063–6069; (e) Gutsche CD, Iqbal M, Alam I (1987) J Am Chem Soc 109:4314–4320; and literature cited there
9. (a) Schatz J (2004) Collect Czech Chem Commun 69:1169–1194; (b) Mendes A, Bonal C, Morel-Desrosiers N, Morel JP, Malfreyt P (2002) J Phys Chem B 106:4516–4524; (c) Baaden M, Burgard M, Boehme C, Wipff G (2001) Phys Chem Chem Phys, 3:1317–1325; (d) Ogden MI, Rohl AL, Gale JD (2001) Chem Comm 1626–1627; (e) Baaden M, Wipff G, Yaftian MR, Burgard M, Matt D (2000) J Chem Soc, Perkin Trans 2 1315–1321; (f) Arnaud-Neu F, Barboso S, Berny F, Casnati A, Muzet N, Pinalli A, Ungaro R, Schwing-Weil MJ, Wipff G (1999) J Chem Soc, Perkin Trans 2 1727–1738; (g) Bernardino RJ, Cost Cabral BJ (1999) J Phys Chem A 103:9080–9085; (h) Bell SEJ, McKervey MA, Fayne D, Kane P, Diamond DJ (1984) Mol Model 4:44–52; (i) Bell SEJ, Browne JK, McKee V, McKervey MA, Malone JF, O´Leary M, Walker A, Arnaud-Neu F, Boulangeot O, Mauprivez O, Schwing-Weill MJ (1998) J Org Chem 63:489–501; and literature cited there
10. see e.g.; Baur M, Frank M, Schatz J, Schildbach F (2001) Tetrahedron 57:6985–6991; and literature cited there
11. see e.g.; Schatz J, Backes ACh, Siehl H-U (2000) J Chem Soc Perkin Trans 2:609–610; and literature cited there
12. see e.g.; Bonai C, Israeli Y, Morel JP, Morel-Desrosiers N (2001) J Chem Soc Perkin Trans 2:1075–1078; and literature cited there
13. see e.g.; (a) Darbost U, Rager M-N, Petit S, Jabin I, Reinaud O (2005) J Am Chem Soc 127:8517–8525; (b) Tsue H, Ishibashi K, Takahashi H, Tamura R (2005) Org Lett 8:2165–2168; and literature cited there
14. (a) Hunter ChA, Sanders JKM (1990) J Am Chem Soc 112:5525–5534; (b) Hobza P, Selzle HL, Schlag EW (1996) J Phys Chem 100:18790–18794; (c) Castro R, Berardi MJ, Córdova E, Ochoa de Olza M, Kaifer AE, Evanseck JD (1996) J Am Chem Soc 118:10257–10268; (d) Chipot Ch, Jaffe R, Maigret B, Pearlman DA, Kollman PA (1996) J Am Chem Soc 118:11217–11224; (e) Sinnokrot MO, Sherill CD (2004) J Am Chem Soc 126:7690–7697;and literature cited there
15. see e.g.; (a) Cametti M, Nissinen M, Cort AD, Mandolini L, Rissanen K (2005) J Am Chem Soc 127:3831–3837; (b) Garau C, Quinonero D, Frontera A, Ballester P, Costa A, Deyà PM (2003) Org Lett 5:2227–2229; and literature cited there
16. see e.g.; (a) Lehn J-M, Sauvage J-P (1975) J Am Chem Soc 97:6700–6707; (b) Lehn J-M (1978) Acc Chem Res 11:49–57
17. Dietrich B, Viout P, Lehn J-M (1993) Macrocyclic Chemistry. VCH, Weinheim
18. see e.g.; (a) Pedersen CJ (1988) Angew Chem 100:1053–1059; (b) Pedersen CJ (1988) Angew Chem, Int Ed Engl 27:1021– 1027; (c) Cram DJ (1988) Angew Chem 100:1041–1052; (d) Cram DJ (1988) Angew Chem, Int Ed Engl 27:1009–1020; (e) Lehn J-M (1988) Angew Chem 100:91–116; (f) Lehn J-M (1988) Angew Chem Int Ed Engl 27:89–112; and literature cited there; (g) Ellermann J, Bauer W, Schütz M, Heinemann FW, Moll M (1998) Monatshefte für Chemie Chemical Monthly 129:547–566; (h) Hausner SH, Striley CAF, Krause-Bauer JA, Zimmer H (2005) J Org Chem 70:5804–5817
19. see e.g.; (a) Saalfrank RW, Dresel A, Seitz V, Trummer S, Hampel F, Teichert M, Stalke D, Stadler C, Daub J, Schünemann V, Trautwein AX (1997) Chem Eur J 3:2058–2061; (b) Saalfrank RW, Löw N, Kareth S, Seitz V, Hampel F, Stalke D, Teichert M (1998) Angew Chem 110:182–184; (c) Saalfrank RW, Löw N, Kareth S, Seitz V, Hampel F, Stalke D, Teichert M (1998) Angew Chem Int Ed Engl 37:172–174; (d) Saalfrank RW, Seitz V, Caulder DL, Raymond KN, Teichert M, Stalke D (1998) Eur J Inorg Chem 1313–1317; (e) Puchta R, Seitz V, van Eikema Hommes NJR, Saalfrank RW (2000) J Mol Model 6:126–132; (f) Saalfrank RW, Seitz V, Heinemann FW, Göbel C, Herbst-Irmer R (2001) J Chem Soc, Dalton Trans 599–603; (g) Saalfrank RW, Bernt I, Uller E, Hampel F (1997) Angew Chem 109:2596–2599; (h) Saalfrank RW, Bernt I, Uller E, Hampel F (1997) Angew Chem Int Ed Engl 36:2482–2485; (i) Waldmann O, Schülein J, Koch R, Müller P, Bernt I, Saalfrank RW, Andres H-P, Güdel HU (1999) Inorg Chem 38:5879–5886; (j) Saalfrank RW, Deutscher Ch, Maid H, Ako AM, Sperner St, Nakajima T, Bauer W, Hampel F, Heß BA, van Eikema Hommes NJR, Puchta R, Heinemann FW (2004) Chem Eur J 10:1899–1905; (k) Saalfrank RW, Nakajima T, Mooren N, Scheurer A, Maid H, Hampel F, Trieflinger Ch, Daub J (2005) Eur J Inorg Chem 8:1149–1153; and literature cited there
20. Dietrich B, Kintzinger JP, Lehn J-M, Metz B, Zahidi A (1987) J Phys Chem 91:6600–6606
21. see e.g.; (a) Radius U (2004) Z Anorg Allg Chemie 630:957–972; (b) Zeller J, Büschel S, Reiser BKH, Begum F, Radius U (2005) Eur J Inorg Chem 2037–2043; (c) Steyer S, Jeunesse C, Harrowfield J, Matt D (2005) Dalton Trans 1301–1309; (d) Frank M, Maas G, Schatz J (2004) Eur J Inorg Chem 607–613; (e) Ishii Y, Onaka K, Hirakawa H, Shiramizu K (2002) Chem Commun 1150–1151; (f) Seitz J, Maas G (2002) Chem Commun 338–339; (g) Cacciapaglia R, Casnati A, Mandolini L, Reinhoudt DN, Salvio R, Sartori A, Ungaro R (2005) J Org Chem 70:5398–5402; and literature cited there
22. (a) Chen Q-Y, Chen C-F (2005) Eur J Inorg Chem 2468–2472; (b) Zeng X, Sun H, Chen L, Leng X, Xu F, Li Q, He X, Zhang W, Zhang Z-Z (2003) Org Biomol Chem 1:1073–1079; (c) Cha NR, Kim MY, Kim YH, Choe J-I, Chang S-K (2002) J Chem Soc, Perkin Trans 1193–1196; (d) Zeng, X, Leng X, Chen L, Sun H, Xu F, Li Q, He X, Zhang Z-Z, Zhang Z-Z (2002) J Chem Soc, Perkin Trans 796–801; (e) Zeng X, Weng L, Chen L, Leng X, Ju H, He X, Zhang Z-Z (2001) J Chem Soc, Perkin Trans 545–549; and literature cited there
23. Casnati A, Della Ca1 N, Fontanella M, Sansone F, Ugozzoli F, Ungaro R, Liger K, Dozol J-F (2005) Eur J Inorg Chem 2338–2348 and literature cited there

24. (a) Talanova GG, Talanov VS, Gorbunova MG, Hwang H-S, Bartsch RA (2002) Chem Soc Perkin Trans 2:2070–2077; (b) Liu Y, Zhao B-T, Zhang H-Y, Wada T, Inoue Y (2001) J Chem Soc Perkin Trans 2:1219–1223; (c) Li J-S, Chen Y-Y, Lu X-R (2000) Eur J Org Chem 485–490; (d) Buschmann H-J, Wenz G, Cleve E, Schollmeyer E (2000) Acta Chim Slov 47:55–61; (e) Arnaud-Neu F, Browne JK, Byrne D, Marrs DJ, McKervey MA, O'Hagan P, Schwing-Weill MJ, Walker A (1999) Chem Eur J 5:175–186; (f) Sachleben RA, Urvoas A, Bryan JC, Haverlock TJ, Hay BP, Moyer BA (1999) Chem Comm 1751–1752; (g) Kakoi T, Toh T, Kubota F, Goto M, Shinkai S, Nakashio F (1984) Analytical Science 14:501–506; (h) Sonoda M, Hayashi K, Nishida M, Ishii D, Yoshida I (1998) Analytical Science 14:493–499; and literature cited there
25. (a) Guo Q-L, Zhu W-X, Liu Y-C, Yuan D-Q, Zhang J, Ma S-L (2004) Polyhedron 23:2055–2061; (b) Rudkevich DM (2004) Angew Chem 116:568–581; (c) Rudkevich DM (2004) Angew Chem Int Ed 43:558–571; (d) Rebeck Jr J (2000) Chem Comm 637–643; (e) Brody MS, Schalley CA, Rudkevich DM, Rebek Jr J (1999) Angew Chem 111:1738–1741; (f) Brody MS, Schalley CA, Rudkevich DM, Rebek Jr J (1999) Angew Chem Int Ed 38:1640–1644; (g) Drljaca A, Hardie MJ, Raston CL, Spiccia L (1999) Chem Eur J 5:2295–2299; and literature cited there
26. (a) Szemes F, Drew MGB, Beer PD (2002) Chem Comm 1228–1229; (b) Roy R, Kim JM (1999) Angew Chem 111:380–384; (c) Roy R, Kim JM (1999) Angew Chem Int Ed 38:369–372; and literature cited there
27. (a) Aime S, Barge A, Botta M, Casnati A, Fragai M, Luchinat C, Ungaro R (2001) Angew Chem 113:4873–4875; (b) Aime S, Barge A, Botta M, Casnati A, Fragai M, Luchinat C, Ungaro R (2001) Angew Chem Int Ed Engl 40:4737–4739; and literature cited there
28. Lippmann T, Mann G (1995) Git Fachz Lab 203–207
29. Soi A, Hirsch A (1998) New J Chem 22:1337–1338
30. (a) Stott K, Stonehouse J, Keeler J, Hwang T-L, Shaka AJ (1995) J Am Chem Soc 117:4199–4200; (b) Stott K, Keeler J, Van QN, Shaka AJ (2000) J Magn Reson 125:302–307
31. (a) Stephens PJ, Devlin FJ, Chabalowski CF, Frisch MJ (1994) J Phys Chem 98:11623–11627; (b) Becke AD (1993) J Chem Phys 97:5648–5652; (c) Lee C, Yang W, Parr RG (1988) Phys Rev B 37:785–789; (d) Hehre WJ, Ditchfeld R, Pople JA (1972) J Chem Phys 56:2257–2261; (e) Francl MM, Pietro WJ, Hehre WJ, Binkley JS, DeFrees DJ, Pople JA,Gordon MS (1982) J Chem Phys 77:3654–3665
32. Frisch MJ, Trucks GW, Schlegel HB, Scuseria GE, Robb MA, Cheeseman JR, Montgomery JA Jr, Vreven T, Kudin KN, Burant JC, Millam JM, Iyengar SS, Tomasi J, Barone V, Mennucci B, Cossi M, Scalmani G, Rega N, Petersson GA, Nakatsuji H, Hada M, Ehara M, Toyota K, Fukuda R, Hasegawa J, Ishida M, Nakajima T, Honda Y, Kitao O, Nakai H, Klene M, Li X, Knox JE, Hratchian HP, Cross JB, Bakken V, Adamo C, Jaramillo J, Gomperts, R, Stratmann RE, Yazyev O, Austin AJ, Cammi R, Pomelli C, Ochterski JW, Ayala PY, Morokuma K, Voth GA, Salvador P, Dannenberg JJ, Zakrzewski VG, Dapprich S, Daniels AD, Strain MC, Farkas O, Malick DK, Rabuck AD, Raghavachari K, Foresman JB, Ortiz JV, Cui Q, Baboul AG, Clifford S, Cioslowski J, Stefanov BB, Liu G, Liashenko A, Piskorz P, Komaromi I, Martin RL, Fox DJ, Keith T, Al-Laham MA, Peng CY, Nanayakkara A, Challacombe M, Gill PMW, Johnson B, Chen W, Wong MW, Gonzalez C, Pople JA (2004) Gaussian 03. Gaussian Inc, Wallingford CT
33. see e.g.; (a) Koch W, Holthausen MC (2000) A chemist's guide to density functional theory. Wiley-VCH, Weinheim; (b) Dosche C, Kumke MU, Ariese F, Bader AN, Gooijer C, Dosa PI, Han S, Miljanic OS, Vollhardt KPC, Puchta R, van Eikema Hommes NJR (2003) Phys Chem Chem Phys 5:4563–4569; (c) Dosche C, Kumke MU, Löhmannsröben H-G, Ariese F, Bader AN, Gooijer C, Miljanic OS, Iwamoto M, Vollhardt KPC, Puchta R, van Eikema Hommes NJR (2004) Phys Chem Chem Phys 6:5476–5483; and literature cited there
34. Backes ACh, Schatz J, Siehl H-U (2002) J Chem Soc Perkin Trans 2:484–488
35. van Eikema Hommes NJR, Timothy Clark T (2005) J Mol Model 11:175–185
36. (a) von Ragué Schleyer P, Maerker C, Dransfeld A, Jiao H, van Eikema Hommes NJR (1996) J Am Chem Soc 118:6317–6318; (b) von Ragué Schleyer P, Jiao H, van Eikema Hommes NJR, Malkin VG, Malkina OLJ (1997) J Am Chem Soc 119;12669–12670; (c) von Ragué Schleyer P, Manoharan M, Wang Z-X, Kiran B, Jiao H, Puchta R, van Eikema Hommes NJR (2001) Org Lett 3:2465–2468; (d) Chen Z, Wannere C S, Corminboeuf C, Puchta R, von Ragué Schleyer P (2005) Chem Rev 105:3842–3888
37. Soi A, Bauer W, Mauser H, Moll C, Hampel F, Hirsch A (1998) J Chem Soc Perkin Trans 2:1471–1478
38. Nachtigall FF, Lazzarotto M, Nome FJ (2002) Brazil Chem Soc 13:295–299; Available on the web: http://www.scielo.br/pdf/jbchs/v13n3/9591.pdf
39. Derome AE (1987) Modern NMR techniques for chemistry research. Pergamon, Oxford, UK
40. Illner P, Zahl A, Puchta R, van Eikema Hommes NJR, Wasserscheidt P, van Eldik R (2005) J Orgmet Chem 690:3567–3576